Plant Physiology under Abiotic Stresses

Plant Physiology under Abiotic Stresses

Editor

Yanyou Wu

MDPI • Basel • Beijing • Wuhan • Barcelona • Belgrade • Manchester • Tokyo • Cluj • Tianjin

Editor
Yanyou Wu
Chinese Academy of Sciences
China

Editorial Office
MDPI
St. Alban-Anlage 66
4052 Basel, Switzerland

This is a reprint of articles from the Special Issue published online in the open access journal *Horticulturae* (ISSN 2311-7524) (available at: https://www.mdpi.com/journal/horticulturae/special_issues/plant_physiology_abiotic).

For citation purposes, cite each article independently as indicated on the article page online and as indicated below:

LastName, A.A.; LastName, B.B.; LastName, C.C. Article Title. *Journal Name* **Year**, *Volume Number*, Page Range.

ISBN 978-3-0365-7218-5 (Hbk)
ISBN 978-3-0365-7219-2 (PDF)

Cover image courtesy of Yanyou Wu

Contents

About the Editor

Yanyou Wu

Yanyou Wu received his Ph.D. degree from Sichuan University, China, in 1994. Now he is a professor at the Research Center for Environmental Bio-Science and Technology, State Key Laboratory of Environmental Geochemistry, Institute of Geochemistry, Chinese Academy of Sciences. His research focuses on interdisciplinary of biology and geology, such as biogeochemical cycle of carbon. His recent interests concentrate on the adaptation mechanism of plants to the environment and plant electrophysiology.

horticulturae

Editorial

Plant Physiology under Abiotic Stresses: Deepening the Connotation and Expanding the Denotation

Yanyou Wu

Research Center for Environmental Bio-Science and Technology, State Key Laboratory of Environmental Geochemistry, Institute of Geochemistry, Chinese Academy of Sciences, Guiyang 550081, China; wuyanyou@mail.gyig.ac.cn

Abstract: Abiotic stress factors influence many aspects of plant physiology. The works collected in the Special Issue deepen plant physiology's connotation (such as plant electrophysiology) under abiotic stress and expand the denotation (such as environmental pollutants as abiotic stress factors). At the same time, the achievements of the selected papers published in the Special Issue also exhibit their potential application value in the production of horticultural plants.

Citation: Wu, Y. Plant Physiology under Abiotic Stresses: Deepening the Connotation and Expanding the Denotation. *Horticulturae* **2023**, *9*, 218. https://doi.org/10.3390/horticulturae9020218

Received: 18 December 2022
Accepted: 20 January 2023
Published: 7 February 2023

1. The Effects of Various Abiotic Stresses

The effects of various abiotic stresses on substance metabolism, energy transformation, and the growth and development of different plants are disparate. Abiotic stress includes several adversities, such as drought, temperature, light, salinity, nutrients, and heavy metals, as well as complex stresses, such as karst environments, saline–alkali soils, and coastal wetland environments. Water is the essential substance necessary for plant survival. A water deficit adversely affects plant morphology, growth and development, and physiological and biochemical processes. Therefore, water stress is also the most common abiotic stress factor. Most of the research works collected in the Special Issue concern the effects of water stress on plant physiology, biochemistry, and growth and development [1–4]; however, these works also focus on some complex multifactorial stresses [5–7].

2. Moderate Drought Stress Has a Positive Effect on Plant Secondary Metabolism

Generally, water stress harms the primitive metabolism of plants [1–4], but appropriate water stress has a positive effect on plant secondary metabolism and adaptation. Honório et al. found that under moderate drought stress, the production of liriodenine and total alkaloids in *Annona crassiflora* Mart. increased, but photosynthesis did not decrease [3]. Al-Quraan et al. demonstrated that drought can promote γ-aminobutyric acid (GABA) accumulation, but it also inhibits the synthesis of chlorophyll a and chlorophyll b in green pea seedlings (*Pisum sativum* L.) [2]. The abovementioned works show that humans can control soil to reach a moderate humidity level to produce secondary metabolites.

3. The Application of Some Exogenous Organic Compounds Can Restore Metabolic Activity Reduced Due to Drought Stress

Drought causes damage to the cell membrane, disrupts cell function, and reduces the physiological vitality of plants. Yeast extract is rich in essential nutrients and is a natural and safe biofertilizer. Chitosan is an environmentally friendly biological polysaccharide obtained by the deacetylation of chitin, which is widely present in nature. However, Abdelaal et al. presented that the application of yeast extract (8 g/L) or chitosan (30 mM) alone or in combination can enhance the antioxidant capacity of plants and restore metabolic activity reduced due to drought stress [8]. We can expect that appropriate drought combined with the application of yeast extract and/or chitosan can increase the synthesis of secondary metabolites in plants.

4. Plant Electrophysiology Can Deepen the Understanding of Plant Physiology under Abiotic Stress

When affected by abiotic stresses, the anatomy, function, and metabolic activities of plant cells change, and these changes are bound to be reflected in the electrical properties of plant tissues and organs. Plant electrical signals can quickly respond to water status in the environment in real-time and characterize intracellular water metabolism. The works collected in the Special Issue demonstrate plant electrophysiology's great potential in the study of plant stress biology. According to the Gibbs free energy equation and Nernst equation, the relationship model between the clamping force and the electrical signal (capacitive reactance, resistance, impedance, and inductive reactance) of plant leaves can be quantitatively reflected in real-time by the intracellular water-holding capacity, water transfer rate, water-holding time, and water use efficiency of plant leaves. *Broussonetia papyrifera* (L.) Vent. and *Morus alba* L. have different intracellular water metabolism mechanisms adapted to soil water deficits [4]. The karst-adaptable plant *Orychophragmus violaceus* (L.) O. E. Schulz adapts to karst drought environments by synergistically regulating intracellular water transport, cell elasticity, and leaf anatomy [7]. The acquisition and analysis of plant electrophysiological information have opened up a broad path for studying plant physiology under abiotic stress.

5. The Environmental Pollutants Become Abiotic Stress Factors

Climate change and environmental pollution caused by human activities have gradually become abiotic stresses, which hugely impact plant life activities. With a background of elevated atmospheric carbon dioxide, the adverse effects of low temperature on plant growth and development are much greater than the adverse effects of high temperature on plant growth and development [5]. Environmental pollutants such as sulfur dioxide and bisulfites have become abiotic stress factors that seriously affect the physiological activities of plants. The work of Li et al. shows that suitable concentrations of $NaHSO_3$ can be used as a source of sulfur for plants to play a positive role in the physiological activities of plants, and high concentrations of $NaHSO_3$ reduce the photosynthetic capacity of plants and adversely affect plant growth and development. The effects of bisulfites as photorespiratory inhibitors are often masked [9].

6. Complexity and Diversity of Salinity Stress

Both salt in the soil and salt spray in the atmosphere can become abiotic stress factors. The influence of salt stress on coastal plants is caused by the superposition of soil salinity and atmospheric salt spray stress, and salt composition is also compounded. The root is the earliest organ to respond to salt stress in the soil. *Tilia cordata* Mill. can only respond to salinity by adjusting the stomatal conductance of leaves and reducing photosynthetic capacity by improving water use efficiency, as its roots do not prevent salt intake. *Pyrus pyraster* L. (Burgsd.) maintains stable photosynthesis under salt stress, as its roots limit the transfer of salt ions to leaves [10]. The first organ to respond to salt spray in the atmosphere is the leaves of plants. *Callistemon citrinus* significantly decreases photosynthetic capacity under salt spray stress [11]. In fact, plants growing in soil salinization areas are not simply affected by single-salinity adversity but by salinity–alkali compound adversity. Wang et al. found that low concentrations of saline–alkali stress ($NaCl:NaHCO_3$ = 3:1, 0~90 mM) can promote glucose metabolism and the synthesis of antioxidant enzymes, while high concentrations of saline–alkali stress ($NaCl:NaHCO_3$ = 3:1, 120~150 mM) reduce the activities of peroxidase and catalase [6]. This is related to the fact that at an appropriate concentration, bicarbonate improves the stress tolerance of plants [12]. In horticultural plant production, saltwater dilution irrigation is an effective measure.

Funding: This research was funded by Support Plan Projects of Science and Technology of Guizhou Province [No. (2021)YB453].

Data Availability Statement: Not applicable.

Acknowledgments: I gratefully acknowledge all the authors that participated in this Special Issue.

Conflicts of Interest: The author declares no conflict of interest.

References

1. Shafqat, W.; Mazrou, Y.S.A.; Sami-ur-Rehman; Nehela, Y.; Ikram, S.; Bibi, S.; Naqvi, S.A.; Hameed, M.; Jaskani, M.J. Effect of Three Water Regimes on the Physiological and Anatomical Structure of Stem and Leaves of Different Citrus Rootstocks with Distinct Degrees of Tolerance to Drought Stress. *Horticulturae* **2021**, *7*, 554. [CrossRef]
2. AL-Quraan, N.A.; Al-Ajlouni, Z.I.; Qawasma, N.F. Physiological and Biochemical Characterization of the GABA Shunt Pathway in Pea (*Pisum sativum* L.) Seedlings under Drought Stress. *Horticulturae* **2021**, *7*, 125. [CrossRef]
3. Honório, A.B.M.; De-la-Cruz-Chacón, I.; Martínez-Vázquez, M.; da Silva, M.R.; Campos, F.G.; Martin, B.C.; da Silva, G.C.; Fernandes Boaro, C.S.; Ferreira, G. Impact of Drought and Flooding on Alkaloid Production in Annona crassiflora Mart. *Horticulturae* **2021**, *7*, 414. [CrossRef]
4. Yu, R.; Wu, Y.; Xing, D. The Differential Response of Intracellular Water Metabolism Derived from Intrinsic Electrophysiological Information in *Morus alba* L. and *Broussonetia papyrifera* (L.) Vent. Subjected to Water Shortage. *Horticulturae* **2022**, *8*, 182. [CrossRef]
5. Barickman, T.C.; Olorunwa, O.J.; Sehgal, A.; Walne, C.H.; Reddy, K.R.; Gao, W. Interactive Impacts of Temperature and Elevated CO_2 on Basil (*Ocimum basilicum* L.) Root and Shoot Morphology and Growth. *Horticulturae* **2021**, *7*, 112. [CrossRef]
6. Wang, Y.; Feng, Y.; Yan, M.; Yu, J.; Zhou, X.; Bao, J.; Zhang, Q.; Wu, C. Effect of Saline–Alkali Stress on Sugar Metabolism of Jujube Fruit. *Horticulturae* **2022**, *8*, 474. [CrossRef]
7. Xing, D.; Wang, W.; Wu, Y.; Qin, X.; Li, M.; Chen, X.; Yu, R. Translocation and Utilization Mechanisms of Leaf Intracellular Water in Karst Plants *Orychophragmus violaceus* (L.) O. E. Schulz and *Brassica napus* L. *Horticulturae* **2022**, *8*, 1082. [CrossRef]
8. Abdelaal, K.; Attia, K.A.; Niedbała, G.; Wojciechowski, T.; Hafez, Y.; Alamery, S.; Alateeq, T.K.; Arafa, S.A. Mitigation of Drought Damages by Exogenous Chitosan and Yeast Extract with Modulating the Photosynthetic Pigments, Antioxidant Defense System and Improving the Productivity of Garlic Plants. *Horticulturae* **2021**, *7*, 510. [CrossRef]
9. Li, Z.; Wu, Y.; Xing, D.; Zhang, K.; Xie, J.; Yu, R.; Chen, T.; Duan, R. Effects of Foliage Spraying with Sodium Bisulfite on the Photosynthesis of *Orychophragmus violaceus*. *Horticulturae* **2021**, *7*, 137. [CrossRef]
10. Paganová, V.; Hus, M.; Lichtnerová, H. Effect of Salt Treatment on the Growth, Water Status, and Gas Exchange of *Pyrus pyraster* L. (Burgsd.) and *Tilia cordata* Mill. Seedlings. *Horticulturae* **2022**, *8*, 519. [CrossRef]
11. Toscano, S.; La Fornara, G.; Romano, D. Salt Spray and Surfactants Induced Morphological, Physiological, and Biochemical Responses in *Callistemon citrinus* (Curtis) Plants. *Horticulturae* **2022**, *8*, 261. [CrossRef]
12. Yao, K.; Wu, Y.Y. Rhizospheric Bicarbonate Improves Glucose Metabolism and Stress Tolerance of *Broussonetia papyrifera* L. seedlings under simulated drought stress. *Russ. J. Plant Physiol.* **2021**, *68*, 126–135. [CrossRef]

horticulturae

MDPI

Article

Effect of Three Water Regimes on the Physiological and Anatomical Structure of Stem and Leaves of Different Citrus Rootstocks with Distinct Degrees of Tolerance to Drought Stress

Waqar Shafqat [1,2], Yasser S. A. Mazrou [3,4], Sami-ur-Rehman [5], Yasser Nehela [6,7,*], Sufian Ikram [5,8], Sana Bibi [9], Summar A. Naqvi [5], Mansoor Hameed [9] and Muhammad Jafar Jaskani [5,*]

1 Indian River Research and Education Center, Horticultural Sciences Department, University of Florida, 2199 South Rock Road, Fort Pierce, FL 34945, USA; waqar.shafqat@ufl.edu
2 Department of Forestry, Mississippi State University, Starkville, MS 39762, USA
3 Business Administration Department, Community College, King Khalid University, Guraiger, Abha 62529, Saudi Arabia; ymazrou@kku.edu.sa or yasser.mazroua@agr.tanta.edu.eg
4 Department of Agriculture Economic, Faculty of Agriculture, Tanta University, Tanta 31527, Egypt
5 Institute of Horticultural Science, University of Agriculture, Faisalabad 38000, Pakistan; samiurrehman96@gmail.com (S.-u.-R.); chsufian1@yahoo.com (S.I.); summar.naqvi@uaf.edu.pk (S.A.N.)
6 Department of Agricultural Botany, Faculty of Agriculture, Tanta University, Tanta 31527, Egypt
7 Citrus Research and Education Center, Department of Plant Pathology, University of Florida, 700 Experiment Station Road, Lake Alfred, FL 33850, USA
8 Institute of Vegetables and Flowers, Chinese Academy of Agricultural Sciences (CAAS), No. 12 Zhongguancun South Street, Haidian District, Beijing 100081, China
9 Department of Botany, University of Agriculture, Faisalabad 38040, Pakistan; sanisana446@gmail.com (S.B.); hameedmansoor@yahoo.com (M.H.)
* Correspondence: yasser.nehela@ufl.edu (Y.N.); jjaskani@uaf.edu.pk (M.J.J.)

Citation: Shafqat, W.; Mazrou, Y.S.A.; Sami-ur-Rehman; Nehela, Y.; Ikram, S.; Bibi, S.; Naqvi, S.A.; Hameed, M.; Jaskani, M.J. Effect of Three Water Regimes on the Physiological and Anatomical Structure of Stem and Leaves of Different Citrus Rootstocks with Distinct Degrees of Tolerance to Drought Stress. *Horticulturae* **2021**, *7*, 554. https://doi.org/10.3390/horticulturae7120554

Academic Editor: Yanyou Wu

Received: 21 October 2021
Accepted: 1 December 2021
Published: 6 December 2021

Abstract: Citrus is grown globally throughout the subtropics and semi-arid to humid tropics. Abiotic stresses such as soil water deficit negatively affect plant growth, physiology, biochemistry, and anatomy. Herein, we investigated the effect(s) of three water regimes (control, moderate drought, and severe drought) on the physiological and anatomical structure of 10 different citrus rootstocks with different degrees of tolerance to drought stress. Brazilian sour orange and Gadha dahi performed well by avoiding desiccation and maintaining plant growth, plant water status, and biochemical characters, while Rangpur Poona nucellar (*C. limonia*) and Sunki × bentake were the most sensitive rootstocks at all stress conditions. At severe water stress, the highest root length (24.33 ± 0.58), shoot length (17.00 ± 1.00), root moisture content (57.67 ± 1.53), shoot moisture content (64.59 ± 1.71), and plant water potential (−1.57 ± 0.03) was observed in tolerant genotype, Brazilian sour orange. Likewise, chlorophyll *a* (2.70 ± 0.06), chlorophyll *b* (0.87 ± 0.06) and carotenoids (0.69 ± 0.08) were higher in the same genotype. The lowest H_2O_2 content (77.00 ± 1.00) and highest proline content (0.51 ± 0.06) were also recorded by Brazilian sour orange. The tolerance mechanism of tolerant genotypes was elucidated by modification in anatomical structures. Stem anatomy at severe drought, 27.5% increase in epidermal cell thickness, 25.4% in vascular bundle length, 30.5% in xylem thickness, 27.7% in the phloem cell area, 8% in the pith cell area, and 43.4% in cortical thickness were also observed in tolerant genotypes. Likewise, leaf anatomy showed an increase of 27.9% in epidermal cell thickness, 11.4% in vascular bundle length, 21% in xylem thickness, and 15% in phloem cell area in tolerant genotypes compared with sensitive ones. These modifications in tolerant genotypes enabled them to maintain steady nutrient transport while reducing the risk of embolisms, increasing water-flow resistance, and constant transport of nutrients across.

Keywords: drought stress; citrus; oxidative stress; proline; photosynthesis; water potential; vascular bundle modifications

1. Introduction

Citrus fruit crops in the family Rutaceae have the largest fruit industry globally [1,2]. Rootstock choice and selection for the citrus scion variety are the most valuable decisions for growers for the implication of better yield and quality with other valuable characters. Citrus rootstock controls the physiological, biochemical, morphological, and genetic characteristics of scion cultivars grafted on selected rootstocks through the rootstock scion interaction pathway [3]. Fruit juice quality and tree productivity of scion cultivars are also affected by rootstock characters. Rootstock has a significant impact on nutrition, horticultural and pathological traits of citrus cultivars, growth, vigor, stress resistance, and fruit quality of the scion [4–7]. Rootstock controls translocation of water and nutrients from the soil and distributes among different up ground plant organs, which were also disturbed negatively by the impact of stresses [8,9].

Citrus plants affected by several biotic (fungal, viral, and bacterial diseases) and abiotic (water deficit, heat, flooding, and salinity) stresses exhibit wide-ranging losses in citrus production [10,11]. Cultivars of genus *Citrus* are grown in a climate of wide range due to evergreens and perennial tree crops [12]. The capacity to tolerate the unfavorable climatic conditions of the citrus crop is high but the estimated yield loss due to environmental abiotic stresses in citrus is up to 82% [13]. Climate change conditioning causes a global rise in temperature and limited the availability of freshwater for the crop. Meanwhile, crop consumption is increasing rapidly resulting in reduced levels of groundwater ultimately limiting productivity [14]. Global warming due to climate change caused frequent extreme water deficit conditions, an important component for global agriculture [15,16]. Abiotic stresses are interlinked with each other in which soil water deficit/drought is a serious environmental factor, which frequently limits the growth and productivity of important crop species [17]. Supply in water restriction can severely limit plant cell division, plant growth, fruit development, and fruit production [18,19]. Soil water deficit for long-term events causes permanent negative changes in the plant which can be in response to previous stress. Water deficit conditions can be developed at different phenological stages which can change physiological and molecular processes of the plant [20].

Plant responses to water stress are mediated by changes in root growth pattern and stomatal closure at moderate stress resulting in reduced CO_2 intake, impairing photosynthesis, and loss in production [21,22]. Extreme water stress conditions alter the physiological and morphological processes of the plant. The molecular and biochemical machinery is also affected under stress conditions. Reactive oxygen species (ROS), i.e., hydrogen peroxide, superoxide, singlet oxygen, and hydroxyl radical are excessively produced and accumulate in plant cells at water deficit conditions. Photoinhibition or photooxidation caused by ROS accumulation leads to uncontrolled photosynthetic components oxidation [23,24]. Water deficit/drought damages photosynthetic apparatus and photosynthetic pigments (chlorophyll *a*, chlorophyll *b*, and carotenoids) which also reduced the photosynthetic rate [25,26]. Photosynthetic pigments are linked to stress tolerance in plants [27,28]. Resistant cultivars had a higher amount of chlorophyll contents and carotenoids when exposed to water deficit conditions [29]. Water deficit conditions decrease contents of total chlorophyll *a* and *b* contents [30]. Water deficit conditions damage thylakoid lamella and reduce active oxygen species' chlorophyll contents [31]. The chlorophyll content is linked inversely with the severity and duration of water deficit conditions [32]. After oxidative stress, accumulation of proline acts as an adaptive stress response. The intracellular levels of proline can increase by >100-fold during stress [33]. Proline accumulation acts as a compatible osmolyte to buffer cytosolic pH and to balance cell redox status. Proline can also function as a molecular chaperone stabilizing the structure of proteins and as a ROS scavenger [24,34].

Plant anatomical structure is also disturbed when the plant is subjected to water deficit which affects the regulation of water through the vascular bundles. The effect of drought stress firstly occurs on the cell structure [35,36]. Increased hydraulic resistance and decreased growth are directly associated with xylem structure [37]. Anatomical changes in citrus rootstocks under abiotic stress influence their ability to survive. In drought conditions, the xylem vessel becomes emboli and dysfunctional. Therefore, plants with narrow and higher number of vessels are considered to be more drought tolerant [38]. The plant cells that face an environment with a shortage of water have generally shown an increase in vessel tissue, thick cell wall, reduction in cell size, and most severe condition cell wall and the cell membrane becomes ruptured [39,40]. The anatomical changes may occur to protect the plants under stress conditions. Multiple changes occur in response to water stress such as alteration of xylem phloem ratio, wall structure, lumen size, and lumen area that resist environmental stress on the plant [41]. The sensitivity of rootstock against drought stress is directly related to vessel dimension [42,43]. The vessel density of root and stem decreases with tree height as vessel diameter increases. The rootstock growth ability is dramatically affected by several xylem traits, xylem phloem ratio, vessel size, and vessel density [43,44]. Thus, traits play an important role in maintaining hydraulic conductance of root and stem and leaves [37,45]. Vessel number and size is the key factor to maintaining hydraulic conductance [46].

Strategies for drought tolerance are highly relevant in the case of rootstock selection and multiplication for ensuring continuous productivity. Rootstocks with a higher root growth ratio, better water use efficiency, higher root hydraulic conductivity, and lower osmatic conductance can withstand drought conditions while maintaining higher growth levels and mass accumulation [47,48]. After a severe drought, the recovery of tolerant rootstocks is much better than other rootstocks. Therefore, screening of drought-tolerant rootstocks is of utmost importance. For drought stress prevention, the plant increases water uptake efficiency either by increasing root density or deepening roots [26].

Climate changes shift the weather conditions by a high degree of unpredictability; water shortage in the soil and plant by a continuous increase in daily average temperature every year are inevitable, which intimidate overall globe agriculture industry stability by the negative effect on plant health and production consumer demand. Survival of the citrus industry against the water deficit needs to evaluate a tolerant/resistant citrus rootstock against the climate change scenario. The objective of the current study was to evaluate 10 citrus rootstocks against drought, based on visual changes, water potential, morphological and biochemical characters. After initial screening, the two most sensitive and tolerant genotypes were used to study the anatomical differences in leaf and stem to elucidate the drought tolerance mechanism in tolerant genotypes.

2. Materials and Methods

2.1. Plant Materials, Experimental Site, and Growth Conditions

Six months old citrus rootstocks potted plants were taken as experimental material. Ten genetically diverse citrus rootstocks were examined (details are given in Table 1). A potted plant experiment was executed at the Institute of Horticultural Science, University of Agriculture Faisalabad, Pakistan. Seeds of selected rootstock were taken from citrus rootstocks progeny block, HIS, UAF, Pakistan.

Table 1. Characteristics of citrus rootstocks used in this study.

Rootstock	Botanical Name	Citrus Category	Leaf Shape	Parentage/Origin
Gabbuchini	*Citrus aurantium* L.	Sour orange hybrid	Unifoliate	*C. aurantium* 'Bittersweet' × *C. sinensis*
Gada dahi	*Citrus maxima* Burm. Merrill/ *Citrus grandis* L. Osbeck/ *Citrus decumana* L.	Pummelo hybrid	Unifoliate	Subcontinent (Indo-Pak), seed selection
Sour orange	*Citrus aurantium* L.	Sour orange	Unifoliate	Subcontinent (Indo-Pak)
Keen sour orange	*Citrus aurantium* L.	Sour orange	Unifoliate	Selection/root sprout
Brazilian sour orange	*Citrus aurantium* L.	Sour orange	Unifoliate	Open-pollinated seed selection
Rough lemon	*Citrus jambhiri*	Lemon	Unifoliate	Open-pollinated seed selection
Sunki × bentake,	*Citrus* spp.	Unknown	Trifoliate	*Citrus sunki* × bentake hybrid

Table 1. *Cont.*

Rootstock	Botanical Name	Citrus Category	Leaf Shape	Parentage/Origin
X639	*Citroncirus* spp.	Mandarin × Poncirus	Trifoliate	Cleopatra mandarin × *Poncirus trifoliata* hybrid
Kirrumakki nucellar	*Citrus limonia* Osbeck	Lime	Unifoliate	Unknow/Subcontinent (Indo-Pak)
Rangpur Poona nucellar	*Citrus limonia* Osbeck	Lime	Unifoliate	Unknow/Subcontinent (Indo-Pak)

The seeds were sown in transparent grow bags (height 12″ and width 6″) and placed in a growth chamber (Model: BST/PGC-175; Bionics Scientific Technologies (p) ltd., Delhi, India) at 32 ± 2 °C, relative humidity oscillating between 80 and 95% and 12–14 h of light. Potted media contained sand, silt, and clay (1:1:1). These plants were placed in the chamber for six months before the treatment application and during growth, pots were regularly irrigated with tap water (75% field capacity) and fertilized weekly with nutrient solution (1.0 g L^{-1} $Ca(NO_3)_2$, 0.4 g L^{-1} KNO_3, 0.6 g L^{-1} $MgSO_4$ and 0.4 g L^{-1} $NH_4H_2PO_4$ (MAP)).

2.2. Water Regimes and Treatments

Citrus rootstocks were subjected to water deficit treatment applications after six months of growth in the controlled growth condition. Plants were exposed to three different groups: controlled condition, moderate water deficit (moderate drought), and severe water deficit (severe drought). Each treatment group consists of 10 citrus rootstocks with three replications. Control plants (sufficient water and optimal temperature ~32 °C) were irrigated once every two days (75% field capacity). Water deficit treatments were moderate (50% field capacity) and severe (25% field capacity) deficit conditions. Water was applied with 2 days interval for control and treated plants. The controlled and water deficit exposed genotypes were kept at 32 ± 2 °C for day and night temperature in a controlled growth room. Thereby, three experimental groups were established as control, moderate water deficit, and severe water deficit conditions.

2.3. Morphological and Biomass Measurements

Leaves were visually observed for leaf necrosis or chlorosis on rootstock seedlings after 15 days of stress applications. Plant height (cm) and root length (cm) were measured with help of a scale after uprooting the plants at the end of the experiment (15 DAS). Leaf water potential was determined to collect the healthy leaf samples at dawn (Shafqat et al., 2021) by using a pressure chamber (Model 3000, Soil moisture Equipment, Santa Barbara, CA, USA). Root and shoot fresh biomass were weighed on electric balance (TS-200 Digital Electronic Scale), oven-dried (70 °C) for 48 h, and weighed again separately [48]. Shoot and root moisture content were calculated using Equation (1):

$$\text{Moisture content } (\%) = \frac{\text{Fresh weight} - \text{dry weight}}{\text{Fresh weight}} \times 100 \tag{1}$$

2.4. Biochemical and Stress-Associated Biomarker Measurements

2.4.1. Leaf Photosynthetic Pigments Contents

Leaf sample of 0.5 g was cut into small pieces and homogenized by adding 80% acetone (v/v). The extract was transferred to a 15 mL tube [48]. Test tubes were placed in the dark to avoid light for 24 h and filtered through filter paper. The absorbance was determined at 663 nm for chlorophyll a, 647 nm for chlorophyll b, and 470 nm for carotenoid. The Chl *a*, Chl *b*, and carotenoids contents were determined using Equations (2)–(4), respectively.

$$\text{Chl } a \text{ (mg/g fresh weight)} = \frac{(12.7\text{OD}663 - 2.69\text{ OD}645) \times \text{V}}{1000 \times \text{W}} \tag{2}$$

$$\text{Chl } b \text{ (mg/g fresh weight)} = \frac{(22.9\text{OD}645 - 4.68 \text{ OD}663) \times \text{V}}{1000 \times \text{W}} \quad (3)$$

$$\text{Carotenoids (mg/g fresh weight)} = \frac{\text{O.D } 480 + 0.114 \text{ (O.D } 663) - 0.638 \text{ (O.D } 645)}{\text{Em } \times 100} \quad (4)$$

where V = volume of the sample, W = weight of fresh tissue, Em = 2500.

2.4.2. Determination of Proline

Proline contents were assessed by following the acid ninhydrin method [49]. Fresh 0.5 g leaf material was extracted using 10 mL of 3% sulfosalicylic acid (Panreac, Barcelona, Spain) for 30 min. Centrifugation at 2000 *g* for 20 min at 4 °C was done. The 2 mL of filtered aqueous extract was mixed with acid ninhydrin reagent (2 mL), and glacial acetic acid (2 mL), and heated at 100 °C for 1 h. The reaction mixture after cooling was segregated against toluene (4 mL) and the absorbance of the organic phase was recorded at 520 nm using a spectrophotometer (IRMECO UV/VIS Model U2020). The resulting values were related with a standard curve plotted using known amounts of proline (Sigma, St Louis, MO, USA).

2.4.3. Determination of Hydrogen Peroxide (H_2O_2)

The leaf tissues of 0.1 g were ground in 1 mL of trichloroacetic acid solution (0.1%) within an ice bath [50]. After preparation in Eppendorf, the samples were centrifuged at 9719× *g* for 15 min. The supernatant of 500 µL was transferred into a new tube having a mixture of 1 M KI (1000 µL) and 10 mM potassium phosphate buffer (500 µL). Absorbance was read at 390 nm in a UV-1900 spectrophotometer (Eppendorf BioSpectrometer® basic). H_2O_2 content was calculated as µmol g^{-1} DW by comparing the absorbance values against the standard curve of commercial H_2O_2.

2.5. Stem and Leaf Anatomical Evaluation

2.5.1. Plant Material and Experiment

Two highly tolerant (Brazilian sour orange and Gadha dahi) and highly sensitive (Rangpur Poona nucellar and Sunki × bentake) genotypes from the first screening study were selected to screen based on the leaf and stem anatomical study. These selected rootstocks were exposed to the same treatments for 15 days then leaf and stem samples were collected for anatomical studies.

2.5.2. Preservation, Sectioning, Staining, and Mounting

Stem and leaf samples (2 cm) were preserved in a formalin acetic alcohol (FAA) solution containing 5% formalin, 10% acetic acid, 50% ethanol, and 35% distilled water. Thereafter, the preserved material was subsequently transferred to an acetic alcohol solution (acetic acid 25% v/v, ethanol 75%) for the long-term preservation of samples. A free-hand sectioning technique was used to prepare permanent slides of stem and leaf transverse sections cut with the razor blade, and some fine sections were carefully picked up on wash glass for staining and dehydration through a series of washings with ethanol (30%, 50%, and then 70% for 15 min each). For staining, the lignified tissues (xylem vessels and sclerenchyma) were transferred to safranin (1.0 g dissolved in 100 mL, 70% alcohol) for 20 min, dehydrated in 90% alcohol for 5 min, and stained with fast green (1.0 g dissolved in 90% ethanol) for one minute. Finally, the tissues were washed three times with absolute alcohol and then transferred to xylene for cleaning the contrast. The sections were mounted in Canada balsam by putting a drop of resin on a slide and placing the sections on the slides and photographed with a digital camera attached to a compound microscope.

2.5.3. Anatomical Traits

The stem and leaf cross-sectional area (mm^2) was measured under a compound light microscope (Olympus MX63, Japan) by recording the maximum length and width of the root sections. Cells present in plants epidermis were measured with the help of ruler or cm scales, and the length of the epidermal cell was multiplied by 41.5 to obtain a value in micrometers. Vascular bundle length was calculated by measuring xylem and phloem which were multiplied by 41.5 to obtain the value in μm. Xylem thickness was calculated by the width of the xylem region. Metaxylem area, phloem cell area, pith cell area, and the cortical area were calculated in μm^2 by measuring the length and width of each trait. Cortical thickness was obtained by measuring the total width of the cortical region in micrometers. By using Equation (5), the areas of the different cells and tissues were calculated (which was modified from the area of the circle).

$$\text{Area} = \frac{\text{Maximum length} \times \text{Maximum width}}{28} \times 22 \quad (5)$$

2.6. Statistical Analysis

Throughout this study, all experiments were laid out using a full factorial split-plot design arranged in randomized complete blocks using rootstocks as main plots and water regimes in the subplots. All experiments were carried out using at least three biological replicates for each treatment. The analysis of variance (ANOVA) was used to test the significant differences among different drought levels (p_{Drought}), rootstocks ($p_{\text{Rootstocks}}$), and their interaction ($p_{\text{Drought} \times \text{rootstocks}}$). Tukey's honestly significant difference (HSD) test was used for post-hoc analysis ($p_{\text{Drought} \times \text{rootstocks}} < 0.05$). Moreover, the data matrix of all dependent variables was used to perform the principal component analysis (PCA). Finally, means of all dependent variables were used for two-way hierarchical cluster analysis (HCA). Similarities and variations between treatments were presented as a heat map.

3. Results

3.1. Drought Negatively Affects the Roots and Shoots Length

Both root and shoot length were significantly affected by varying drought levels ($p_{\text{Drought}} < 0.0001$) and rootstocks ($p_{\text{Rootstock}} < 0.0001$) (Table 2). Briefly, the root length of different citrus rootstocks was significantly reduced when plants were stressed with different levels of drought. Although control Brazilian sour orange had the highest root length (27.67 ± 0.58 cm; $p_{\text{Drought} \times \text{Rootstock}} < 0.0001$), shortest root lengths were recorded by Sunki × bentake when stressed with moderate (10.33 ± 0.58 cm) or severe drought (9.33 ± 0.58 cm) and Rangpur Poona nucellar rootstock under severe drought conditions (10.33 ± 0.58 cm) (Table 2). On the other hand, while there were no significant differences in shoot length of all tested rootstock, except Gabbuchini, under normal growth conditions, shoot lengths were significantly reduced under various drought levels. Rangpur Poona nucellar rootstock had the shortest shoots (5.33 ± 0.58 cm), followed by Sunki × bentake (6.00 ± 1.00 cm) when stressed under severe drought ($p_{\text{Drought} \times \text{Rootstock}} < 0.0001$) (Table 2).

3.2. Drought Stress Disrupts the Water Relations of Citrus Rootstocks

As strong wilt phenotype was visually observed on citrus seedlings after 15 days post applications (DPA; Figure 1), the water relations (water potential, root moisture content, and shoot moisture content) were assessed (Table 2). For instance, water potential was reduced significantly by water-deficient ($p_{\text{Drought}} < 0.0001$) and rootstocks ($p_{\text{Rootstock}} = 0.0018$). Kirrumakki nucellar rootstock had the highest water potential (−0.11 ± 0.66 Mpa), whereas, Keen sour orange, Rough lemon, Brazilian sour orange, Sunki × bentake, X639, Kirrumakki nucellar, and Rangpur Poona nucellar rootstocks had the lowest water potential when severely stressed with water deficiency, without significant differences between them ($p_{\text{Drought} \times \text{Rootstock}} = 0.0250$; Table 2). Likewise, the moisture content of both root and shoot was significantly reduced by drought levels ($p_{\text{Drought}} < 0.0001$) and rootstocks ($p_{\text{Rootstock}} < 0.0001$) (Table 2). Normal irrigated Brazilian sour orange had the highest moisture content of root (75.67 ± 0.59%) and shoot (84.75 ± 0.65%), whereas Rangpur Poona nucellar rootstock had the lowest moisture content of root (36.33 ± 0.58%) and shoot (40.69 ± 0.65%) under severe drought conditions (Table 2).

Table 2. Effect of different water regimes on the root length (cm), shoot length (cm), and water relations of 10 differentcitrus rootstocks with distinct degrees of tolerance to drought stress.

	Rootstock	Root Length (cm)	Shoot Length (cm)	Water Potential (Mpa)	Moisture Content	
					Root (%)	Shoot (%)
Control	Gabbuchini	25.33 ± 0.58 ab	21.00 ± 1.00 b	−0.32 ± 0.02 a–e	66.00 ± 1.00 cde	73.92 ± 1.12
	Gada dahi	24.33 ± 0.58 bc	24.33 ± 0.58 a	−0.23 ± 0.02 a–d	69.00 ± 1.00 b	77.28 ± 1.12 b
	Sour orange	24.33 ± 0.58 bc	23.67 ± 0.58 a	−0.27 ± 0.06 a–e	64.67 ± 0.58 e	72.43 ± 0.65 e
	Keen sour orange	25.33 ± 0.58 ab	25.33 ± 0.58 a	−0.19 ± 0.02 abc	67.33 ± 0.58 bcd	75.41 ± 0.65 bcd
	Rough lemon	26.00 ± 1.00 ab	25.00 ± 1.00 a	−0.19 ± 0.03 abc	64.67 ± 0.58 e	72.43 ± 0.65 e
	Brazilian sour orange	27.67 ± 0.58 a	25.00 ± 1.00 a	−0.23 ± 0.02 a–d	75.67 ± 0.58 a	84.75 ± 0.65 a
	Sunki × bentake	12.33 ± 0.58 jkl	24.67 ± 0.58 a	−0.28 ± 0.05 a–e	65.33 ± 0.58 de	73.17 ± 0.65 de
	X639	24.33 ± 0.58 bc	24.33 ± 0.58 a	−0.26 ± 0.06 a–e	67.00 ± 1.00 b–e	75.04 ± 1.12 b–e
	Kirrumakki nucellar	24.00 ± 1.00 bc	24.00 ± 1.00 a	−0.23 ± 0.02 a–d	68.00 ± 1.00 bc	76.16 ± 1.12 bc
	Rangpur poona nucellar	24.67 ± 0.58 bc	24.67 ± 0.58 a	−0.23 ± 0.02 a–d	66.00 ± 1.00 cde	73.92 ± 1.12 cde
Moderate drought	Gabbuchini	15.67 ± 0.58 ghi	12.00 ± 1.00 ghi	−0.60 ± 0.05 a–f	55.67 ± 0.58 fg	62.35 ± 0.65 fg
	Gada dahi	22.67 ± 0.58 cd	18.33 ± 0.58 c	−0.15 ± 0.58 ab	64.67 ± 0.58 e	72.43 ± 0.65 e
	Sour orange	19.00 ± 1.00 ef	15.67 ± 0.58 de	−0.61 ± 0.06 a–f	54.00 ± 1.00 g	60.48 ± 1.12 g
	Keen sour orange	21.00 ± 1.00 de	18.33 ± 0.58 c	−0.62 ± 0.14 a–f	58.00 ± 1.00 f	64.96 ± 1.12 f
	Rough lemon	17.33 ± 0.58 fg	14.33 ± 0.58 efg	−0.68 ± 0.03 b–f	54.33 ± 0.58 g	60.85 ± 0.65 g
	Brazilian sour orange	25.67 ± 0.58 ab	19.00 ± 1.00 bc	−0.70 ± 0.03 c–f	65.00 ± 1.00 de	72.80 ± 1.12 de
	Sunki × bentake	10.33 ± 0.58 lm	14.67 ± 0.58 def	−0.66 ± 0.04 b–f	50.33 ± 0.58 h	56.37 ± 0.65 h
	X639	16.33 ± 1.53 gh	13.00 ± 1.00 fgh	−0.74 ± 0.06 def	54.33 ± 0.58 g	60.85 ± 0.65 g
	Kirrumakki nucellar	15.33 ± 0.58 ghi	12.00 ± 1.00 ghi	−0.11 ± 0.66 a	54.00 ± 1.00 g	60.48 ± 1.12 g
	Rangpur poona nucellar	14.67 ± 0.58 hij	11.00 ± 1.00 hij	−0.78 ± 0.03 efg	48.67 ± 0.58 hi	54.51 ± 0.65 hi
Severe drought	Gabbuchini	11.67 ± 1.15 klm	08.33 ± 0.58 kl	−1.30 ± 0.06 gh	46.33 ± 0.58 ij	51.89 ± 0.65 ij
	Gada dahi	19.33 ± 0.58 ef	14.67 ± 0.58 def	−1.14 ± 0.01 fgh	54.67 ± 0.58 g	61.23 ± 0.65 g
	Sour orange	14.67 ± 0.58 hij	11.67 ± 0.58 hi	−1.30 ± 0.05 gh	46.67 ± 0.58 ij	52.27 ± 0.65 ij
	Keen sour orange	17.33 ± 0.58 fg	13.00 ± 1.00 fgh	−1.39 ± 0.07 h	50.00 ± 1.00 h	56.00 ± 1.12 h
	Rough lemon	13.67 ± 0.58 ijk	10.33 ± 0.58 ijk	−1.39 ± 0.04 h	45.00 ± 1.00 jk	50.40 ± 1.12 jk
	Brazilian sour orange	24.33 ± 0.58 bc	17.00 ± 1.00 cd	−1.57 ± 0.03 h	57.67 ± 1.53 f	64.59 ± 1.71 f
	Sunki × bentake	09.33 ± 0.58 lm	06.00 ± 1.00 lm	−1.39 ± 0.03 h	39.67 ± 0.58 l	44.43 ± 0.65 l
	X639	14.00 ± 1.00 hijk	08.33 ± 0.58 kl	−1.36 ± 0.06 h	43.33 ± 0.58 k	48.53 ± 0.65 k
	Kirrumakki nucellar	13.33 ± 0.58 ijk	08.67 ± 0.58 jk	−1.38 ± 0.03 h	45.00 ± 1.00 jk	50.40 ± 1.12 jk
	Rangpur poona nucellar	10.33 ± 0.58 lm	05.33 ± 0.58 m	−1.40 ± 0.03 h	36.33 ± 0.58 m	40.69 ± 0.65 m
	***p*-value**					
	$p_{Drought}$	<0.0001	<0.0001	<0.0001	<0.0001	<0.0001
	$p_{Rootstock}$	<0.0001	<0.0001	=0.0018	<0.0001	<0.0001
	$p_{Drought \times Rootstock}$	<0.0001	<0.0001	=0.0250	<0.0001	<0.0001

Data presented are means ± standard deviation (mean ± SD) of three replicates. Different letters indicate statistically significant differences among treatments, while "ns" signifies no significant differences between them according to Tukey's honestly significant difference test ($p < 0.05$).

Figure 1. Effect of different water regimes on the leaf phenotype of 10 different citrus rootstocks with distinct degrees of tolerance to drought stress.

3.3. *Water Deficiency Interrupts the Photosynthetic Pigments of Citrus Rootstocks*

Drought stress considerably lessened chlorophyll *a*, chlorophyll *b*, and carotenoids content of different rootstocks ($p_{Drought} < 0.0001$; Table 3). Normal irrigated Brazilian sour orange had the highest chlorophyll *a* content (3.74 ± 0.12 mg g^{-1} FW), while Rangpur Poona nucellar rootstock had the lowest chlorophyll *a* content (1.57 ± 0.04% mg g^{-1} FW; $p_{Drought \times Rootstock} < 0.0001$). Furthermore, there was no significant difference in chlorophyll *b* content among all tested rootstocks, except Gabbuchini and Gada dahi, under normal irrigation conditions. However, chlorophyll *b* content was significantly decreased under water deficiency conditions (Table 3). Likewise, there were no significant differences in carotenoids content of all studied citrus rootstocks, except Gabbuchini, under normal irrigation conditions. Nevertheless, carotenoids content was significantly reduced when citrus rootstocks were stressed with different drought levels. It is worth mentioning that Rangpur Poona nucellar rootstock had the lowest carotenoids content (0.14 ± 0.03 mg g^{-1} FW), followed by Sunki × bentake (0.23 ± 0.02 mg g^{-1} FW) under severe drought conditions (Table 3).

Table 3. Effect of different water regimes on the photosynthetic pigments, H_2O_2, and proline content of 10 different citrus rootstocks with distinct degrees of tolerance to drought stress.

	Rootstock	Chlorophyll *a* (mg g^{-1} FW)	Chlorophyll *b* (mg g^{-1} FW)	Carotenoids (mg g^{-1} FW)	H_2O_2 (μmol g^{-1} FW)	Proline (μmol g^{-1} FW)
Control	Gabbuchini	3.52 ± 0.03 b	1.43 ± 0.06 abc	1.05 ± 0.04 ab	43.00 ± 2.65 d	0.33 ± 0.03 c–h
	Gada dahi	3.54 ± 0.01 ab	1.47 ± 0.06 ab	1.13 ± 0.04 a	44.33 ± 5.03 d	0.35 ± 0.01 c–g
	Sour orange	3.53 ± 0.02 b	1.53 ± 0.06 a	1.18 ± 0.04 a	42.33 ± 1.15 d	0.38 ± 0.02 c–f
	Keen sour orange	3.57 ± 0.05 ab	1.50 ± 0.10 a	1.15 ± 0.08 a	46.00 ± 2.00 d	0.29 ± 0.00 fgh
	Rough lemon	3.60 ± 0.08 ab	1.50 ± 0.10 a	1.15 ± 0.08 a	45.33 ± 2.52 d	0.32 ± 0.03 c–h
	Brazilian sour orange	3.74 ± 0.12 a	1.57 ± 0.06 a	1.26 ± 0.12 a	45.00 ± 2.65 d	0.38 ± 0.02 c–f
	Sunki × bentake	3.50 ± 0.05 b	1.53 ± 0.06 a	1.18 ± 0.04 a	44.67 ± 0.58 d	0.25 ± 0.01 ghi
	X639	3.51 ± 0.06 b	1.57 ± 0.06 a	1.21 ± 0.04 a	39.67 ± 6.03 d	0.26 ± 0.00 ghi
	Kirrumakki nucellar	3.57 ± 0.08 ab	1.57 ± 0.06 a	1.26 ± 0.04 a	42.67 ± 1.53 d	0.41 ± 0.01 bcd
	Rangpur poona nucellar	3.68 ± 0.04 ab	1.53 ± 0.06 a	1.18 ± 0.04 a	44.67 ± 2.08 d	0.17 ± 0.02 i–l
Moderate drought	Gabbuchini	2.95 ± 0.04 de	1.23 ± 0.06 cde	0.77 ± 0.08 c–f	64.33 ± 2.08 c	0.18 ± 0.03 i–l
	Gada dahi	3.10 ± 0.01 cde	1.27 ± 0.06 bcd	0.85 ± 0.05 bcd	68.33 ± 3.06 bc	0.29 ± 0.01 fgh
	Sour orange	2.99 ± 0.04 cde	1.03 ± 0.12 efg	0.79 ± 0.09 cde	64.67 ± 1.53 c	0.26 ± 0.04 ghi
	Keen sour orange	3.12 ± 0.02 cde	1.23 ± 0.06 cde	0.82 ± 0.12 cde	62.67 ± 2.08 c	0.23 ± 0.06 hij
	Rough lemon	3.14 ± 0.03 cd	0.93 ± 0.15 fgh	0.72 ± 0.12 c–f	62.00 ± 1.00 c	0.25 ± 0.05 ghi
	Brazilian sour orange	3.19 ± 0.04 c	1.37 ± 0.06 abc	0.87 ± 0.04 bcd	66.00 ± 2.65 c	0.31 ± 0.02 d–h
	Sunki × bentake	2.66 ± 0.05 g	0.80 ± 0.10 hi	0.61 ± 0.05 e–h	66.00 ± 1.73 c	0.14 ± 0.05 jkl
	X639	3.11 ± 0.03 cde	1.13 ± 0.06 def	0.79 ± 0.09 cde	60.67 ± 1.53 c	0.25 ± 0.02 ghi
	Kirrumakki nucellar	2.92 ± 0.03 ef	1.13 ± 0.06 def	0.90 ± 0.12 bc	64.67 ± 1.53 c	0.41 ± 0.00 bcd
	Rangpur poona nucellar	2.54 ± 0.02 gh	0.73 ± 0.06 h–k	0.57 ± 0.07 f–i	64.67 ± 2.52 c	0.08 ± 0.06 l
Severe drought	Gabbuchini	2.20 ± 0.03 jk	0.77 ± 0.06 hij	0.44 ± 0.04 c–f	83.00 ± 2.00 a	0.41 ± 0.01 abc
	Gada dahi	2.45 ± 0.04 hi	0.83 ± 0.06 ghi	0.66 ± 0.01 d–g	83.33 ± 2.08 a	0.40 ± 0.00 cde
	Sour orange	2.31 ± 0.03 ij	0.63 ± 0.06 i–l	0.41 ± 0.04 hij	84.00 ± 2.65 a	0.38 ± 0.06 c–f
	Keen sour orange	2.24 ± 0.06 ijk	0.63 ± 0.06 i–l	0.49 ± 0.04 ghi	82.33 ± 3.06 a	0.34 ± 0.03 c–g
	Rough lemon	2.71 ± 0.25 fg	0.57 ± 0.06 jkl	0.38 ± 0.08 ij	84.00 ± 1.00 a	0.29 ± 0.04 fgh
	Brazilian sour orange	2.70 ± 0.06 g	0.87 ± 0.06 gh	0.69 ± 0.08 c–g	77.00 ± 1.00 ab	0.51 ± 0.06 a
	Sunki × bentake	1.87 ± 0.04 l	0.47 ± 0.06 l	0.23 ± 0.02 jk	85.67 ± 4.16 a	0.18 ± 0.01 ijk
	X639	2.10 ± 0.06 k	0.57 ± 0.06 jkl	0.36 ± 0.04 ij	81.00 ± 6.08 a	0.30 ± 0.03 e–h
	Kirrumakki nucellar	2.10 ± 0.04 k	0.53 ± 0.06 kl	0.41 ± 0.04 hij	82.33 ± 2.31 a	0.51 ± 0.01 ab
	Rangpur poona nucellar	1.57 ± 0.04 m	0.43 ± 0.06 l	0.14 ± 0.03 k	82.00 ± 1.73 a	0.10 ± 0.00 kl
	***p*-value**					
	$p_{Drought}$	<0.0001	<0.0001	<0.0001	<0.0001	<0.0001
	$p_{Rootstock}$	<0.0001	<0.0001	<0.0001	=0.0261	<0.0001
	$p_{Drought \times Rootstock}$	<0.0001	<0.0001	<0.0001	=0.0978	<0.0001

Data presented are means ± standard deviation (mean ± SD) of three replicates. Different letters indicate statistically significant differences among treatments, while "ns" signifies no significant differences between them according to Tukey's honestly significant difference test ($p < 0.05$).

3.4. *Drought Stress Induced the Accumulation of Stress-Associated Biomarkers in Citrus Rootstocks*

Two major stress-associated biomarkers, including H_2O_2 and endogenous proline content, were assessed (Table 3). Generally, both drought levels ($p_{Drought} < 0.0001$) and rootstocks ($p_{Rootstock} = 0.0261$) induced the H_2O_2 accumulation. Plants under drought stress conditions (moderate or severe) had significantly higher H_2O_2 levels, compared with non-stressed rootstocks. It is worth mentioning that all tested rootstocks had their highest H_2O_2 levels under severe drought conditions with no noticeable difference between them (Table 3). Similarly, endogenous proline content was significantly affected by water-deficient ($p_{Drought} < 0.0001$) and citrus rootstock ($p_{Rootstock} < 0.0001$), and its levels were boosted with raising the severity of drought stress. Brazilian sour orange had the highest proline levels (0.51 ± 0.06 μmol g^{-1} FW) under severe drought conditions. Nevertheless, there was no significant difference in proline content of different citrus rootstocks under regular irrigation conditions (Table 3).

3.5. *Principal Component Analysis (PCA) and Two-Way Hierarchical Cluster Analysis (HCA) Revealed the Differences among Water Treatments and Citrus Rootstocks*

To better understand the water deficiency and citrus rootstocks interactions, principal component analysis (PCA) and two-way hierarchical cluster analysis (HCA) were carried out (Figure 2). PCA-associated scatterplot revealed a clear separation among water deficit treatment (control, moderate drought, and severe drought), as well as all studied rootstocks with respect to PC1 (approximately 81.28%) and PC2 (about 11.94%) (Figure 2A). Moreover, the PCA-associated loading plot showed that while root length, root moisture content, shoot moisture content, shoot length, carotenoids, chlorophyll a, chlorophyll b, and water potential were positively correlated with non-stressed control plants, H_2O_2, and proline content were positively associated with water deficit treatments (Figure 2B).

In harmony with PCA findings, the HCA and its associated heatmap revealed the differences among water-deficient treatments (Figure 2C). For example, HCA-associated dendrogram among rootstocks revealed that all studied rootstocks were clustered separately into three distinct clusters. Cluster (a) included all non-stressed rootstocks, Cluster (b) included all moderate drought-stressed rootstocks except Rangpur Poona nucellar and Sunki × bentake which were clustered with severely stressed rootstocks within Cluster (c). Additionally, the HCA-associated dendrogram among investigated variables revealed that they were clustered into two separate clusters. Cluster 'I' included root length, root moisture content, shoot moisture content, shoot length, carotenoids, chlorophyll a, chlorophyll b, and water potential which all were higher in regularly irrigated non-stressed control rootstocks. On the other hand, Cluster 'II' included only H_2O_2 and proline contents which were higher in citrus rootstocks grown under severe deficient water stress (Figure 2C).

3.6. *Water Deficiency Alters the Anatomical Structure of Citrus Rootstocks*

To better understand the mechanism of drought resistance in citrus, the effect of different drought levels on the anatomical structure of stem and leaves of two highly tolerant (Brazilian sour orange and Gadha dahi) and two highly sensitive genotypes (Rangpur Poona nucellar and Sunki × bentake) from the first screening study was investigated. As we mentioned above, under drought stress conditions, the moisture content of both roots and shoots was reduced, resulting in yellow, curled, wilted leaves, and some other adverse wilt-associated symptoms (Figure 1). Briefly, under severe drought stress, the length of epidermal cells and vascular bundles was reduced, as well as the thickness of the xylem and cortical was thinner.

3.6.1. Effect of Drought Stress on Stem Anatomy of Citrus Rootstocks

Microscopic observation of stem cross-section showed that the highly tolerant genotypes (Brazilian sour orange and Gadha dahi) had some features such as water-filled cells, small cell gaps, tight and round cells under normal water conditions, as well as under drought stress conditions (Figure 3).

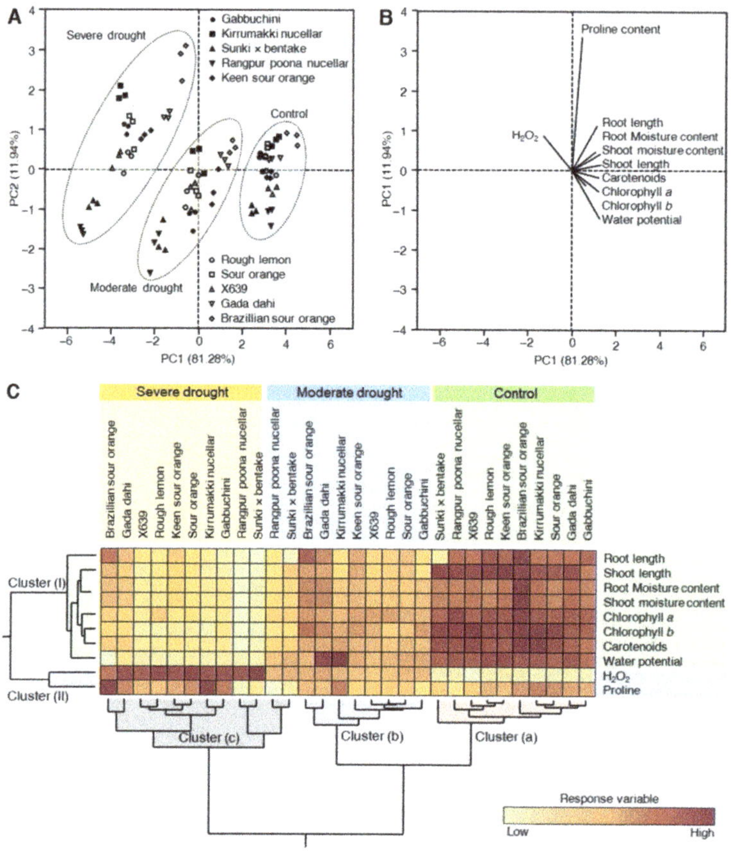

Figure 2. Principal component analysis (PCA) and two-way hierarchical cluster analysis (HCA) of individual morphological and physiological parameters were assessed in 10citrus rootstocks with distinct degrees of tolerance to drought stress under three water regimes. (**A**) PCA-associated scatterplots, (**B**) PCA-associated loading plots, and (**C**) two-way HCA. Variations in the dependent variables among studied treatments are visualized as a heat map. Rows correspond to dependent variables, whereas columns correspond to different treatments. Low numerical values are light-yellow-colored, while high numerical values are colored dark red (see the scale at the right bottom corner of the heat map).

However, both drought levels and rootstocks ($p_{\text{Drought}} < 0.0001$ and $p_{\text{Rootstock}} = 0.0002$, respectively) significantly affected the length of the epidermal cell and vascular bundle (Figure 4A,B, respectively), xylem thickness (Figure 4C), area of metaxylem cell (Figure 4D), phloem cell (Figure 4E), pith cell (Figure 4F), pith thickness (Figure 4G), and cortical cell (Figure 4H), and cortical thickness (Figure 4I). Under all water conditions, the anatomical structures of the highly tolerant genotypes Brazilian sour orange and Gadha dahi performed better than two highly sensitive genotypes Rangpur Poona nucellar and Sunki × bentake. For instance, under regular irrigation, both Brazilian sour orange and Gadha dahi rootstocks had the highest epidermal cell length ($p_{\text{Drought} \times \text{Rootstock}} = 0.0255$), vascular bundle length ($p_{\text{Drought} \times \text{Rootstock}} < 0.0001$), xylem thickness ($p_{\text{Drought} \times \text{Rootstock}} = 0.488$), metaxylem cell area ($p_{\text{Drought} \times \text{Rootstock}} = 0.0140$), phloem cell area ($p_{\text{Drought} \times \text{Rootstock}} = 0.0346$), pith cell area ($p_{\text{Drought} \times \text{Rootstock}} = 0.0316$), pith thickness area ($p_{\text{Drought} \times \text{Rootstock}} = 0.0112$), cortical cell area ($p_{\text{Drought} \times \text{Rootstock}} = 0.0180$), and cortical thickness ($p_{\text{Drought} \times \text{Rootstock}} = 0.0319$) with no significant differences between them. Although, both moderate and severe drought stresses significantly reduced most, if not all, of these anatomical attributes, both highly tolerant genotypes did not show noticeable changes under drought stress. Nevertheless, the mesophyll cells of drought-sensitive Rangpur Poona nucellar and Sunki × bentake genotypes were slightly deformed and had shorter epidermal cell and vascular bundle length, as well as narrower metaxylem, phloem, pith, and cortical area (Figure 4).

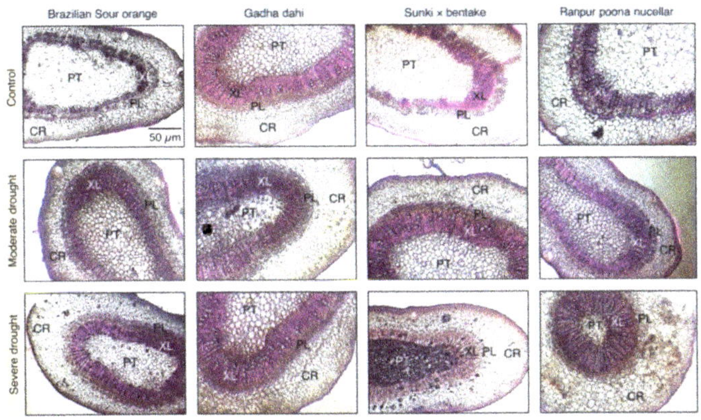

Figure 3. Stem transverse section of two highly tolerant (Brazilian sour orange and Gadha dahi) and two highly sensitive genotypes (Rangpur Poona nucellar and Sunki × bentake). PT: pith; CR: cortex; XL: xylem; PL: phloem.

3.6.2. PCA and Two-Way HCA Divulged the Variations in Stem Anatomy of Different Citrus Rootstocks

Briefly, the PCA-associated scatter plot showed a clear separation among water deficit treatment (control, moderate drought, and severe drought), as well as all studied rootstocks with respect to PC1 (approximately 87.29%) and PC2 (about 8.41%) (Figure 5A). It is worth mentioning that the drought-tolerant rootstocks Brazilian sour orange and Gadha dahi were grouped close to each other and separately from the two sensitive genotypes Rangpur Poona nucellar and Sunki × bentake under normal irrigation and moderate drought, but not severe drought conditions. Furthermore, the PCA-associated loading plot showed that all studied stem anatomical features were positively associated with normal water application (Figure 5B). In agreement with PCA results, the HCA and its associated heatmap uncovered the differences among different rootstocks under water-deficient treatments (Figure 5C). For example, HCA-associated dendrogram among rootstocks revealed that all studied rootstocks separated into two distinct clusters. Cluster (a) included all non-stressed rootstocks and the highly tolerant (Brazilian sour orange and Gadha dahi), grown under moderate drought, whereas cluster (b) included all severely stressed rootstocks and the two highly sensitive genotypes (Rangpur Poona nucellar and Sunki × bentake) that were moderately stressed with drought (Figure 5C).

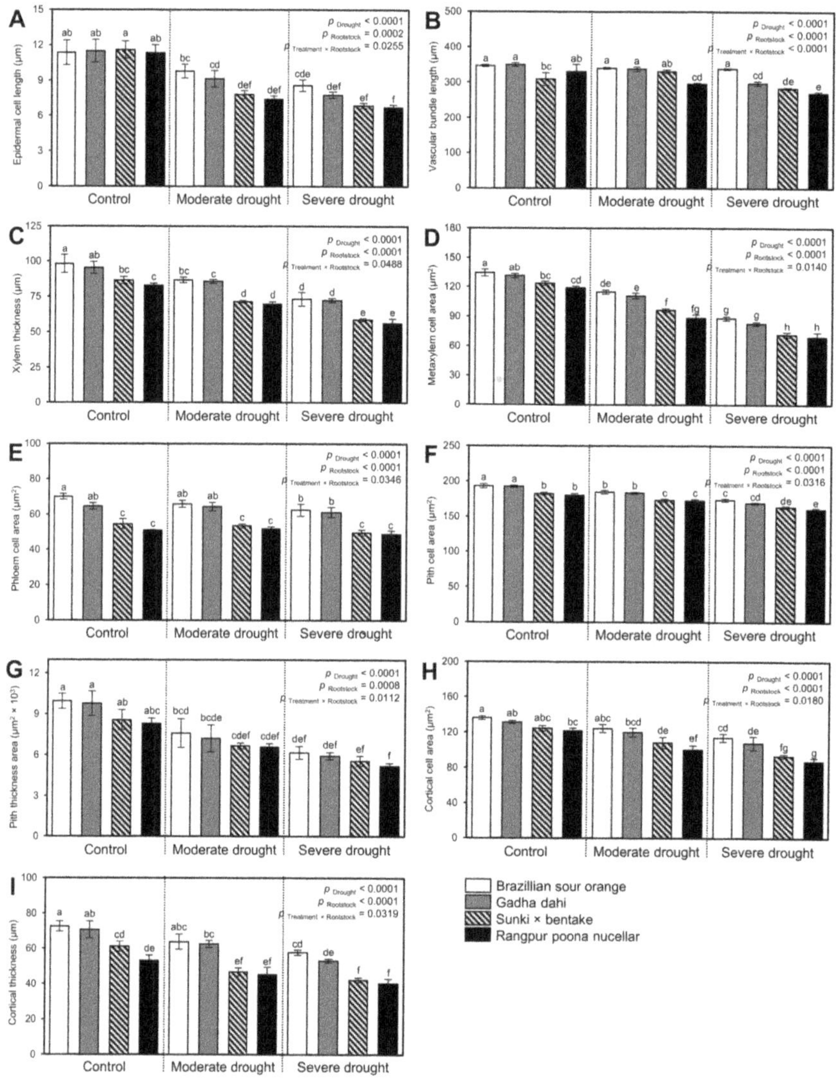

Figure 4. Effect of different water regimes on the stem anatomical features of two highly tolerant (Brazilian sour orange and Gadha dahi) and two highly sensitive genotypes (Rangpur Poona nucellar and Sunki × bentake). (**A**) Epidermal cell length (µm), (**B**) Vascular bundle length (µm), (**C**) Xylem thickness (µm), (**D**) Metaxylem cell area (µm^2), (**E**) Phloem cell area (µm^2), (**F**) Pith cell area (µm^2), (**G**) Pith thickness area (µm^2 × 10^3), (**H**) Cortical cell area (µm^2), and (**I**) Cortical thickness (µm). Data presented are means ± standard deviation (mean ± SD) of three biological replicates. Different letters indicate statistically significant differences among treatments, while "ns" signifies no significant differences between them according to Tukey's honestly significant difference test ($p < 0.05$).

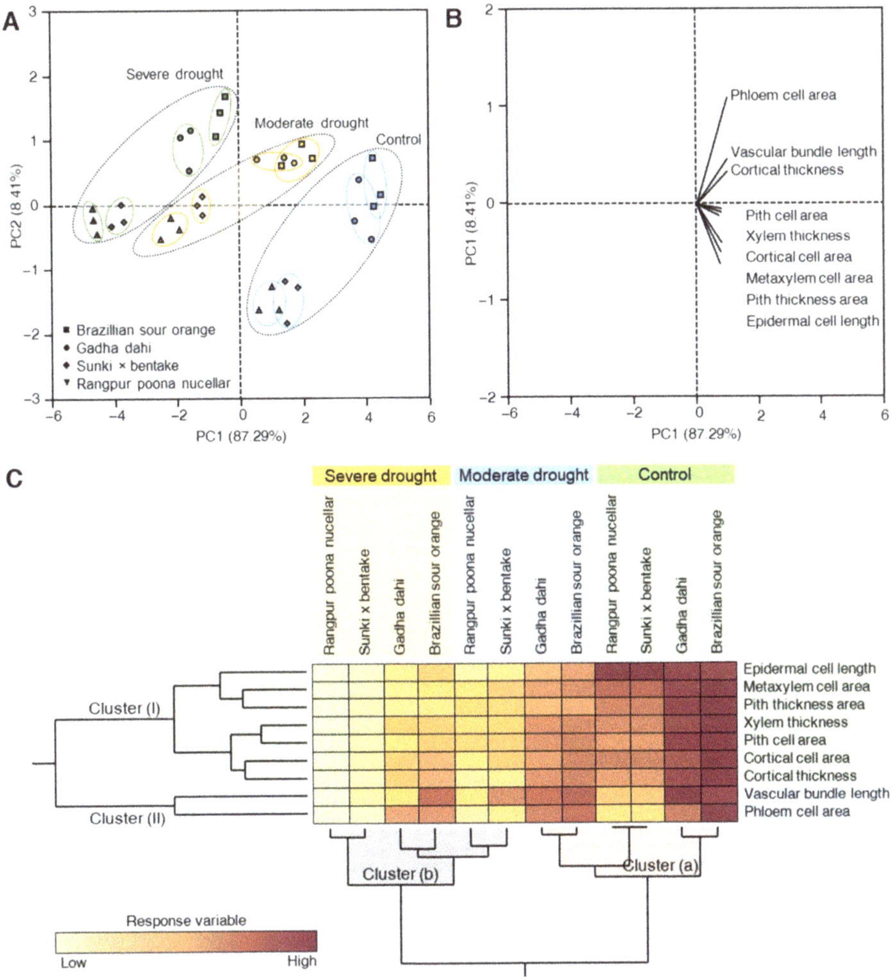

Figure 5. Principal component analysis (PCA) and two-way hierarchical cluster analysis (HCA) of individual stem anatomical features of two highly tolerant (Brazilian sour orange and Gadha dahi) and two highly sensitive genotypes (Rangpur Poona nucellar and Sunki × bentake) under three water regimes. (**A**) PCA-associated scatter plots, (**B**) PCA-associated loading plots, and (**C**) two-way HCA. Variations in the dependent variables among studied treatments are visualized as a heat map. Rows correspond to dependent variables, whereas columns correspond to different treatments. Low numerical values are light-yellow-colored, while high numerical values are colored dark red (see the scale at the right bottom corner of the heat map).

3.6.3. Effect of Drought Stress on Leaf Tissue Structure of Citrus Rootstocks

Microscopic observation of citrus leaves cross-section showed that it had an asymmetric heterogeneous structure that was characterized by two unequal palisade parenchyma (Figure 6).

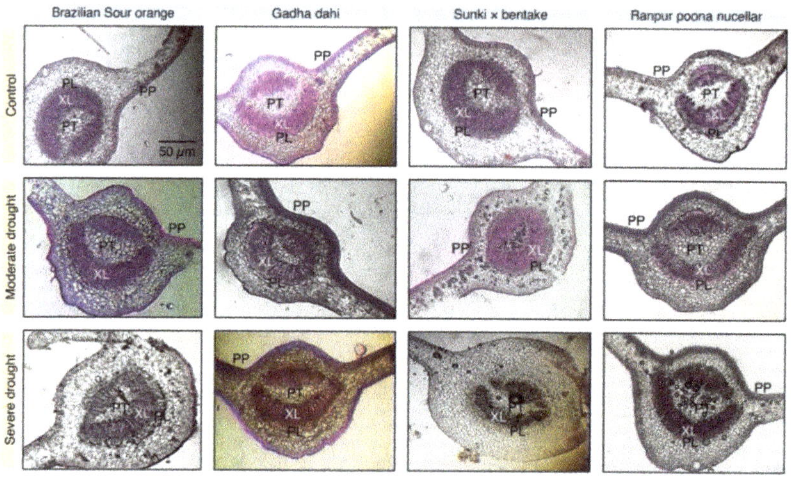

Figure 6. Leaf transverse section of two highly tolerant (Brazilian sour orange and Gadha dahi) and two highly sensitive genotypes (Rangpur Poona nucellar and Sunki × bentake). PT: pith; CR: cortex; XL: xylem; PL: phloem; PP: palisade parenchyma.

Like stem anatomy, citrus leaf anatomical attributes were significantly altered by both drought levels and rootstocks. However, the anatomical changes in leaf tissue structure were less significant in highly tolerant rootstocks (Brazilian sour orange and Gadha dahi) than sensitive genotypes (Rangpur Poona nucellar and Sunki × bentake). Interestingly, differences in all studied anatomical features of citrus leaves under different water regimes were cultivar-dependent. These features included epidermal cell length ($p_{\text{Drought} \times \text{Rootstock}}$ = 0.0315; Figure 7A), vascular bundle length ($p_{\text{Drought} \times \text{Rootstock}}$ = 0.0009; Figure 7B), xylem thickness ($p_{\text{Drought} \times \text{Rootstock}}$ = 0.0251; Figure 7C), metaxylem cell area ($p_{\text{Drought} \times \text{Rootstock}}$ = 0.0349; Figure 7D), phloem cell area ($p_{\text{Drought} \times \text{Rootstock}}$ = 0.0197; Figure 7E), pith cell area ($p_{\text{Drought} \times \text{Rootstock}}$ = 0.0163; Figure 7F), pith thickness area ($p_{\text{Drought} \times \text{Rootstock}}$ = 0.0335; Figure 7G), cortical cell area ($p_{\text{Drought} \times \text{Rootstock}}$ = 0.0250; Figure 7H), and cortical thickness ($p_{\text{Drought} \times \text{Rootstock}}$ = 0.0357; Figure 7I). Under all tested water regimes, highly tolerant rootstocks Brazilian sour orange and Gadha dahi had thicker epidermal and vascular bundle, as well as wider pith and cortical areas compared with sensitive genotypes (Rangpur Poona nucellar and Sunki × bentake). Additionally, severe drought stress significantly reduced the thickness of all leaf tissues, particularly in sensitive genotypes (Figure 7).

3.6.4. PCA and Two-Way HCA Revealed the Differences in Leaf Tissue Structure of Different Citrus Rootstocks

In brief, the PCA-associated scatter plot showed a clear separation among water deficit treatment (control, moderate drought, and severe drought), as well as all studied rootstocks with respect to PC1 (approximately 93.24%) and PC2 (about 3.95%) (Figure 8A). It is worth mentioning that the drought-tolerant rootstocks Brazilian sour orange and Gadha dahi were grouped close to each other and separately from the two sensitive genotypes Rangpur Poona nucellar and Sunki × bentake under all investigated water regimes. Moreover, the PCA-associated loading plot showed that all studied anatomical features of citrus leaves were positively associated with normal water application (Figure 8B). Like PCA, the HCA and its associated heatmap revealed the differences among different rootstocks under water-deficient treatments (Figure 8C). For example, HCA-associated dendrogram among rootstocks revealed that all studied rootstocks separated into two distinct clusters. Cluster (a) included all regularly irrigated genotypes and the two highly tolerant rootstocks (Brazilian sour orange and Gadha dahi) that were grown under moderate drought. On the other hand, cluster (b) included all severely stressed genotypes and the two sensitive rootstocks (Rangpur Poona nucellar and Sunki × bentake) that were moderately stressed with drought (Figure 8C).

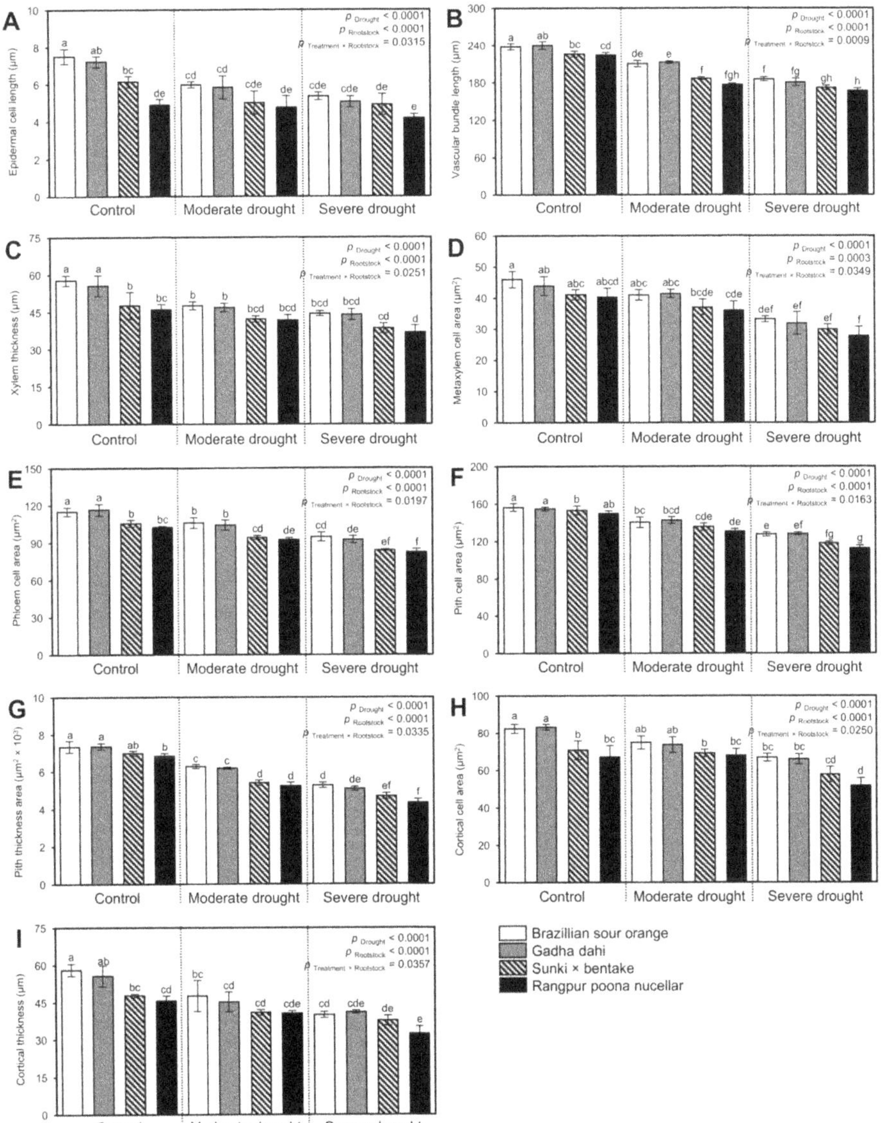

Figure 7. Effect of different water regimes on the leaf anatomical features of different two highly tolerant (Brazilian sour orange and Gadha dahi) and two highly sensitive genotypes (Rangpur Poona nucellar and Sunki × bentake). (**A**) Epidermal cell length (μm), (**B**) Vascular bundle length (μm), (**C**) Xylem thickness (μm), (**D**) Metaxylem cell area (μm^2), (**E**) Phloem cell area (μm^2), (**F**) Pith cell area (μm^2), (**G**) Pith thickness area (μm^2 × 10^3), (**H**) Cortical cell area (μm^2), and (**I**) Cortical thickness (μm). Data presented are means ± standard deviation (mean ± SD) of three biological replicates. Different letters indicate statistically significant differences among treatments, while "ns" signifies no significant differences between them according to Tukey's honestly significant difference test ($p < 0.05$).

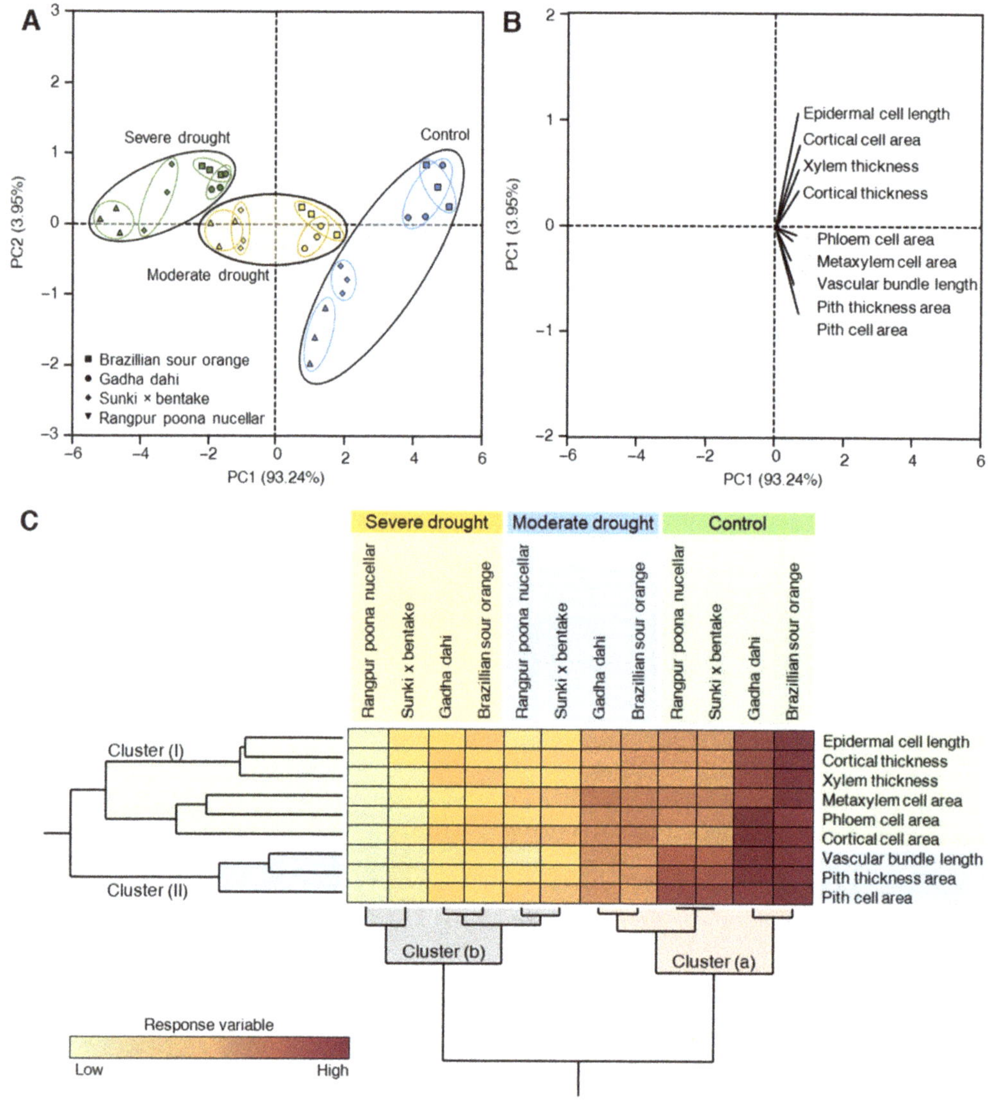

Figure 8. Principal component analysis (PCA) and two-way hierarchical cluster analysis (HCA) of individual leaf anatomical features of two highly tolerant (Brazilian sour orange and Gadha dahi) and two highly sensitive genotypes (Rangpur Poona nucellar and Sunki × bentake) under three water regimes. (**A**) PCA-associated scatter plots, (**B**) PCA-associated loading plots, and (**C**) two-way HCA. Variations in the dependent variables among studied treatments are visualized as a heat map. Rows correspond to dependent variables, whereas columns correspond to different treatments. Low numerical values are light-yellow-colored, while high numerical values are colored dark red (see the scale at the right bottom corner of the heat map).

4. Discussion

Water deficit conditions are a major environmental factor, which frequently limits the growth and productivity of important crop species [17]. Restriction of water supply can severely limit plant growth, development, and production [18,19]. Choice of rootstock is among the most important decisions a grower makes, and implications for yield and quality are enormous. Rootstock in citrus trees influences the morphological, biochemical, physiological, and genetic characteristics of grafted scion cultivars through the rootstock scion interaction pathway [51]. Citrus rootstocks with better drought tolerance ability can greatly reduce production losses [52]. In this study, plant material consisted of 10 genetically diverse citrus rootstocks belonging to different citrus categories, i.e., oranges, pummelo, lemon, lime, their hybrids, and originating from diverse localities. The leaf shape and size of these rootstocks also varied. These rootstocks are reported to have different tolerance towards some biotic and abiotic stresses. The drought tolerance of these rootstocks was studied in this investigation. Leaf water potential in plants is directly related to water availability [53]. Leaf water potential indicates the whole plant water status, and maintenance of high leaf water potential is found to be associated with dehydration avoidance mechanisms. Our results demonstrated the decrease in leaf water potential as drought conditions become severe compared with control. The maintenance of water potential in leaves is a direct indicator of a plant dehydration avoidance mechanism, as genotypes with better water potential at stress conditions are regarded as drought tolerant [23,54,55]. In our studies, Brazilian sour orange showed drought tolerance by performing best at severe drought and Rangpur Poona nucellar at moderate drought. For insensitive genotypes, the decrease in leaf water potential indicated the mechanical injury of leaf chloroplasts caused by stress conditions which result in reduced transpiration rate and oxidative stress [16].

Citrus rootstocks with high chlorophyll *a* and *b*, and carotenoid contents against the stresses, especially the water stress, are regarded as tolerant rootstocks [56]. The normal functioning of photosynthetic machinery is affected by drought stress, the degradation, and photo-oxidation of chlorophyll caused by transpirational imbalance at water stress hamper the plant's ability to harvest light reducing total photosynthetic output [21,57]. Results showed that photosynthetic pigments chlorophyll *a*, chlorophyll *b*, and carotenoids reduced significantly at elevated stress conditions and overall genotype Brazilian sour orange and Gada dahi had the highest chlorophyll contents at drought conditions highlighting their ability to tolerate drought stress. Plants with dark green leaves (chlorophyll) under drought stress are considered tolerant. Visual assessments indicated the Brazilian sour orange as a tolerant rootstock without changes in leaf green color and leaf necrosis; while, Savage citrange emerged as the most sensitive, with maximum plant death and leaf shedding during stress treatments. The compromised photosynthetic machinery reduces carbohydrate transport and as a result plant growth is also reduced. The ability of plants to maintain growth under limited water supply reveals their tolerance ability [43,58]. Results showed that among citrus rootstocks, Brazilian sour orange and Gada dahi at water stress conditions maintained steady root and shoot growth. The root and shoot moisture content of these rootstocks were also high at stress conditions showing their tolerant nature. While, Rangpur Poona nucellar had the lowest shoot and root growth and moisture content emerged as the most sensitive.

Metabolic imbalances triggered by drought stress cause oxidative stress and as a result ROS are produced and accumulated [59]. The increased oxidation greatly reduces metabolic activities and the normal functioning of cell organelles. To combat oxidative stress, plants also have a built-in antioxidant defense mechanism. Proline is an osmoprotectant that is trigged as a result of ROS production in the cell, its production and accumulation work as ROS scavenging, redox balance, and reduce cell damage which normalizes the functionality of plant cells [60]. In drought-tolerant genotypes, the ROS production is reduced, and proline concentration is increased with increasing severity of stress. The results show that Brazilian sour orange has minimal ROS production and the highest proline concentration at severe drought stress.

In citrus under drought stress, leaves are observed to be shorter with thick epidermal cells which facilitates reducing the transpirational rate and oxidative stress [61,62]. Modifications in vascular anatomy are important for plant acclimation potential. Vascular bundles present in mid rid of leaves serve as a source to distribute nutrients and water. At stress conditions with reduced leaf size, the reduction in the vascular bundle is an indicator of plants' abilities to modify their anatomy under stress. In the vascular bundle, the xylem acts as a source of water transport. The plants with greater xylem vessel diameter are unable to survive harsh environmental conditions [63–65]. At the onset of drought, stress transpiration, water uptake from roots, and stem hydraulic capacitance begin to decline. That reduces growth, the vascular bundles in the stem are observed to be reduced along with pith cell area and cortical thickness [66,67]. In this study, the results showed that with increasing drought intensity the anatomy of both sensitive and tolerant genotypes was modified, interestingly the two tolerant genotypes Brazilian sour orange and Gada dahi had greater values for all the leaf and stem anatomy parameters at severe drought stress than sensitive genotypes. This could be because of the continuous adaptability of tolerant genotypes which enabled them to maintain growth and function as the amount of water became limited, in response to sensitive genotypes in which the response could have been triggered at very later stages which abruptly affect their growth. These modifications in tolerant genotypes enabled them to maintain steady nutrient transport while reducing the risk of embolisms, increasing water-flow resistance, and constant transport of nutrients across [64].

5. Conclusions

Drought stress adversely affected plant water status, photosynthetic machinery, biochemical balance, and anatomical structure of all the citrus rootstocks studied. The intensifying drought reduced leaf water potential, and compromised the photosynthetic apparatus by damaging photosynthetic pigments (chlorophyll "*a*", "*b*", and carotenoid) apparent from lighter green color. Oxidative stress caused by ROS production which triggered production of proline. Alteration in anatomical structures of leaf and stem were observed. Citrus rootstocks Brazilian sour orange and Gada dahi performed best under drought stress, mitigated damage at molecular biochemical and anatomical levels, while rootstocks Sunki × bentake and Rangpur Poona nucellar were the most sensitive rootstocks.

Author Contributions: Conceptualization, M.J.J., S.A.N., M.H. and Y.N., methodology, W.S., M.J.J., Y.S.A.M. and M.H.; software, W.S., S.A.N., Y.N. and S.I.; formal analysis, Y.S.A.M., W.S., S.-u.-R. and Y.N.; resources, M.J.J., M.H., S.A.N. and Y.S.A.M.; data curation, S.-u.-R., S.B., S.I. and W.S.; writing—original draft preparation, W.S., S.I., S.-u.-R. and Y.N.; writing—review and editing, M.J.J., Y.S.A.M., S.A.N., Y.N. and M.H.; supervision, M.J.J. All authors have read and agreed to the published version of the manuscript.

Funding: We would like to acknowledge the Higher Education Commission, Islamabad and Endowment Fund Secretariat, University of Agriculture, Faisalabad, Pakistan for providing funding. The authors extend their appreciation to the Deanship of Scientific Research at King Khalid University for funding this work through the Program of Research Groups under grant number (RGP 2/165/42).

Institutional Review Board Statement: Not applicable.

Informed Consent Statement: Not applicable.

Data Availability Statement: The data that supports the findings of this study are contained within the article and available from the corresponding author upon reasonable request.

Acknowledgments: Y.N. would like to extend his appreciation to the Graduate Student and Research Affairs Sector of Tanta University, Egypt. We also thank all staff members of our laboratories for their helpful discussions and comments. The authors extend their appreciation to the Deanship of Scientific Research at King Khalid University for funding this work through the Program of Research Groups under grant number (RGP 2/165/42).

Conflicts of Interest: The authors declare that there is no conflict of interest, and they have no known competing financial interests or personal relationships that could have appeared to influence the work reported in this paper.

References

1. Mehdi, M.; Ashfaq, M.; Hassan, S.; Abid, M. Effect of marketing channel choice on the profitability of citrus farmers: Evidence form Punjab-Pakistan. *Pak. J. Agric. Sci.* **2019**, *56*, 1003–1011.
2. Talat, H.; Shafqat, W.; Qureshi, M.A.; Sharif, N.; Raza, M.K.; Din, S.; Ikram, S.; Jaskani, M.J. Effect of gibberellic acid on fruit quality of Kinnow mandarin. *J. Glob. Innov. Agric. Soc. Sci.* **2020**, *8*, 59–63. [CrossRef]
3. Silva, S.F.; Miranda, M.T.; Costa, V.E.; Machado, E.C.; Ribeiro, R.V. Sink strength of citrus rootstocks under water deficit. *Tree Physiol.* **2021**, *41*, 1372–1383. [CrossRef] [PubMed]
4. Qureshi, M.A.; Jaskani, M.J.; Khan, A.S.; Ahmad, R. Influence of endogenous plant hormones on physiological and growth attributes of Kinnow mandarin grafted on nine rootstocks. *J. Plant Growth Regul.* **2021**, 1–11. [CrossRef]
5. Qureshi, M.A.; Jaskani, M.J.; Khan, A.S.; Haider, M.S.; Shafqat, W.; Asif, M.; Mehmood, A. Influence of different rootstocks on physico-chemical quality attributes of Kinnow mandarin. *Pak. J. Agric. Sci.* **2021**, *58*, 929–935.
6. Khan, K.; Ikram, S.; Ashfaq, M.; Jaskani, M.J.; Shafqat, W. Citrus rootstock characterization against citrus canker and evaluation of antibiotics effect against *Xanthomonas axonopodis* pv. Citri. *J. Innov. Agric.* **2021**, *8*, 1–8.
7. Shafqat, W.; Tahir, T.; Khurshid, T.; Ur-Rahman, H.; Saqib, M.; Jaskani, M.J. Effect of rootstock types on leaf nutrient composition in three commercial citrus scion cultivars of Pakistan under the ASLP Citrus Project. In Proceedings of the XXIX International Horticultural Congress on Horticulture: Sustaining Lives, Livelihoods and Landscapes (IHC2014): 1128, Brisbane, Australia, 25 November 2016.
8. Bowman, K.D.; Joubert, J. Citrus rootstocks. In *The Genus Citrus*; Woodhead Publishing Books—Elsevier: Sawston, UK, 2020; pp. 105–127.
9. Morianou, G.; Ziogas, V.; Kourgialas, N.N.; Karatzas, G.P. Effect of irrigation practices upon yield and fruit quality of four grapefruit (*Citrus paradisi* Mac.) cultivars. *Water Supply* **2021**, *21*, 2735–2747. [CrossRef]
10. Koshita, Y.; Takahara, T. Effect of water stress on flower-bud formation and plant hormone content of satsuma mandarin (*Citrus unshiu* Marc.). *Sci. Hortic.* **2004**, *99*, 301–307. [CrossRef]
11. Syvertsen, J.; Hanlon, E.A. Citrus tree stresses: Effects on growth and yield. *EDIS* **2008**, *2008*, 1–6.
12. Zaman, L.; Shafqat, W.; Qureshi, A.; Sharif, N.; Raza, K.; ud Din, S.; Ikram, S.; Jaskani, M.J. Effect of foliar spray of zinc sulphate and calcium carbonate on fruit quality of Kinnow mandarin (*Citrus reticulata* Blanco). *J. Glob. Innov. Agric. Soc. Sci.* **2019**, *7*, 157–161. [CrossRef]
13. Bray, E.A.; Bailey-Serres, J.; Weretilnyk, E. Responses to abiotic stress. In *Biochemistry & Molecular Biology of Plants*; Gruissem, W., Jones, R., Eds.; American Society of Plant Physiologists: Rockville, MD, USA, 2000; pp. 1158–1203.
14. Bates, B.; Kundzewicz, Z.; Wu, S. *Climate Change and Water*; Intergovernmental Panel on Climate Change Secretariat: Geneva, Switzerland, 2008; ISBN 9291691232.
15. Ramírez, D.A.; Rolando, J.L.; Yactayo, W.; Monneveux, P.; Mares, V.; Quiroz, R. Improving potato drought tolerance through the induction of long-term water stress memory. *Plant Sci.* **2015**, *238*, 26–32. [CrossRef] [PubMed]
16. Zandalinas, S.I.; Rivero, R.M.; Martínez, V.; Gómez-Cadenas, A.; Arbona, V. Tolerance of citrus plants to the combination of high temperatures and drought is associated to the increase in transpiration modulated by a reduction in abscisic acid levels. *BMC Plant Biol.* **2016**, *16*, 105. [CrossRef] [PubMed]
17. Barnabás, B.; Jäger, K.; Fehér, A. The effect of drought and heat stress on reproductive processes in cereals. *Plant Cell Environ.* **2008**, *31*, 11–38. [CrossRef] [PubMed]
18. Chen, Y.; Müller, F.; Rieu, I.; Winter, P. Epigenetic events in plant male germ cell heat stress responses. *Plant Reprod.* **2016**, *29*, 21–29. [CrossRef] [PubMed]
19. Li, X.; Liu, F. Drought stress memory and drought stress tolerance in plants: Biochemical and molecular basis. In *Drought Stress Tolerance in Plants*; Springer: New York, NY, USA, 2016; Volume 1, pp. 17–44.
20. Kinoshita, T.; Seki, M. Epigenetic memory for stress response and adaptation in plants. *Plant Cell Physiol.* **2014**, *55*, 1859–1863. [CrossRef]
21. Shafqat, W.; Jaskani, M.; Maqbool, R.; Sattar Khan, A. Evaluation of citrus rootstocks against drought, heat and their combined stress based on growth and photosynthetic pigments fingerprinting of Jamun (*Syzygium cumini*) genetic resources of Punjab view project. *Int. J. Agri. Biol.* **2019**, *22*, 1001–1009.
22. Chaves, M.M.; Flexas, J.; Pinheiro, C. Photosynthesis under drought and salt stress: Regulation mechanisms from whole plant to cell. *Ann. Bot.* **2009**, *103*, 551–560. [CrossRef]
23. Xiong, B.; Wang, Y.; Zhang, Y.; Ma, M.; Gao, Y.; Zhou, Z.; Wang, B.; Wang, T.; Lv, X.; Wang, X. Alleviation of drought stress and the physiological mechanisms in Citrus cultivar (Huangguogan) treated with methyl jasmonate. *Biosci. Biotechnol. Biochem.* **2020**, *84*, 1958–1965. [CrossRef]
24. Liu, S.; Yang, R. Regulations of reactive oxygen species in plants abiotic stress: An integrated overview. In *Plant Life under Changing Environment: Responses and Management*; Tripathi, D.K., Singh, V.P., Chauhan, D.K., Sharma, S., Prasad, S.M., Dubey, N.K., Ramawat, N., Eds.; Academic Press-Elsevier: Cambridge, MA, USA, 2020; pp. 323–353.
25. Noctor, G.; Veljovic-Jovanovic, S.; Driscoll, S.; Novitskaya, L.; Foyer, C.H. Drought and oxidative load in the leaves of C3 plants: A predominant role for photorespiration? *Ann. Bot.* **2002**, *89*, 841–850. [CrossRef] [PubMed]

26. Guo, Y.Y.; Yu, H.Y.; Yang, M.M.; Kong, D.S.; Zhang, Y.J. Effect of drought stress on lipid peroxidation, osmotic adjustment and antioxidant enzyme activity of leaves and roots of *Lycium ruthenicum* Murr. seedling. *Russ. J. Plant Physiol.* **2018**, *65*, 244–250. [CrossRef]
27. Baker, N.R.; Rosenqvist, E. Applications of chlorophyll fluorescence can improve crop production strategies: An examination of future possibilities. *J. Exp. Bot.* **2004**, *55*, 1607–1621. [CrossRef] [PubMed]
28. de Matos Nunes, J.; Bertodo, L.O.O.; Da Rosa, L.M.G.; Von Poser, G.L.; Rech, S.B. Stress induction of valuable secondary metabolites in *Hypericum polyanthemum* acclimatized plants. *S. Afr. J. Bot.* **2014**, *94*, 182–189. [CrossRef]
29. Zaefyzadeh, M.; Quliyev, R.A.; Babayeva, S.M.; Abbasov, M.A. The effect of the interaction between genotypes and drought stress on the superoxide dismutase and chlorophyll content in durum wheat landraces. *Turk. J. Biol.* **2009**, *33*, 1–7.
30. Osmolovskaya, N.; Shumilina, J.; Kim, A.; Didio, A.; Grishina, T.; Bilova, T.; Keltsieva, O.A.; Zhukov, V.; Tikhonovich, I.; Tarakhovskaya, E. Methodology of drought stress research: Experimental setup and physiological characterization. *Int. J. Mol. Sci.* **2018**, *19*, 4089. [CrossRef]
31. Nyachiro, J.M.; Briggs, K.G.; Hoddinott, J.; Johnson-Flanagan, A.M. Chlorophyll content, chlorophyll fluorescence and water deficit in spring wheat. *Cereal Res. Commun.* **2001**, *29*, 135–142. [CrossRef]
32. Kyparissis, A.; Grammatikopoulos, G.; Manetas, Y. Leaf morphological and physiological adjustments to the spectrally selective shade imposed by anthocyanins in *Prunus cerasifera*. *Tree Physiol.* **2007**, *27*, 849–857. [CrossRef] [PubMed]
33. Liang, X.; Zhang, L.; Natarajan, S.K.; Becker, D.F. Proline mechanisms of stress survival. *Antioxid. Redox Signal.* **2013**, *19*, 998–1011. [CrossRef] [PubMed]
34. Verbruggen, N.; Hermans, C. Proline accumulation in plants: A review. *Amino Acids* **2008**, *35*, 753–759. [CrossRef]
35. Matsuda, K.; Rayan, A. Anatomy: A key factor regulating plant tissue response to water stress. In *Environment Injury to Plants*; Kafternan, F., Ed.; Academic Press-Elsevier: Cambridge, MA, USA, 1990; p. 290.
36. Olmos, E.; Sánchez-Blanco, M.J.; Ferrández, T.; Alarcon, J.J. Subcellular effects of drought stress in *Rosmarinus officinalis*. *Plant Biol.* **2007**, *9*, 77–84. [CrossRef]
37. Tombesi, S.; Johnson, R.S.; Day, K.R.; DeJong, T.M. Relationships between xylem vessel characteristics, calculated axial hydraulic conductance and size-controlling capacity of peach rootstocks. *Ann. Bot.* **2009**, *105*, 327–331. [CrossRef] [PubMed]
38. Carlquist, S. *Comparative Wood Anatomy: Systematic, Ecological, and Evolutionary Aspects of Dicotyledon Wood*; Springer Science & Business Media: Berlin/Heidelberg, Germany, 2013; ISBN 3662217147.
39. Pitman, W.D.; Holt, E.C.; Conrad, B.E.; Bashaw, E.C. Histological differences in moisture-stressed and nonstressed Kleingrass forage 1. *Crop. Sci.* **1983**, *23*, 793–795. [CrossRef]
40. Guerfel, M.; Baccouri, O.; Boujnah, D.; Chaïbi, W.; Zarrouk, M. Impacts of water stress on gas exchange, water relations, chlorophyll content and leaf structure in the two main Tunisian olive (*Olea europaea* L.) cultivars. *Sci. Hortic.* **2009**, *119*, 257–263. [CrossRef]
41. Child, R.D.; Summers, J.E.; Babij, J.; Farrent, J.W.; Bruce, D.M. Increased resistance to pod shatter is associated with changes in the vascular structure in pods of a resynthesized *Brassica napus* line. *J. Exp. Bot.* **2003**, *54*, 1919–1930. [CrossRef]
42. Lo Gullo, M.A.; Salleo, S.; Piaceri, E.C.; Rosso, R. Relations between vulnerability to xylem embolism and xylem conduit dimensions in young trees of *Quercus corris*. *Plant Cell Environ.* **1995**, *18*, 661–669. [CrossRef]
43. Trifilo, P.; Lo Gullo, M.A.; Nardini, A.; Pernice, F.; Salleo, S. Rootstock effects on xylem conduit dimensions and vulnerability to cavitation of *Olea europaea* L. *Trees* **2007**, *21*, 549–556. [CrossRef]
44. Meland, M.; Moe, M.E.; Frøynes, O. Differences in growth and development of functional xylem of grafted and budded sweet cherry trees. In Proceedings of the VIII International Symposium on Canopy, Rootstocks and Environmental Physiology in Orchard Systems, Budapest, Hungary, 13–18 June 2004; pp. 311–316.
45. Zach, A.; Schuldt, B.; Brix, S.; Horna, V.; Culmsee, H.; Leuschner, C. Vessel diameter and xylem hydraulic conductivity increase with tree height in tropical rainforest trees in Sulawesi, Indonesia. *Flora-Morphol. Distrib. Funct. Ecol. Plants* **2010**, *205*, 506–512. [CrossRef]
46. Tyree, M.T.; Ewers, F.W. The hydraulic architecture of trees and other woody plants. *New Phytol.* **1991**, *119*, 345–360. [CrossRef]
47. Şahin-Çevik, M.; Çevik, B.; Topkaya-Kütük, B.; Yazıcı, K. Identification of drought-induced genes from the leaves of Rangpur lime (*Citrus limon* (L) Osbeck). *J. Hortic. Sci. Biotechnol.* **2017**, *92*, 636–645. [CrossRef]
48. Shafqat, W.; Jaskani, M.J.; Maqbool, R.; Chattha, W.S.; Ali, Z.; Naqvi, S.A.; Haider, M.S.; Khan, I.A.; Vincent, C.I. Heat shock protein and aquaporin expression enhance water conserving behavior of citrus under water deficits and high temperature conditions. *Environ. Exp. Bot.* **2021**, *181*, 104270. [CrossRef]
49. Ábrahám, E.; Hourton-Cabassa, C.; Erdei, L.; Szabados, L. Methods for determination of proline in plants. In *Plant Stress Tolerance*; Humana Press-Springer: Totowa, NJ, USA, 2010; pp. 317–331.
50. Velikova, V.; Yordanov, I.; Edreva, A. Oxidative stress and some antioxidant systems in acid rain-treated bean plants: Protective role of exogenous polyamines. *Plant Sci.* **2000**, *151*, 59–66. [CrossRef]
51. Hussain, S.; Khalid, M.F.; Saqib, M.; Ahmad, S.; Zafar, W.; Rao, M.J.; Morillon, R.; Anjum, M.A. Drought tolerance in citrus rootstocks is associated with better antioxidant defense mechanism. *Acta Physiol. Plant* **2018**, *40*, 135. [CrossRef]
52. Santana-Vieira, D.D.S.; Freschi, L.; da Hora Almeida, L.A.; de Moraes, D.H.S.; Neves, D.M.; Dos Santos, L.M.; Bertolde, F.Z.; dos Santos Soares Filho, W.; Coelho Filho, M.A.; da Silva Gesteira, A. Survival strategies of citrus rootstocks subjected to drought. *Sci. Rep.* **2016**, *6*, 38775. [CrossRef]

53. Aroca, R. Plant responses to drought stress. In *From Morphological to Molecular Features*; Springer: Berlin/Heidelberg, Germany, 2012.
54. Zhang, J.; Kirkham, M.B. Water status of drought-resistant and drought-sensitive sorghum treated with ethephon. *J. Plant Growth Regul.* **1990**, *9*, 189–194. [CrossRef]
55. Savé, R.; Biel, C.; Domingo, R.; Ruiz-Sánchez, M.C.; Torrecillas, A. Some physiological and morphological characteristics of citrus plants for drought resistance. *Plant Sci.* **1995**, *110*, 167–172. [CrossRef]
56. Homayoun, H.; Daliri, M.S.; Mehrabi, P. Effect of drought stress on leaf chlorophyll in corn cultivars (*Zea mays*). *Middle East. J. Sci. Res.* **2011**, *9*, 418–420.
57. Mafakheri, A.; Siosemardeh, A.F.; Bahramnejad, B.; Struik, P.C.; Sohrabi, Y. Effect of drought stress on yield, proline and chlorophyll contents in three chickpea cultivars. *Aust. J. Crop. Sci.* **2010**, *4*, 580–585.
58. Nawazish, S.; Hameed, M.; Naurin, S. Leaf anatomical adaptations of *Cenchrus ciliaris* L. from the Salt Range, Pakistan against drought stress. *Pak. J. Bot.* **2006**, *38*, 1723–1730.
59. Molassiotis, A.; Job, D.; Ziogas, V.; Tanou, G. Citrus plants: A model system for unlocking the secrets of NO and ROS-inspired priming against salinity and drought. *Front. Plant Sci.* **2016**, *7*, 229. [CrossRef] [PubMed]
60. Hayat, S.; Hayat, Q.; Alyemeni, M.N.; Wani, A.S.; Pichtel, J.; Ahmad, A. Role of proline under changing environments: A review. *Plant Signal. Behav.* **2012**, *7*, 1456–1466. [CrossRef]
61. Chartzoulakis, K.; Patakas, A.; Bosabalidis, A.M. Changes in water relations, photosynthesis and leaf anatomy induced by intermittent drought in two olive cultivars. *Environ. Exp. Bot.* **1999**, *42*, 113–120. [CrossRef]
62. Xiao, J.; Chen, J.; Zhang, H.; Xu, H.; Wang, H.; Xie, M. Gene expression profiling in response to drought stress in citrus leaves by cDNA-AFLP. *Acta Hortic. Sin.* **2011**, *38*, 417–424.
63. Wu, Q.-S.; Srivastava, A.K.; Zou, Y.-N. AMF-induced tolerance to drought stress in citrus: A review. *Sci. Hortic.* **2013**, *164*, 77–87. [CrossRef]
64. Qaderi, M.M.; Martel, A.B.; Dixon, S.L. Environmental factors influence plant vascular system and water regulation. *Plants* **2019**, *8*, 65. [CrossRef] [PubMed]
65. Balfagón, D.; Terán, F.; de Oliveira, T.; Santa-Catarina, C.; Gómez-Cadenas, A. Citrus rootstocks modify scion antioxidant system under drought and heat stress combination. *Plant Cell Rep.* **2021**, 1–10. [CrossRef]
66. Mansoor, U.; Fatima, S.; Hameed, M.; Naseer, M.; Ahmad, M.S.A.; Ashraf, M.; Ahmad, F.; Waseem, M. Structural modifications for drought tolerance in stem and leaves of *Cenchrus ciliaris* L. ecotypes from the Cholistan Desert. *Flora* **2019**, *261*, 151485. [CrossRef]
67. Crous, C.J.; Greyling, I.; Wingfield, M.J. Dissimilar stem and leaf hydraulic traits suggest varying drought tolerance among co-occurring *Eucalyptus grandis* × *E. urophylla* clones. *South. For. J. For. Sci.* **2018**, *80*, 175–184. [CrossRef]

horticulturae

MDPI

Article

Physiological and Biochemical Characterization of the GABA Shunt Pathway in Pea (*Pisum sativum* L.) Seedlings under Drought Stress

Nisreen A. AL-Quraan [1,*], Zakaria I. Al-Ajlouni [2] and Nima F. Qawasma [1]

1 Department of Biotechnology and Genetic Engineering, Faculty of Science and Arts, Jordan University of Science and Technology, Irbid 22110, Jordan; qawasmehnana@gmail.com
2 Department of Plant Production, Faculty of Agriculture, Jordan University of Science and Technology, Irbid 22110, Jordan; ziajlouni@just.edu.jo
* Correspondence: naquraan@just.edu.jo; Tel.: +962-2-7201000 (ext. 23460); Fax: +962-2-7201071

Abstract: The physiological and biochemical role of the γ-aminobutyric acid (GABA) shunt pathway in green pea seedlings (*Pisum sativum* L.) was studied in response to soil water holding capacity levels: 80%, 60%, 40%, 20%, and 10% grown under continuous light at 25 °C for 7 days and 14 days, separately. Characterization of seeds germination pattern, seedlings growth (plant height, fresh and dry weight, and chlorophyll contents), GABA shunt metabolite (GABA, glutamate, and alanine) levels, total protein and carbohydrate levels, and oxidative damage (MDA level) were examined. Data showed a significant effect of drought stress on seed germination, plant growth, GABA shunt metabolites level, total protein and carbohydrate contents, and MDA level. A significant decline in seed germination percentage was recorded at a 20% drought level, which indicated that 20% of soil water holding capacity is the threshold value of water availability for normal germination after 14 days. Seedling fresh weight, dry weight, and plant height were significantly reduced with a positive correlation as water availability was decreased. There was a significant decrease with a positive correlation in Chl *a* and Chl *b* contents in response to 7 days and 14 days of drought. GABA shunt metabolites were significantly increased with a negative correlation as water availability decreased. Pea seedlings showed a significant increase in protein content as drought stress was increased. Total carbohydrate levels increased significantly when the amount of water availability decreased. MDA content increased slightly but significantly after 7 days and sharply after 14 days under all water stress levels. The maximum increase in MDA content was observed at 20% and 10% water levels. Overall, the significant increases in GABA, protein and carbohydrate contents were to cope with the physiological impact of drought stress on *Pisum sativum* L. seedlings by maintaining cellular osmotic adjustment, protecting plants from oxidative stress, balancing carbon and nitrogen (C:N) metabolism, and maintaining cell metabolic homeostasis and cell turgor. The results presented in this study indicated that severe (less than 40% water content of the holding capacity) and long-term drought stress should be avoided during the germination stage to ensure proper seedling growth and metabolism in *Pisum sativum* L.

Keywords: drought stress; GABA: gamma aminobutyric acid; metabolism; pea; *Pisum sativum* L.; seedling growth; water deficit

Citation: AL-Quraan, N.A.; Al-Ajlouni, Z.I.; Qawasma, N.F. Physiological and Biochemical Characterization of the GABA Shunt Pathway in Pea (*Pisum sativum* L.) Seedlings under Drought Stress. *Horticulturae* **2021**, *7*, 125. https://doi.org/10.3390/horticulturae7060125

Academic Editor: Yanyou Wu

Received: 24 April 2021
Accepted: 25 May 2021
Published: 27 May 2021

1. Introduction

Pea (*Pisum sativum* L.) was used by Mendel to lay the foundation of modern genetics [1]. It is one of the major food legumes that can grow in different regions and rich in proteins, vitamins, minerals, carbohydrates, and seed oil [2]. Pea is predominantly a self-pollinated crop with limited variation in the number of flowers per node [3]. Most garden pea germplasm/varieties lines have either one or two flowers per node [4]. The agro-ecological importance of pea is connected with its capability to form symbiotic nitrogen fixation with rhizobial bacteria that promotes and enhances seed germination, seedling vigor and emergence, root and shoot growth, plant stand and biomass, and seed weight [5].

Drought is a major abiotic stress that affects plant growth and yield [6]. Plant species and genotype adaptation and level of tolerance vary according to the type of drought stress (meteorological, agricultural, hydrological, or socioeconomic drought) [7]. Drought limits the productivity of many crops, especially during the seedling stage [8]. Water scarcity negatively affects vegetative growth by direct influence on its metabolic sink strength and fruit composition [9]. It causes reduced quality and quantity of the crop yield, growth rate, leaf expansion, stem elongation [8,10], stomatal conductance [11], and grain filling) [12]. Increasing the level of CO_2 in plants mitigates the impact of drought stress in many legume species [13].

Physiological, biochemical, and morphological processes in plants can occur as a result of drought stress causing changes in the expression of genes that lead to alteration in protein production [8]. In addition, respiration, photosynthesis, enzyme activities, and mineral nutrition, Redox (oxidation/reduction) homeostasis, and chloroplast metabolism are influenced by drought [12,14]. Analysis of proteomics data in plant leaves showed that many drought-responsive proteins that are involved in osmotic regulation, cell structure modulation, ROS scavenging, drought signal transduction, as well as carbohydrate metabolism, were upregulated under drought stress [15].

Drought stress reduces the quality and yield of many crops. Extended periods of water deficit results in a reduction in plant growth, photosynthesis efficiency and initiates a series of actions to maintain plant survival. These actions include regulation of stomatal conductance, osmotic adjustment, cell turgor maintenance, and protection of cellular membranes, enzymes, and macromolecules from oxidative damage [16]. During drought stress, plants close their stomata which leads to plasma membrane damage; consequently, the internal CO_2 concentration will be decreased, and excessive generation of reactive oxygen species will be increased, leading to a reduction in the photosynthetic rate and plant growth [17,18]. Water deficiency in maize caused a reduction in seedling survival rate, acceleration of the post-pollination embryo abortion rate, and ultimately, yield loss by postponing silking through the increase in the anthesis to the silking interval that leads to a reduction in ear and kernel number per plant [19–22]. Furthermore, water scarcity decreased photosynthetic efficiency, chlorophyll content, and CO_2 exchange in maize seedlings [21,23].

Drought stress decreased total barley (*Hordeum vulgare* L.) grain yield through a drop in tillers, spikes, and grains number per individual plant [24]. In soybean, drought stress decreased branch seed and total crop yields [25]. Cell elongation in higher plants can be inhibited through water movement interruption to the surrounding elongating cells from the xylem under severe water deficiency [26].

The development of candidate drought-tolerant associated genes is based on precise screening for germplasm and breeding materials in limited water environments using bioinformatics [27]. Various studies on some plant genes subjected to drought stress described the biochemical pathways that are involved in drought acclimation. Proteins and metabolites that were involved in protective mechanisms against drought conditions were identified. These mechanisms include detoxification enzymes, redox status regulation, signaling pathways, protein folding and degradation, photosynthesis stability, and primary metabolism [18]. In addition, drought and low water content reduction lead to an increase in sugar and amino acid concentrations in plants [28].

GABA is a four-carbon non-protein amino acid that increases in plant tissues under stress [29]. The GABA shunt has a functional role in biotic and abiotic stress in plants through the improvement in the antioxidant activity to restrict ROS species production, plant cell signaling, and metabolism under stress [29,30]. The GABA shunt pathway is composed of three enzymatic reactions. GABA is largely produced through glutamate decarboxylation that is catalyzed by glutamate decarboxylase enzyme, then GABA is catabolized inside the mitochondrial matrix to succinic semialdehyde (SSA) by GABA transaminase enzyme. Then, succinate is produced from SSA oxidization by the mitochondrial succinic semialdehyde dehydrogenase enzyme or reduced to gamma-hydroxybutyrate (GHB) [30].

GABA is made from glutamate by glutamate decarboxylase that is present in symbiotic rhizobia bacteroids [31]. Furthermore, GABA was shown to be the second most plentiful amino acid in detached pea (*Pisum sativum* L.) nodules through the nuclear magnetic resonance analysis, which, in turn, reflects the vital role of GABA in amino acid cycling during bacteroid metabolism [31]. Data showed that large amounts of GABA is accumulated in root nodules, while GABA that is newly made appeared to be limited to pea nodules that are metabolically active [32]. Bound forms of GABA accumulated in nodules of many legume species, reaching 20% of the total nitrogen content [32].

Drought stress could increase the activation of GABA transporters (ProTs and AAP_3) that regulate the entry of GABA across the cell membrane. The entrance of GABA decreased drought leaf wilting and improved membrane stability via the reduction in oxidative damage in plants [33]. Endogenous GABA enhanced drought tolerance through the inhibition of lipid peroxidation and photosynthesis [33]. In addition, GABA stabilized the intracellular pH during drought stress and provided a source of nitrogen and carbon during the Krebs cycle and carbon–nitrogen metabolism in the maturation of green peas [34]. Drought stress directly influences metabolic sink strength during vegetative growth, which triggers an imbalance in redox homeostasis that affects overall plant growth and development.

In this study, the effects of drought stress on the physiological and biochemical characterization of the GABA shunt pathway in green pea (*Pisum sativum* L.) were investigated. Characterization of seeds germination pattern, seedling growth (plant height, fresh weight, dry weight, and chlorophyll content), oxidative damage (Malondialdehyde level), GABA shunt metabolite levels (GABA, Glutamate and Alanine), total proteins, and total carbohydrate levels were determined.

2. Materials and Methods

2.1. Plant Materials and Growth Conditions

The green pea seeds (*Pisum sativum* L.) that were used in this study were harvested in 2018 and obtained from local pea growers in Irbid/Jordan. All experiments were performed in the laboratory using a plant growth medium. The growth medium was a mixture of peat moss and perlite growth soil with a ratio of (2:1) wt/wt, later referred to as "soil" in a pot (7 g). Seeds were surface disinfected with 6% (*v*/*v*) sodium hypochlorite for 10 min and then washed five times with sterile distilled water. Surface sterilized seeds were planted in soil. Tap water was used to irrigate the seeds at a full water holding capacity level (80–100%) for two weeks until the appearance of seedlings) [35]. The two-week-old seedlings were then subjected to drought treatments according to soil water holding capacity levels: 80%, 60%, 40%, 20%, and 10% separately for 7 days and another set for 14 days. Plants were grown at 25 °C under continuous cool white fluorescent lamp illumination (40 µmol m^{-2} s^{-1}). For assays and parameters determinations, three replicates of 6 pea (*Pisum sativum* L.) seeds for each replicate were used.

2.2. Seed Sensitivity to Drought Stress Assay

Three replicates of 15 pea (*Pisum sativum* L.) surface-sterilized seeds were placed on three filter papers as artificial growth surface in sealed Petri dishes and irrigated with tap water in drought treatments according to full water holding capacity levels: 80%, 60%, 40%, 20%, and 10% separately for 14 days and allowed to grow at 25 °C. Radicle emergence from the seeds was recorded daily for 14 days. Drought treatment effect on seed germination was calculated.

2.3. Seedling Physiological Growth Parameters Assay

After each drought treatment for the 7 and 14 days separately, plant height (cm), fresh weight (g), and dry weight (g) were determined. Shoot fresh weight (g) was determined by directly weighing the collected seedlings' shoot tissues. Shoot dry weight (g) was determined following weighing the seedlings' shoot tissues after oven drying at 70 °C for 48 h.

2.4. Chlorophyll Content Determination

After each drought treatment for the 7 and 14 days separately, fresh shoots tissues were harvested. Chlorophyll pigments (Chl *a* and Chl *b*) extraction and determination were performed according to Jeffrey et al. protocol [36] with the following adjustments: 300 mg fresh leaves mixed with one ml of ice-cold 90% acetone were ground in an Eppendorf tube using a micro centrifuge tube pestle to prepare the chlorophyll extract. The extracted liquid was placed in a new Eppendorf tube and centrifuged at 11,000× *g* for 5 min. The supernatant was used for the determination of both chlorophyll pigments. The resulted supernatant absorbance was measured spectrophotometrically at 647 nm, 664 nm, and 750 nm wavelengths. Equation and extension coefficients [36] were used to calculate the concentration of Chl *a* and Chl *b* (μg/mL). The average of three replicates for each treatment was calculated.

2.5. Metabolites Extraction

After each drought treatment for the 7 and 14 days separately, fresh shoot tissues were harvested. GABA shunt metabolites were extracted and determined according to Zhang and Bown [37] with the following adjustments: 500 mg of fresh shoot tissues were ground in an Eppendorf tube using a micro centrifuge tube pestle until a fine powder was obtained, and then placed in 1.5 mL Eppendorf tubes and vortexed with 400 μL methanol for 10 min. Methanol from samples was removed by overnight evaporation at room temperature. To each tube, 500 μL of 70 mM lanthanum chloride was added and vortexed for 15 min. Then tubes were centrifuged at 12,400× *g* for 5 min. One hundred and sixty microliters of 1 M KOH was added to the collected supernatants and vortexed for 10 min. Tubes were centrifuged at 12,400× *g* for 5 min. The resulted supernatant (metabolites extract) was used for the determination of GABA shunt metabolites (GABA, alanine, and glutamate). The average of three replicates was used for each drought treatment.

2.6. GABA (γ-Aminobutyric Acid) Level Determination

The protocol of Zhang and Bown [37] was used for GABA determination with the following adjustments: 50 μL of metabolites extract, 14 μL of 4 mM $NADP^+$, 19 μL of 0.5 M potassium pyrophosphate, pH (8.6), 10 μL of (2 u/μL) GABASE enzyme (GABASE enzyme (Sigma-Aldrich Corp., St. Louis, MO, USA) was suspended in 0.1 M potassium pyrophosphate, pH (7.2), containing 12.5% Glycerol and 5 mM β-mercaptoethanol), and 10 μL of α-ketoglutarate were mixed in an Eppendorf tube to prepare the reaction mixture. The change in absorbance at 340 nm after the addition of α-ketoglutarate was measured after 90 min incubation at 25 °C using a microplate reader. The GABA level (nmol/mg FW) was determined using the NADPH standard curve. For each drought treatment, the average of three replicates was used.

2.7. Alanine Level Determination

The Bergmeyer protocol [38] was used for alanine determination with the following adjustments: 180 μL of 0.05 M Na-carbonate buffer, pH (10), 7 μL of 30 mM β-NAD^+, 10 μL of sample extract, and 1 μL of 0.3 u/μL alanine dehydrogenase (Sigma-Aldrich, St. Louis, MO, USA) enzyme suspension were mixed in an Eppendorf tube to prepare the reaction mixture. The change in absorbance at 340 nm after the addition of alanine dehydrogenase was measured after 60 min incubation at 25 °C using a microplate reader. The alanine level (nmol/mg FW) was determined using the NADH standard curve. For each drought treatment, the average of three replicates was used.

2.8. Glutamate Level Determination

The Bergmeyer protocol [38] was used for glutamate determination with the following adjustments: 180 µL of 0.1 M Tris-HCl pH (8.3), 8 µL of 7.5 mM β-NAD^+, 10 µL of sample extract, and 1 µL of 0.8 u/µL glutamate dehydrogenase enzyme suspension (Sigma-Aldrich, St. Louis, MO, USA) were mixed in an Eppendorf tube to prepare the deamination reaction mixture. The change in absorbance at 340 nm after the addition of glutamate dehydrogenase was measured after 60 min incubation at 25 °C using a microplate reader. The glutamate level (nmol/mg FW) was determined using the NADH standard curve. For each drought treatment, the average of three replicates was used.

2.9. Total Protein Content Determination

After each drought treatment for the 7 and 14 days separately, fresh shoot tissues were harvested. Total protein content was determined using a SMART BCA Protein Assay Kit (Intron Biotechnology, Gangnam-gu, Seoul, Korea) according to the manufacturer's instructions. Five hundred milligrams of fresh shoots tissues mixed with 100 µL of distilled water were ground in an Eppendorf tube using a micro centrifuge tube pestle. Fifty microliters of the extracted liquid was obtained and mixed well with 1 mL of the kit working solution. All tubes were incubated at 37 °C for 30 min and then kept at room temperature for 5 min. The absorbance of the samples was measured at 562 nm using a microplate reader. The BSA standard curve was used to determine the total protein concentration (µg/mL). For each drought treatment, the average of three replicates was used.

2.10. Total Carbohydrates Content Determination

After each drought treatment for the 7 and 14 days separately, fresh shoot tissues were harvested. Total carbohydrates content was determined using the Total Carbohydrate Quantification Assay Kit (catalog number: ab155891, Abcam, Cambridge, MA, USA) according to the manufacturer's instructions. Fifty milligrams of fresh shoot tissues mixed with 200 µL ice-cold kit assay buffer was ground in an Eppendorf tube using a micro centrifuge tube pestle. Tubes were then centrifuged at 12,400× *g* for 5 min. The resulted supernatant was collected. In a 96-well microplate, 30 µL of collected supernatant from each sample and the kit standard solution wells were mixed with 150 µL of concentrated H_2SO_4. The microplate was mixed well for 5 min on a shaker and then incubated at 90 °C for 15 min. The microplate was cooled at room temperature for 15 min. Thirty microliters of the kit developer solution was added to all treated samples and kit standard solution wells, mixed on a shaker at room temperature for 15 min, and measured OD at 490 nm. A glucose standard curve was used to determine the total carbohydrate concentration (µg/mg FW). For each drought treatment, the average of three replicates was used.

2.11. Oxidative Damage Assay

After each drought treatment for the 7 and 14 days separately, fresh shoot tissues were harvested. Lipid peroxidation as malondialdehyde (MDA) equivalent was determined by estimating the TBA reactive substances (TBARS) as described by Heath and Packer [39] with some adjustments: 20 mg of fresh shoots tissues was mixed with 0.25 mL of 0.5% (*w/v*) thiobarbituric acid in 20% (*w/v*) trichloroacetic acid and 0.25 mL of 175 mM NaCl in 50 mM Tris-HCl (pH 8) and ground in an Eppendorf tube using a micro centrifuge tube pestle. Tubes were placed in a water bath at 90 °C for 25 min. Tubes were then centrifuged at 12,400× *g* for 20 min and the supernatant was collected. A microplate reader was used to measure the absorbance of the supernatant at 532 nm. A standard curve of MDA was used to calculate the MDA level (nmol/mg FW). For each drought treatment, the average of three replicates was used.

2.12. Experimental Design and Data Analysis

A completely randomized design (CRD) was used for all experiments in this study. All treatments and assays were replicated three times. For all assays parameters, mean and standard deviation (SD) values were determined. Results were expressed as mean. Data analysis was performed by one-way analysis of variance (ANOVA) using the Least Significant Difference (LSD) multiple comparison tests on the means using the SPSS version 16.0 software. For all data analyses, p-value ≤ 0.05–0.01 was considered significant, and p-value < 0.01 was considered highly significant.

3. Results and Discussion

3.1. Seed Sensitivity to Drought Stress

Germination percentage of *Pisum sativum* L. was affected significantly ($p < 0.05$) by drought treatments to all saturation levels (80%, 60%, 40%, 20%, and 10%) in pea seeds for 14 days. In general, the germination percentage significantly decreased with a positive correlation as water content decreased (Table 1). The highest seed germination percentage was observed at the 80% saturation level ($p < 0.05$). A significant decline was recorded at 40% water saturation level. After 14 days, the germination percentage was 100, 80, and 30 at 60, 40, and 20% soil water saturation levels, respectively. No germination was observed at 10% drought treatment point (Table 1). These results indicated that the 40% water saturation level is the threshold value for normal pea seed germination after 14 days ($p = 0.0001$, $r = 0.929$).

Table 1. Germination percentage of pea (*Pisum sativum* L.) in response to soil water holding capacity levels: 80%, 60%, 40%, 20%, and 10% grown under continuous light at 25 °C. Means followed by different letters are significantly different ($p \leq 0.05$) by LSD; r = correlation coefficient.

Treatment	Day 1	Day 2	Day 3	Day 4	Day 5	Day 6	Day 8	Day 10	Day 12	Day 14
80%	0.0	40.0 ^{a}	59.0 ^{a}	75.0 ^{a}	75.0 ^{a}	95.0 ^{a}	100.0 ^{a}	100.0 ^{a}	100.0 ^{a}	100.0 ^{a}
60%	0.0	30.0 ^{b}	60.0 ^{a}	72.5 ^{a}	75.0 ^{a}	85.0 ^{b}	100.0 ^{a}	100.0 ^{a}	100.0 ^{a}	100.0 ^{a}
40%	0.0	12.5 ^{c}	45.0 ^{b}	60.0 ^{b}	65.0 ^{b}	70.0 ^{c}	75.0 ^{b}	75.0 ^{b}	80.0 ^{b}	80.0 ^{b}
20%	0.0	0.0 ^{d}	0.0 ^{c}	7.5 ^{c}	15.0 ^{c}	18.0 ^{d}	20.0 ^{c}	25.0 ^{c}	30.0 ^{c}	30.0 ^{c}
10%	0.0	0.0 ^{d}	0.0 ^{c}	0.0 ^{cd}	0.0 ^{d}	0.0 ^{e}	0.0 ^{d}	0.0 ^{d}	0.0 ^{d}	0.0 ^{d}
r *		0.987	0.924	0.929	0.913	0.953	0.942	0.942	0.929	0.929

* means $\alpha < 0.05$.

Seed germination is a critical stage for plant survival. Drought stress greatly affects seed germination, but the response intensity and harmful effects of stress depend on the plant species [40]. A decrease in the germination of *Eremosparton songoricum* was observed under drought stress [40]. The results of the current study indicated that as water saturation decreased, a significant drop in the germination percentage in pea (*Pisum sativum* L.) was observed. Similar to our finding, Liu et al.'s [41] study revealed that seed germination of two maize cultivars (Liansheng15 cultivar and Zhengdan 958 cultivar) was reduced under drought stress. Increased water deficit minimized absorption of water by seeds, therefore preventing their germination [42]. However, it has been shown that failure of the emergence of the root was due to a reduction in water level between the soil and the seeds [43]. In addition, a reduced water gradient affected enzymatic reactions, which caused a delay in seed germination [44].

Muscolo et al. [45] found significant differences between two lentil cultivars, *Ustica* and *Pantelleria*, collected from Sicilian islands subjected to drought stress with a remarkable decrease and delay in seed germination. Inhibition of seed germination was directly related to energy production through respiration, enzyme and hormonal activity, reserve mobilization, and dilution of the protoplasm to increase metabolism for successful embryonic growth [45]. Water availability is a limiting factor for non-dormant seed germination. It also affected the speed, uniformity, and percentage of seed emergence [46]. Dornbos et al. [47] showed that the germination percentage was decreased during the seed filling stage in soybean (*Glycine max* L.c.v Merr.) under severe drought stress. Inhibition of starch catabolism under drought and salt stress decreased seed germination [48,49]. AL-Quraan et al. [50] showed a significant decrease in seed germination of five wheat cultivars (*Triticum aes-*

tivum L.) under salt and osmotic stress. Furthermore, AL-Quraan et al. [51] reported that cold, heat, salt, and osmotic stresses significantly reduced seed germination in two lentil (*Lens culinaris* Medik) cultivars. The results of this study showed that reduction in water content had a severe impact on seed germination and early seedling growth in pea (*Pisum sativum* L.) that might be caused by a reduction in water absorption by seeds during metabolic processes and enzymatic activity in germinating pea seeds.

3.2. Seedling Physiological Growth in Response to Drought Stress

Seedling fresh weight, dry weight, and seedling height of *Pisum sativum* L. were measured in response to drought treatments according to soil water holding capacity levels of 80%, 60%, 40%, 20%, and 10% for 7 and 14 days, separately. Seedling fresh weight and seedlings height were significantly reduced with a positive correlation as water availability decreased (Table 2). The seedling fresh weight ($p = 0.001, 0.05$, $r = 0.957, 0.979$) and seedling height ($p = 0.001, 0.001$, $r = 0.867, 0.992$) were reduced significantly after 7 and 14 days, respectively. Dry weight reduction was not significant under all drought treatment after 7 ($p = 0.096$) and 14 days ($p = 0.228$), respectively.

Table 2. Seedling height (cm), fresh weight, and dry weight (gm), and water content (%) of pea (*Pisum sativum* L.) in response to soil water holding capacity levels: 80%, 60%, 40%, 20%, and 10% grown under continuous light at 25 °C for 7 days and 14 days, separately. Means followed by different letters are significantly different ($p \leq 0.05$) by LSD; r = correlation coefficient.

	Day 7			Day 14		
Treatment	Seedlings Height (cm)	Fresh Weight (gm)	Dry Weight (gm)	Seedlings Height (cm)	Fresh Weight (gm)	Dry Weight (gm)
80%	21.0 ^{a}	0.51 ^{a}	0.051 ^{a}	25.5 ^{a}	0.57 ^{a}	0.052 ^{a}
60%	15.0 ^{b}	0.355 ^{b}	0.046 ^{a}	14.5 ^{b}	0.415 ^{a}	0.046 ^{a}
40%	10.4 ^{bcd}	0.25 ^{c}	0.035 ^{a}	13.5 ^{bc}	0.3 ^{ab}	0.04 ^{a}
20%	6.36 ^{d}	0.185 ^{cd}	0.031 ^{ab}	11.5 ^{bd}	0.205 ^{ac}	0.031 ^{a}
10%	5.4 ^{de}	0.2 ^{ce}	0.03 ^{ac}	11.5 ^{bd}	0.21 ^{ad}	0.03 ^{a}
r *	0.992	0.957	0.979	0.867	0.979	0.995

* means $\alpha \leq 0.05$.

In agreement with our study, Embiale et al. [52] reported that unavailability of water resulted in a significant decline in pea's basal diameter increment, area, width, and expansion length of leaf, number of branches and leaves, and total seedlings length. Plant cells save water by avoiding active growth. Stomatal closure, diminished leaf water potential, turgor loss, reduction in cell enlargement and growth [53], and inhibition of shoot and root growth [54] are common plant responses due to water stress. In general, reduction in plant biomass is positively linked with prolonged water deficiency and inhibition of cell expansion per cell growth due to low turgor pressure [55]. Reduction in plant dry weight under water stress could be due to imbalanced stomatal conductance that leads to a reduction in carbon assimilation per unit leaf area and low biomass production [56,57]. Drought stress was associated with reduced cellular division and elongation during germination, causing a reduction in root length [58].

Khorasaninejad et al. [59] reported a reduction in shoot fresh and dry weight, root dry weight, internodes, and internodes length of peppermint (*Mentha piperita* L.) under drought stress. A decrease in root and shoot fresh and dry weights and shoot length was observed in sensitive bean genotypes under water stress [60]. Zhang et al. [61] reported that the length and width of maize leaves on seedlings were shorter under water stress due to a reduction in chlorophyll content. A reduction in water content decreased plant height, stem diameter, stem and root biomass, and total biomass of *Populus nigra* (poli) [62]. In agreement with our study, Riad et al. [63] reported that decreasing water content resulted in a reduction in plant growth parameters (root length, root, and leaves fresh and dry weights, and plant length), biomass accumulation, and vegetative growth of green peas as compared with 100% water availability.

3.3. The Effect of Drought on Chlorophyll Content

Table 3 shows that the chlorophyll content were significantly decreased after 7 and 14 days as water content decreased. *Pisum sativum* L. seedlings possessed the highest total Chl *a* and Chl *b* contents at 80% water saturation level (p = 0.020, 0.003) with positive correlation (r = 0.938, 0.986) after 7 and 14 days, respectively. There was a significant decrease with a positive correlation in Chl *a* and Chl *b* contents in response to 7 days and 14 days of drought treatments.

Table 3. Chlorophyll content (Chl *a*, Chl *b*, total chlorophyll) (μg/mL) of pea (*Pisum sativum* L.) seedlings in response to soil water holding capacity levels: 80%, 60%, 40%, 20%, and 10% grown under continuous light at 25 °C for 7 days and 14 days, separately. Means followed by different letters are significantly different ($p \leq 0.05$) by LSD; r = correlation coefficient.

	Day 7			Day 14		
Treatment	Chl *a*	Chl *b*	Total	Chl *a*	Chl *b*	Total
80%	7.670 ^{a}	5.350 ^{a}	13.020 ^{a}	10.470 ^{a}	9.950 ^{a}	20.420 ^{a}
60%	3.552 ^{b}	3.870 ^{b}	7.422 ^{b}	8.200 ^{b}	8.220 ^{a}	16.420 ^{b}
40%	2.500 ^{c}	1.420 ^{c}	3.920 ^{c}	6.460 ^{c}	7.040 ^{ab}	13.500 ^{c}
20%	2.300 ^{cd}	1.220 ^{cd}	3.520 ^{cd}	5.740 ^{cd}	5.380 ^{bc}	11.120 ^{d}
10%	1.140 ^{e}	0.699 ^{e}	1.839 ^{e}	3.210 ^{e}	4.870 ^{bd}	8.080 ^{e}
r *	0.913	0.963	0.938	0.975	0.998	0.986

* means $\alpha < 0.05$.

Chlorophyll is a major chloroplast component and has a positive relationship with photosynthetic rate. Our results agreed with Nyachiro et al.'s [64] study, which found a significant reduction in Chl *a* and Chl *b* caused by drought stress in six wheat (*Triticum aestivum* L.) cultivars. Shinde and Thakur [65] reported that drought stress significantly reduced Chl *a* and Chl *b* contents in three varieties of chickpeas during vegetative growth or anthesis. Our data indicated that decreased water availability caused a significant decrease in Chl *a* and Chl *b* contents in *Pisum sativum* L. after 7 and 14 days. This reduction in total chlorophyll content leads to a reduction in photosynthesis and plant growth which was correlated with a significant reduction in seedling growth under all water stress levels. Depending on the duration and severity of drought stress, plant genotypes, and environmental conditions, a reduction in Chl *a* and *b* contents has been reported in other plant species [66,67]. Previous studies on *Triticum aestivum* and *Zea mays* [68], *Gossypium hirsutum* [69], and *Tectona grandis* [70] found that the content of photosynthetic pigments reduced as water availability decreased. It has been reported that metabolic imbalance occurred as a result of a decrease in chlorophyll synthesis and an increase in chlorophyll degradation under drought stress [71–73].

Nitrogen is a component of the chlorophyll structure in plant tissues. Under stress conditions, a decline in chlorophyll content was associated with changes in nitrogen metabolism as a result of proline synthesis to maintain osmotic adjustment [74]. Low nitrogen levels reduced photosynthetic rates due to a reduction in chlorophyll synthesis [75]. Severe water stress decreased stomatal conductance, transpiration, and photosynthetic rate [56,76]. As a drought response mechanism, chloroplasts reduced light absorption by decreasing the chlorophyll content [77]. Studies revealed that the inhibition of photosynthesis efficiency as a result of an imbalance between light capture and usage under water shortage enhances oxidative stress. Furthermore, under water stress, stomatal closure and variation in photosynthetic metabolism lead to a reduction in CO_2 availability and directly impacted the photosynthetic rate [78]. Under severe drought conditions, chlorophyll content in wheat seedlings was reduced as a result of the increased activity of chlorophillase and peroxidase enzymes [79]. In response to drought stress, chlorophyll content was reduced in the vegetative and flowering stages of the *Avena* species [80]. Alberte et al. [81] reported a reduction in Chl *a*/*b* contents occurred in lamellar content in response to water deficit. Patro et al. [82] and Ghotbi-Ravandi et al. [83] reported that stomatal closure due to water deficit is a major factor inhibiting photosynthesis in *Arabidopsis thaliana* and barley (*Hordeum vulgare* L.) seedlings, respectively. Moreover, drought stress significantly decreased the chlorophyll content in *Pisum sativum* L. seedlings [84] and three wheat varieties (*Triticum aestivum* L.) [85]. The results of our study indicated that a water imbalance due to drought stress caused a reduction in chlorophyll content that might be attributed to the photosynthetic apparatus activity disturbance.

3.4. GABA Shunt Activation in Response to Drought Stress

The GABA shunt metabolites increased significantly as a result of drought stress caused by different water saturation levels (Table 4). There were significant differences in GABA, alanine, and glutamate content at 80%, 60%, 40%, 20%, and 10% soil water content between the 7- and 14-day treatments. In general, there was a significant increase in GABA, alanine, and glutamate (p = 0.001, 0.0001, 0.002, respectively) with a negative correlation (r = −0.987, −0.968, −0.978, respectively) with water availability after the 7-day treatment. A significant increase was also found in GABA, alanine, and glutamate (p = 0.0001, 0.003, 0.870, respectively) with a negative correlation (r = −0.906, −0.989, −0.939, respectively) after the 14-day treatment. The increase in GABA content was about 4 and 5 times higher at the 20% and 10% water levels, respectively, after the 14-day treatment when compared with GABA content at the same water levels after the 7 day treatment. The increase in GABA levels was accompanied by a slight but significant ($p \leq 0.01$–0.05) increase in the levels of alanine and glutamate after 7- and 14-day treatments under all water saturation levels.

Table 4. Levels of GABA shunt metabolites (GABA, alanine, and glutamate) of pea (*Pisum sativum* L.) seedlings in response to soil water holding capacity levels: 80%, 60%, 40%, 20%, and 10% grown under continuous light at 25 °C for 7 days and 14 days, separately. Metabolite levels were calculated as nmol/mgFW. GABA (γ-Aminobutyric acid), Ala (alanine), Glu (glutamate). Means followed by different letters are significantly different ($p \leq 0.05$) by LSD; r = correlation coefficient.

	Day 7			Day 14		
Treatment	**GABA**	**Ala**	**Glu**	**GABA**	**Ala**	**Glu**
80%	3.966 ^{a}	0.185 ^{a}	0.166 ^{a}	22.291 ^{a}	0.535 ^{a}	0.429 ^{a}
60%	6.999 ^{b}	0.237 ^{b}	0.198 ^{b}	25.211 ^{a}	0.603 ^{a}	0.446 ^{a}
40%	8.170 ^{bc}	0.268 ^{c}	0.221 ^{bc}	25.648 ^{ab}	0.666 ^{a}	0.584 ^{a}
20%	10.106 ^{cd}	0.275 ^{cd}	0.228 ^{bd}	40.973 ^{c}	0.797 ^{ab}	0.590 ^{a}
10%	12.111 ^{de}	0.302 ^{e}	0.252 ^{be}	49.969 ^{d}	0.838 ^{bc}	0.722 ^{a}
r *	−0.987	−0.968	−0.978	−0.906	−0.989	−0.939

* means $\alpha < 0.05$.

The GABA shunt has been associated with physiological responses, such as cytosolic pH regulation [86], nitrogen metabolism, and carbon fluxes into the Krebs cycle [87], protection against ROS production [88], and osmoregulation and signaling [89]. The GABA shunt is also involved in carbon and nitrogen metabolism, maintenance of cell membrane integrity [90], and minimizing the negative effects of abiotic stresses on plant metabolism [91].

Many studies reported that various stresses caused an increase in endogenous GABA accumulation in plants [92]. The current study showed a significant increase in GABA content as a result of a reduction in water availability to ease the metabolic damage in pea seedlings during vegetative growth. Studies showed that GABA effectively decreased leaf wilting and improved membrane stability induced by drought stress in perennial ryegrass (*Lolium perenne*) [93], black pepper [94], and white clover (*Trifolium repens*) [33], which confirmed the beneficial effect of GABA in protecting plants from oxidative stress. Insufficient availability of oxygen resulted in significant suppression of melon seedling growth. However, seedling growth was significantly improved when exogenous GABA was applied [95]. Several amino acid syntheses are regulated by the TCA cycle [96]. Glutamate, as a precursor amino acid of GABA production, is produced from the α-ketoglutaric acid (an intermediate of the TCA cycle) transamination [90]. Glutamate metabolic pathway activation via the TCA cycle was confirmed by GABA production and defense of creeping bentgrass against drought stress [97]. An increase in endogenous GABA levels due to an increase in glutamate content occurred in white clover in response to drought stress [33]. Increased content of glutamic acid and GABA were found in six winter wheat genotypes due to drought stress [98]. In addition, alanine was increased at the beginning of drought stress due to an increase in glutamate levels [99]. In this study, the increase in GABA content as water availability decreased mitigated the damaging effects of drought stress on pea (*Pisum sativum* L.) seedlings by supplying enough carbon/nitrogen source to the TCA cycle and amino acid synthesis.

3.5. The Effect of Drought on Seedling Proteins Level

In general, the total protein level in *Pisum sativum* L. seedlings increased with increasing water deficit. A significant difference was found in the total protein level of pea seedlings between the 7-day (p = 0.0001, r = −0.984) and 14-day (p = 0.0001, r = −0.982) treatments (Figure 1). The total protein level was 2 to 3 times higher after 7-day treatments when compared with the 14-day treatments under all water saturation levels. The significant increase in protein content after 7 days indicated that *Pisum sativum* L. seedlings could accumulate protein to lower the osmotic potential to maintain cellular structures and metabolic stability under drought stress. However, the steady-state level of protein content after 14 days indicated the ability of pea seedlings to tolerate prolonged water deficit by keeping stable protein metabolism to adjust cellular osmolarity and provide protection from further dehydration damage.

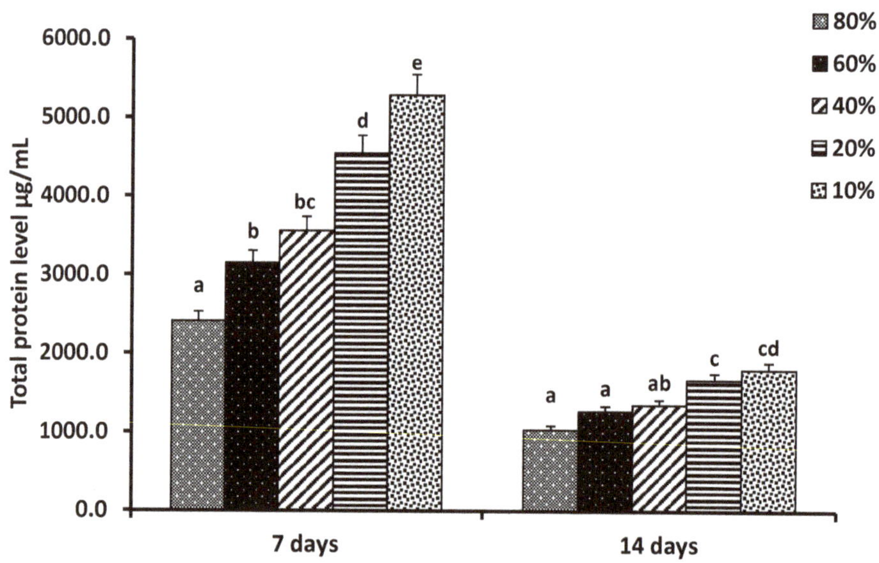

Figure 1. Total proteins level of pea (*Pisum sativum* L.) seedlings in response to soil water holding capacity levels: 80%, 60%, 40%, 20%, and 10% grown under continuous light at 25 °C for 7 days and 14 days, separately. Protein levels were calculated as µg/mL. Means (columns) with different letter scripts are significantly different ($p \leq 0.05$) by LSD.

Under environmental stress factors, plants stimulated specific changes in protein synthesis [100]. In agreement with our study, Guttieri et al. [101] reported increased protein concentration under water deficit due to higher rates of nitrogen accumulation in hard wheat (*Triticum aestivum* L.) grains. Ozturk and Aydin [102] also reported an increase in protein and gluten content in winter wheat cultivars grain in response to drought stress when compared with the fully irrigated treatment. During a combination of heat and drought stress, glutamine, ornithine, tyrosine, valine, and tryptophan were accumulated in *Arabidopsis thaliana* and purslane plants to maintain cellular osmotic adjustment and keep leaf turgor in response to such stress combination [103,104]. Drought stress induced the accumulation of protein by enhancing the protein biosynthesis in wheat and barley [105,106]. However, a reserve of available substrates for protein synthesis could be associated with amino acid accumulation to facilitate quick retrieval of osmotic adjustment and plant metabolism in response to water deficiency [107]. In a study that used three Australian bread wheat (*Triticum aestivum* L.) cultivars, Ford et al. [108] suggested that the cultivar RAC875 had the highest capacity to withstand drought stress by increasing the cellular protein synthesis. Although there was an increase in total protein levels which were involved in ROS scavenging, they observed a decrease in proteins involved in the Calvin cycle and photosynthesis. In contrast to our results, Gołębiowska et al. [109] observed a decrease in the number of differential proteins in leaves of winter barley subjected to drought stress for three weeks. Mohammadkhani and Heidar [110] reported an initial increase followed by a decrease in protein concentration in maize varieties (*Zea mays* L.) subjected to drought stress. The initial increase in total soluble proteins might be due to the expression of new proteins involved in stress adaptation. However, the decrease in protein concentration might be due to a severe reduction in photosynthesis. Under water stress, proline amino acid accumulated in plant cells prevents cellular oxidation through scavenge ROS [111] and regulates plant osmotic pressure for efficient water absorption [112]. Free proline level was reported to be increased under water deficit in wheat [113,114]. The amount of proline was increased significantly in ginger (*Zingiber officinale)* in response to a reduction in water availability and prolonged duration of drought stress [112].

3.6. The Effect of Drought on Seedling Carbohydrates Content

In the current study, data showed that the total carbohydrate level in *Pisum sativum* L. seedlings increased as water availability decreased (Figure 2)., the total carbohydrate level increased significantly ($p = 0.0001$) with a negative correlation ($r = -0.970$) in response to drought stress for 7 days. Similar results were obtained under drought stress for 14 days ($p = 0.0001$, $r = -0.0980$).

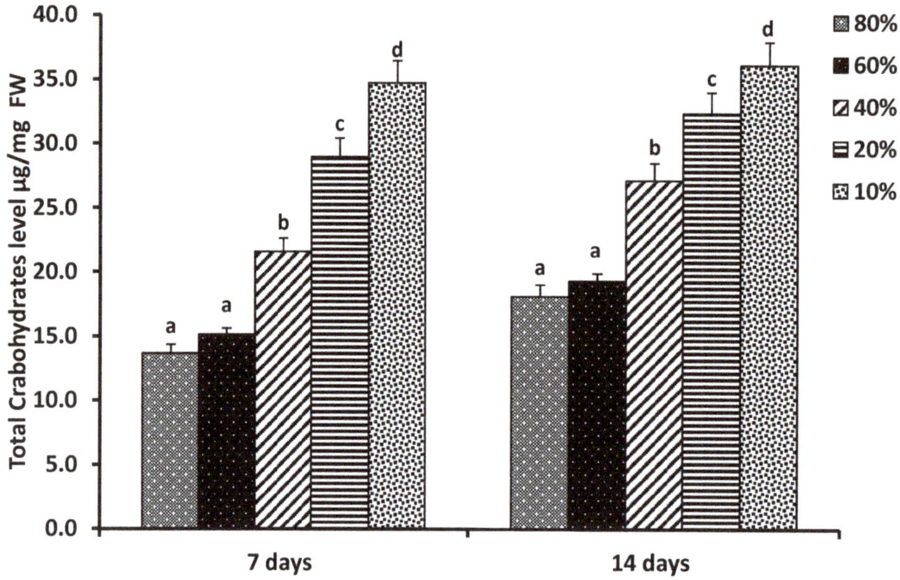

Figure 2. Total carbohydrates level of pea (*Pisum sativum* L.) seedlings in response to soil water holding capacity levels: 80%, 60%, 40%, 20%, and 10% grown under continuous light at 25 °C for 7 days and 14 days, separately. Total carbohydrate level was calculated as µg/mg FW. Means (columns) with different letter scripts are significantly different ($p \leq 0.05$) by LSD.

Water stress induces carbohydrate accumulation and osmolytes synthesis to maintain the water potential of plants [115,116]. Soluble sugar levels in plants also increased in response to drought stress. Plants respond to water shortage by balancing their potential osmotic proportion with the external environment by increasing their soluble sugars at the cellular level, reducing activities in roots, reducing the metabolism of carbohydrates as a result of severe pressure, and reducing the transfer of sugars in rinsing vessels [117]. In agreement with our study, an increase in total soluble sugars in durum (*Triticum durum* L.) wheat [118] and oligosaccharides in two sesame cultivars [119] were reported. In contrast, Akinci and Losel [120] reported a decrease in total sugar in cucumber cultivars under water stress. However, a major reserve of carbohydrates was detected in the leaves of cucumber seedlings. Under drought stress, sugar accumulation might be due to the fact that sucrose content increased because the enzyme activities involved in sucrose breakdown were diminished [121]. Furthermore, sugars protect the cells during drought by two physiological mechanisms. The first mechanism involves the hydroxyl group of sugars, which substitute the water to maintain hydrophilic interactions and hydrogen bonding in membranes and proteins during dehydration to prevent protein denaturation [122]. The second mechanism involves sugar vitrification in dehydrated cells through the formation of a biological glass in the cytoplasm to reduce water permeability [123]. Mohammadkhani and Heidari [110] reported an increase in soluble sugar concentrations as a result of starch degradation in two maize (*Zea mays* L.) cultivars. Soluble sugars accumulated only in roots of eucalyptus trees under drought stress [124]. Regier et al. [62] reported an increase in soluble sugar concentrations in the roots of two divergent clones of *Populus nigra* to maintain osmotic adjustment under water deficit. Soluble sugars were also reported to be increased and participated in plant metabolic regulation and stress signaling in response to osmotic stress [125]. Additionally, increased soluble sugars may act as osmoprotectants to stabilize the membranes and sustain cell turgor in response to water deficit [126]. Li and Li [127] showed an increase in the glucose, fructose, and sucrose content in micro-propagated apple plants (*Malus domestica* Borkh) in response to water stress. It is concluded from these previous studies that fluctuations of sugar concentrations in different plants under drought stress might be due to variations in CO_2 assimilation, partitioning of carbon source-sink, activity of enzymes involved in sucrose–starch partitioning, and inhibition of Calvin cycle enzymes.

3.7. Oxidative Damage in Response to Drought Stress

The accumulation of MDA, which is a byproduct of membrane lipids' oxidative damage, can be used as a marker for oxidative stress [128]. In this study, the MDA levels in the pea seedlings were determined to evaluate the lipid peroxidation in response to drought stress. Figure 3 shows that the malondialdehyde (MDA) level of pea (*Pisum sativum* L.) seedlings was slightly but significantly increased after 7 days ($p = 0.0001$, $r = -0.967$). However, the MDA level was increased sharply and significantly under all water stress levels reaching the highest level (140 nmol/mgFW at 10% water saturation level) ($p = 0.0001$, $r = -0.975$) after 14 days. Our results showed that the pea seedlings suffered minor damage under mild (60% water saturation level) drought stress, but severe lipid peroxidation and oxidative damage of the cell membrane occurred with increased (40%, 20%, and 10% water saturation levels) water deficit (Figure 3).

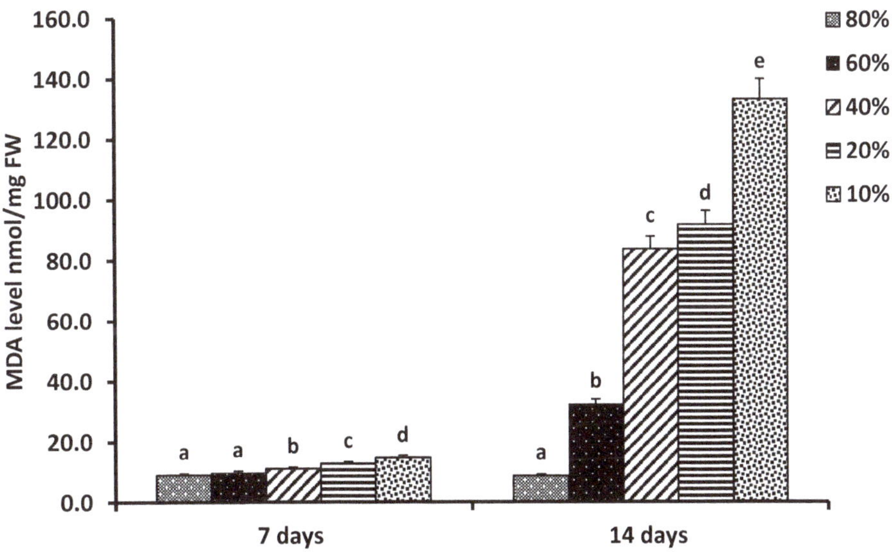

Figure 3. Malondialdehyde (MDA) level of pea (*Pisum sativum* L.) seedlings in response to soil water holding capacity levels: 80%, 60%, 40%, 20%, and 10% grown under continuous light at 25 °C for 7 days and 14 days, separately. The MDA level was calculated as nmol/mg FW. Means (columns) with different letter scripts are significantly different ($p \leq 0.05$) by LSD.

Many abiotic stresses lead to the accumulation of MDA in plant tissues due to lipid peroxidation of cellular membranes and overproduction of ROS [39]. Severe water stress caused agitation and instability in the metabolic processes in the mitochondria and chloroplasts, leading to high ROS production [129]. ROS resulting from oxidative stress are toxic and highly reactive molecules that can damage cellular macromolecules and subcellular structures [130,131]. The levels of MDA and H_2O_2 are indicators of the free radical reactions occurring in plant-stressed tissues [132]. Similar to our findings, Pandey et al. [133] reported that the MDA level was increased in *Avena* species leaves under drought stress, increasing membrane leakage. In addition, Zlatev et al. [134] reported an increase in membrane damage and ROS production in three bean (*Phaseolus vulgaris* L.) cultivars as a result of water deficit. Similarly, Tatar and Gevrek [135] reported an increase in the level of MDA in wheat as the duration of drought stress increased. Leaves of growing carrots (*Daucus carota* L.) accumulated high levels of MDA content under water deficit stress [91]. Morabito and Guerrier [136] reported that oxidative stress occurred 12 h after the application of drought stress. However, after a longer period of drought stress, an antioxidants defense system was initiated. Abid et al. [137] showed that a higher MDA concentration in wheat (*Triticum aestivum* L.) was associated with higher H_2O_2 content and greater rate of $O_2^{\bullet-}$ generation. Hernandez et al. [138] reported that salinity stress in plant tissues could lead to cell membrane rupture due to the accumulation of ROS and lipid peroxidation.

3.8. Correlation between GABA Level and All Physiological and Metabolic Parameters

Under all water treatment levels used in this study, the GABA level was negatively correlated with seed germination, seedlings height, fresh and dry weight, and chlorophyll content (Table 5). On the other hand, GABA level was positively correlated with protein and carbohydrate contents and MDA level under all water treatments. The elevated level of GABA metabolites (GABA, alanine, and glutamate), total proteins, and total carbohydrates content might be involved in cellular osmotic adjustment, protecting plants from oxidative stress, balancing of carbon and nitrogen (C:N) metabolism, and maintaining cell metabolic homeostasis and cell turgor under water stress [9,22].

Table 5. Correlation analysis between the GABA level and all physiological and metabolic parameters that were measured in this study under drought stress treatments after 7 and14 days, separately of pea (*Pisum sativum* L.) seedlings. G% = germination percentage; *r* = correlation coefficient.

r	**Drought Treatments Duration** **7 Days**	**14 Days**
GABA vs. G%	−0.897	−0.989
GABA vs. Seedling height	−0.886	−0.861
GABA vs. Fresh weight	−0.937	−0.812
GABA vs. Dry weight	−0.941	−0.911
GABA vs. Chl *a*	−0.936	−0.894
GABA vs. Chl *b*	−0.935	−0.901
GABA vs. Protein	0.988	0.956
GABA vs. Carbohydrates	0.955	0.923
GABA vs. MDA	0.967	0.877

4. Conclusions

Our study showed that water deficit had a suppressive effect on *Pisum sativum* L. growth and metabolism. Drought stress significantly inhibited seed germination, decreased fresh and dry weight and plant height due to a reduction in water absorption by seeds during metabolic processes and enzymatic activities in germinating pea seeds and early seedling growth. In addition, water deficit caused a significant decrease in chlorophyll pigments that might be attributed to disturbances in the activity of the photosynthetic apparatus. Pea seedlings suffered minor cellular damage under mild drought stress. However, the oxidative damage and ROS production were more severe with increased duration and levels of water deficit. The increase in GABA content as water availability decreased mitigated the damaging effects of drought stress on pea (*Pisum sativum* L.) seedlings by supplying enough carbon/nitrogen sources to the TCA cycle and amino acid synthesis. The significant increase in protein content might lower the osmotic potential to maintain cellular structures and metabolic stability under drought stress. Pea seedlings tolerate prolonged water deficit by keeping stable protein metabolism to adjust cellular osmolarity and to provide protection from further dehydration damage. Water deficit induced carbohydrate accumulation to maintain water potential and osmotic adjustment via the activation of osmolytes synthesis. Overall, the significant increases in GABA, protein and carbohydrate contents were to cope with the physiological impact of drought stress on

Pisum sativum L. seedlings by maintaining cellular osmotic adjustment, protecting plants from oxidative stress, balancing carbon and nitrogen (C:N) metabolism and maintaining cell metabolic homeostasis and cell turgor. According to data presented in the current study, a sufficient water supply is vital for normal growth and metabolism in pea seedlings. Severe (less than 40% water content of the holding capacity) and long-term drought stress should be avoided during the germination stage to ensure proper seedling growth.

Author Contributions: N.A.A.-Q.: Study design, statistical analysis, result interpretation, writing and revision of the manuscript. Z.I.A.-A.: Methodology assistance, data analysis, and result discussion. N.F.Q.: Performing the experimental work, data collection and writing the first draft of the manuscript. All authors have read and agreed to the published version of the manuscript.

Funding: This research was financially supported by the Deanship of Research, Jordan University of Science and Technology, Jordan, grant number [118/2019].

Institutional Review Board Statement: Not applicable.

Informed Consent Statement: Not applicable.

Data Availability Statement: Raw data is available on request.

Conflicts of Interest: No conflict of interest to declare.

References

1. Yang, T.; Fang, L.; Zhang, X.; Hu, J.; Bao, S.; Hao, J.; Li, L.; He, Y.; Jiang, J.; Wang, F.; et al. High-throughput development of SSR markers from pea (*Pisum sativum* L.) based on next generation sequencing of a purified Chinese commercial variety. *PLoS ONE* **2015**, *10*, e0139775. [CrossRef]
2. Rungruangmaitree, R.; Jiraungkoorskul, W. Pea, *Pisum sativum*, and its anticancer activity. *Pharmacogn. Rev.* **2017**, *11*, 39.
3. El-Esawi, M.; Al-Ghamdi, A.; Ali, H.; Alayafi, A.; Witczak, J.; Ahmad, M. Analysis of genetic variation and enhancement of salt tolerance in French pea (*Pisum Sativum* L.). *Int. J. Mol. Sci.* **2018**, *19*, 2433. [CrossRef]
4. Devi, J.; Mishra, G.P.; Sanwal, S.K.; Dubey, R.K.; Singh, P.M.; Singh, B. Development and characterization of penta-flowering and triple-flowering genotypes in garden pea (*Pisum sativum* L. var. hortense). *PLoS ONE* **2018**, *13*, e0201235. [CrossRef] [PubMed]
5. Ranjbar Sistani, N.; Kaul, H.P.; Desalegn, G.; Wienkoop, S. Rhizobium impacts on seed productivity, quality, and protection of *Pisum sativum* upon disease stress caused by didymella pinodes: Phenotypic, proteomic, and metabolomic traits. *Front. Plant Sci.* **2017**, *8*, 1961. [CrossRef]
6. Idrissi, O.; Udupa, S.M.; De Keyser, E.; McGee, R.J.; Coyne, C.J.; Saha, G.C.; Muehlbauer, F.J.; Van Damme, P.; De Riek, J. Identification of quantitative trait loci controlling root and shoot traits associated with drought tolerance in a lentil (*Lens culinaris Medik.*) recombinant inbred line population. *Front. Plant Sci.* **2016**, *7*, 1174. [CrossRef] [PubMed]
7. Muscolo, A.; Junker, A.; Klukas, C.; Weigelt-Fischer, K.; Riewe, D.; Altmann, T. Phenotypic and metabolic responses to drought and salinity of four contrasting lentil accessions. *J. Exp. Bot.* **2015**, *66*, 5467–5480. [CrossRef] [PubMed]
8. Li, P.; Zhang, Y.; Wu, X.; Liu, Y. Drought stress impact on leaf proteome variations of faba bean (*Vicia faba* L.) in the Qinghai–Tibet Plateau of China. *3 Biotech* **2018**, *8*, 110. [CrossRef] [PubMed]
9. Rahmati, M.; Mirás-Avalos, J.M.; Valsesia, P.; Lescourret, F.; Génard, M.; Davarynejad, G.H.; Bannayan, M.; Azizi, M.; Vercambre, G. Disentangling the effects of water stress on carbon acquisition, vegetative growth, and fruit quality of peach trees by means of the QualiTree model. *Front. Plant Sci.* **2018**, *9*, 3. [CrossRef]
10. Alexieva, V.; Sergiev, I.; Mapelli, S.; Karanov, E. The effect of drought and ultraviolet radiation on growth and stress markers in pea and wheat. *Plant Cell Environ.* **2001**, *24*, 1337–1344. [CrossRef]
11. Biswas, D.K.; Jiang, G.M. Differential drought-induced modulation of ozone tolerance in winter wheat species. *J. Exp. Bot.* **2011**, *62*, 4153–4162. [CrossRef]
12. Maiti, R.K.; Satya, P. Research advances in major cereal crops for adaptation to abiotic stresses. *GM Crop. Food* **2014**, *5*, 259–279. [CrossRef]
13. Jin, J.; Lauricella, D.; Armstrong, R.; Sale, P.; Tang, C. Phosphorus application and elevated CO2 enhance drought tolerance in field pea grown in a phosphorus-deficient vertisol. *Ann. Bot.* **2014**, *116*, 975–985. [CrossRef] [PubMed]
14. Szalonek, M.; Sierpien, B.; Rymaszewski, W.; Gieczewska, K.; Garstka, M.; Lichocka, M.; Sass, L.; Paul, K.; Vass, I.; Vankova, R.; et al. Potato annexin STANN1 promotes drought tolerance and mitigates light stress in transgenic *Solanum tuberosum* L. plants. *PLoS ONE* **2015**, *10*, e0132683. [CrossRef]
15. Wang, X.; Cai, X.; Xu, C.; Wang, Q.; Dai, S. Drought-responsive mechanisms in plant leaves revealed by proteomics. *Int. J. Mol. Sci.* **2016**, *17*, 1706. [CrossRef] [PubMed]
16. Charlton, A.J.; Donarski, J.A.; Harrison, M.; Jones, S.A.; Godward, J.; Oehlschlager, S.; Arques, J.L.; Ambrose, M.; Chinoy, C.; Mullineaux, P.M.; et al. Responses of the pea (*Pisum sativum* L.) leaf metabolome to drought stress assessed by nuclear magnetic resonance spectroscopy. *Metabolomics* **2008**, *4*, 312. [CrossRef]

17. Mo, Y.; Wang, Y.; Yang, R.; Zheng, J.; Liu, C.; Li, H.; Ma, J.; Zhang, Y.; Wei, C.; Zhang, X. Regulation of plant growth, photosynthesis, antioxidation and osmosis by an arbuscular mycorrhizal fungus in watermelon seedlings under well-watered and drought conditions. *Front. Plant Sci.* **2016**, *7*, 644. [CrossRef]
18. Chmielnicka, A.; Żabka, A.; Winnicki, K.; Maszewski, J.; Polit, J.T. Endoreplication and its consequences in the suspensor of *Pisum sativum*. *Plant Cell Rep.* **2018**, *37*, 1639–1651. [CrossRef] [PubMed]
19. Cattivelli, L.; Rizza, F.; Badeck, F.W.; Mazzucotelli, E.; Mastrangelo, A.M.; Francia, E.; Marè, C.; Tondelli, A.; Stanca, A.M. Drought tolerance improvement in crop plants: An integrated view from breeding to genomics. *Field Crop. Res.* **2008**, *105*, 1–14. [CrossRef]
20. Kakumanu, A.; Ambavaram, M.M.; Klumas, C.; Krishnan, A.; Batlang, U.; Myers, E.; Grene, R.; Pereira, A. Effects of drought on gene expression in maize reproductive and leaf meristem tissue revealed by RNA-Seq. *Plant Physiol.* **2012**, *160*, 846–867. [CrossRef]
21. Mao, H.; Wang, H.; Liu, S.; Li, Z.; Yang, X.; Yan, J.; Li, J.; Tran, L.S.P.; Qin, F. A transposable element in a NAC gene is associated with drought tolerance in maize seedlings. *Nat. Commun.* **2015**, *6*, 8326. [CrossRef]
22. Chen, D.; Wang, S.; Cao, B.; Cao, D.; Leng, G.; Li, H.; Yin, L.; Shan, L.; Deng, X. Genotypic variation in growth and physiological response to drought stress and re-watering reveals the critical role of recovery in drought adaptation in maize seedlings. *Front. Plant Sci.* **2016**, *6*, 1241. [CrossRef]
23. Bunce, J.A. Leaf transpiration efficiency of some drought-resistant maize lines. *Crop Sci.* **2010**, *50*, 1409–1413. [CrossRef]
24. Alghabari, F.; Ihsan, M.Z. Effects of drought stress on growth, grain filling duration, yield and quality attributes of barley (*Hordeum vulgare* L.). *Bangladesh J. Bot.* **2018**, *47*, 421–428. [CrossRef]
25. Hussain, M.; Farooq, S.; Hasan, W.; Ul-Allah, S.; Tanveer, M.; Farooq, M.; Nawaz, A. Drought stress in sunflower: Physiological effects and its management through breeding and agronomic alternatives. *Agric. Water Manag.* **2018**, *201*, 152–166. [CrossRef]
26. Nonami, H. Plant water relations and control of cell elongation at low water potentials. *J. Plant Res.* **1998**, *111*, 373–382. [CrossRef]
27. Zhang, J.; Mason, A.S.; Wu, J.; Liu, S.; Zhang, X.; Luo, T.; Redden, R.; Batley, J.; Hu, L.; Yan, G. Identification of putative candidate genes for water stress tolerance in canola (Brassica napus). *Front. Plant Sci.* **2015**, *6*, 1058. [CrossRef]
28. Sorensen, J.N.; Edelenbos, M.; Wienberg, L. Drought effects on green pea texture and related physical-chemical properties at comparable maturity. *J. Am. Soc. Hortic. Sci.* **2003**, *128*, 128–135. [CrossRef]
29. Kumar, N.; Dubey, A.K.; Upadhyay, A.K.; Gautam, A.; Ranjan, R.; Srikishna, S.; Sahu, N.; Behera, S.K.; Mallick, S. GABA accretion reduces Lsi-1 and Lsi-2 gene expressions and modulates physiological responses in *Oryza sativa* to provide tolerance towards arsenic. *Sci. Rep.* **2017**, *7*, 8786. [CrossRef]
30. Carillo, P. GABA shunt in durum wheat. *Front. Plant Sci.* **2018**, *9*, 100. [CrossRef] [PubMed]
31. White, J.P.; Prell, J.; Ramachandran, V.K.; Poole, P.S. Characterization of a γ-aminobutyric acid transport system of Rhizobium leguminosarum bv. viciae 3841. *J. Bacteriol.* **2009**, *191*, 1547–1555. [CrossRef]
32. Scharff, A.M.; Egsgaard, H.; Hansen, P.E.; Rosendahl, L. Exploring symbiotic nitrogen fixation and assimilation in pea root nodules by in vivo 15N nuclear magnetic resonance spectroscopy and liquid chromatography-mass spectrometry. *Plant Physiol.* **2003**, *131*, 367–378. [CrossRef]
33. Yong, B.; Xie, H.; Li, Z.; Li, Y.P.; Zhang, Y.; Nie, G.; Zhang, X.Q.; Ma, X.; Huang, L.K.; Yan, Y.H.; et al. Exogenous application of GABA improves PEG-induced drought tolerance positively associated with GABA-shunt, polyamines, and proline metabolism in white clover. *Front. Physiol.* **2017**, *8*, 1107. [CrossRef]
34. Xiang, L.; Hu, L.; Xu, W.; Zhen, A.; Zhang, L.; Hu, X. Exogenous γ-aminobutyric acid improves the structure and function of photosystem II in muskmelon seedlings exposed to salinity-alkalinity stress. *PLoS ONE* **2016**, *11*, e0164847. [CrossRef] [PubMed]
35. Samarah, N.H.; Mullen, R.E.; Cianzio, S.R.; Scott, P. Dehydrin-like proteins in soybean seeds in response to drought stress during seed filling. *Crop Sci.* **2006**, *46*, 2141–2150. [CrossRef]
36. Jeffrey, S.T.; Humphrey, G.F. New spectrophotometric equations for determining chlorophylls a, b, c1 and c2 in higher plants, algae and natural phytoplankton. *Biochem. Und Physiol. Pflanz.* **1975**, *167*, 191–194. [CrossRef]
37. Zhang, G.; Bown, A.W. The rapid determination of γ-aminobutyric acid. *Phytochemistry* **1997**, *44*, 1007–1009. [CrossRef]
38. Bergmeyer, H.U. *Methods of Enzymatic Analysis*, 2nd ed.; Verlag Chemie: Weinheim, Germany, 1983; Volume I, p. 42.
39. Heath, R.L.; Packer, L. Photoperoxidation in isolated chloroplasts: I. Kinetics and stoichiometry of fatty acid peroxidation. *Arch. Biochem. Biophys.* **1968**, *125*, 189–198. [CrossRef]
40. Li, H.; Li, X.; Zhang, D.; Liu, H.; Guan, K. Effects of drought stress on the seed germination and early seedling growth of the endemic desert plant *Eremosparton songoricum* (Fabaceae). *Excli J.* **2013**, *12*, 89.
41. Liu, M.; Li, M.; Liu, K.; Sui, N. Effects of drought stress on seed germination and seedling growth of different maize varieties. *J. Agric. Sci.* **2015**, *7*, 231. [CrossRef]
42. Pirdashti, H.; Sarvestani, Z.T.; Nematzadeh, G.H.; Ismail, A. Effect of water stress on seed germination and seedling growth of rice (*Oryza sativa* L.) genotypes. *Pak. J. Agron.* **2003**, *2*, 217–222. [CrossRef]
43. Murillo-Amador, B.; López-Aguilar, R.; Kaya, C.; Larrinaga-Mayoral, J.; Flores-Hernández, A. Comparative effects of NaCl and polyethylene glycol on germination, emergence and seedling growth of cowpea. *J. Agron. Crop Sci.* **2002**, *188*, 235–247. [CrossRef]
44. Hadas, A. Water uptake and germination of leguniinous seeds under changing external water potential in osmotic solutions. *J. Exp. Bot.* **1976**, *27*, 480–489. [CrossRef]
45. Muscolo, A.; Sidari, M.; Anastasi, U.; Santonoceto, C.; Maggio, A. Effect of PEG-induced drought stress on seed germination of four lentil genotypes. *J. Plant Interact.* **2014**, *9*, 354–363. [CrossRef]

46. Bentsink, L.; Koornneef, M. Seed dormancy and germination. In *The Arabidopsis Book/American Society of Plant Biologists*; American Society of Plant Biologists: Rockville, MD, USA, 2008; Volume 6.
47. Dornbos, D.L.; Mullen, R.E.; Shibles, R.E. Drought stress effects during seed fill on soybean seed germination and vigor. *Crop Sci.* **1989**, *29*, 476–480. [CrossRef]
48. Kim, S.K.; Son, T.K.; Park, S.Y.; Lee, I.J.; Lee, B.H.; Kim, H.Y.; Lee, S.C. Influences of gibberellin and auxin on endogenous plant hormone and starch mobilization during rice seed germination under salt stress. *J. Environ. Biol.* **2006**, *27*, 181.
49. Li, Z.; Peng, Y.; Zhang, X.Q.; Ma, X.; Huang, L.K.; Yan, Y.H. Exogenous spermidine improves seed germination of white clover under water stress via involvement in starch metabolism, antioxidant defenses and relevant gene expression. *Molecules* **2014**, *19*, 18003–18024. [CrossRef] [PubMed]
50. AL-Quraan, N.A.; Sartawe, F.A.B.; Qaryouti, M.M. Characterization of γ-aminobutyric acid metabolism and oxidative damage in wheat (*Triticum aestivum* L.) seedlings under salt and osmotic stress. *J. Plant Physiol.* **2013**, *170*, 1003–1009. [CrossRef]
51. AL-Quraan, N.A.; Al-Sharbati, M.; Dababneh, Y.; Al-Olabi, M. Effect of temperature, salt and osmotic stresses on seed germination and chlorophyll contents in lentil (*Lens culinaris Medik*). *Acta Hortic.* **2014**, *1054*, 47–54. [CrossRef]
52. Embiale, A.; Hussein, M.; Husen, A.; Sahile, S.; Mohammed, K. Differential sensitivity of *Pisum sativum* L. cultivars to water-deficit stress: Changes in growth, water status, chlorophyll fluorescence and gas exchange attributes. *J. Agron.* **2016**, *15*, 45–57. [CrossRef]
53. Morgan, J.M. Osmoregulation and water stress in higher plants. *Annu. Rev. Plant Physiol.* **1984**, *35*, 299–319. [CrossRef]
54. Jaleel, C.A.; Manivannan, P.A.; Wahid, A.; Farooq, M.; Al-Juburi, H.J.; Somasundaram, R.A.; Panneerselvam, R. Drought stress in plants: A review on morphological characteristics and pigments composition. *Int. J. Agric. Biol.* **2009**, *11*, 100–105.
55. Jaleel, C.A.; Manivannan, P.; Lakshmanan, G.M.A.; Gomathinayagam, M.; Panneerselvam, R. Alterations in morphological parameters and photosynthetic pigment responses of *Catharanthus roseus* under soil water deficits. *Colloids Surf. B Biointerfaces* **2008**, *61*, 298–303. [CrossRef] [PubMed]
56. Medrano, H.; Escalona, J.M.; Bota, J.; Gulías, J.; Flexas, J. Regulation of photosynthesis of C3 plants in response to progressive drought: Stomatal conductance as a reference parameter. *Ann. Bot.* **2002**, *89*, 895–905. [CrossRef]
57. Fahad, S.; Bajwa, A.A.; Nazir, U.; Anjum, S.A.; Farooq, A.; Zohaib, A.; Sadia, S.; Nasim, W.; Adkins, S.; Saud, S.; et al. Crop production under drought and heat stress: Plant responses and management options. *Front. Plant Sci.* **2017**, *8*, 1147. [CrossRef] [PubMed]
58. Fraser, T.E.; Silk, W.K.; Rost, T.L. Effects of low water potential on cortical cell length in growing regions of maize roots. *Plant Physiol.* **1990**, *93*, 648–651. [CrossRef] [PubMed]
59. Khorasaninejad, S.; Mousavi, A.; Soltanloo, H.; Hemmati, K.; Khalighi, A. The effect of salinity stress on growth parameters, essential oil yield and constituent of peppermint (*Mentha piperita* L.). *World Appl. Sci. J.* **2010**, *11*, 1403–1407.
60. Kabay, T.; Erdinc, C.; Sensoy, S. Effects of drought stress on plant growth parameters, membrane damage index and nutrient content in common bean genotypes. *J. Anim. Plant Sci.* **2017**, *27*, 940–952.
61. Zhang, X.; Lei, L.; Lai, J.; Zhao, H.; Song, W. Effects of drought stress and water recovery on physiological responses and gene expression in maize seedlings. *Biomedcentral Plant Biol.* **2018**, *18*, 68. [CrossRef]
62. Regier, N.; Streb, S.; Cocozza, C.; Schaub, M.; Cherubini, P.; Zeeman, S.C.; Frey, B. Drought tolerance of two black poplar (*Populus nigra* L.) clones: Contribution of carbohydrates and oxidative stress defense. *Plant Cell Environ.* **2009**, *32*, 1724–1736. [CrossRef]
63. Riad, G.S.; Youssef, S.M.; El-Azm, N.A.A.; Ahmed, E.M. Amending Sandy Soil with Biochar or/and Superabsorbent Polymer Mitigates the Adverse Effects of Drought Stress on Green Pea. *Egypt. J. Hortic.* **2018**, *45*, 169–183.
64. Nyachiro, J.M.; Briggs, K.G.; Hoddinott, J.; Johnson-Flanagan, A.M. Chlorophyll content, chlorophyll fluorescence and water deficit in spring wheat. *Cereal Res. Commun.* **2001**, *29*, 135–142. [CrossRef]
65. Shinde, B.P.; Thakur, J. Influence of Arbuscular mycorrhizal fungi on chlorophyll, proteins, proline and total carbohydrates content of the pea plant under water stress condition. *Int. J. Curr. Microbiol. Appl. Sci.* **2015**, *4*, 809–821.
66. Kyparissis, A.; Petropoulou, Y.; Manetas, Y. Summer survival of leaves in a soft-leaved shrub (*Phlomis fruticosa L., Labiatae*) under Mediterranean field conditions: Avoidance of photoinhibitory damage through decreased chlorophyll contents. *J. Exp. Bot.* **1995**, *46*, 1825–1831. [CrossRef]
67. Sepehri, A.; Golparvar, A.R. The effect of drought stress on water relations, chlorophyll content and leaf area in canola cultivars (*Brassica napus* L.). *Electron. J. Biol.* **2011**, *7*, 49–53.
68. Nayyar, H.; Gupta, D. Differential sensitivity of C3 and C4 plants to water deficit stress: Association with oxidative stress and antioxidants. *Environ. Exp. Bot.* **2006**, *58*, 106–113. [CrossRef]
69. Massacci, A.; Nabiev, S.M.; Pietrosanti, L.; Nematov, S.K.; Chernikova, T.N.; Thor, K.; Leipner, J. Response of the photosynthetic apparatus of cotton (*Gossypium hirsutum*) to the onset of drought stress under field conditions studied by gas-exchange analysis and chlorophyll fluorescence imaging. *Plant Physiol. Biochem.* **2008**, *46*, 189–195. [CrossRef] [PubMed]
70. Husen, A. Growth characteristics, physiological and metabolic responses of teak (*Tectona grandis Linn.* f.) clones differing in rejuvenation capacity subjected to drought stress. *Silvae Genet.* **2010**, *59*, 124–136. [CrossRef]
71. Ashraf, M.Y.; Azmi, A.R.; Khan, A.H.; Ala, S.A. Effect of water stress on total phenols, peroxidase activity and chlorophyll content in wheat (*Triticum aestivum* L.). *Acta Physiol. Plant.* **1994**, *16*, 3.
72. Santos, M.G.D.; Ribeiro, R.V.; Oliveira, R.F.D.; Pimentel, C. Gas exchange and yield response to foliar phosphorus application in *Phaseolus vulgaris* L. under drought. *Braz. J. Plant Physiol.* **2004**, *16*, 171–179. [CrossRef]

73. Srivastava, A.K.; Lokhande, V.H.; Patade, V.Y.; Suprasanna, P.; Sjahril, R.; D'Souza, S.F. Comparative evaluation of hydro-, chemo-, and hormonal-priming methods for imparting salt and PEG stress tolerance in Indian mustard (*Brassicajuncea* L.). *Acta Physiol. Plant.* **2010**, *32*, 1135–1144. [CrossRef]
74. Afshari, M.; Shekari, F.; Azimkhani, R.; Habibi, H.; Fotokian, M.H. Effects of foliar application of salicylic acid on growth and physiological attributes of cowpea under water stress conditions. *Iran Agric. Res.* **2013**, *32*, 55–70.
75. Tóth, V.R.; Mészáros, I.; Veres, S.; Nagy, J. Effects of the available nitrogen on the photosynthetic activity and xanthophyll cycle pool of maize in field. *J. Plant Physiol.* **2002**, *159*, 627–634. [CrossRef]
76. Fariduddin, Q.; Khanam, S.; Hasan, S.A.; Ali, B.; Hayat, S.; Ahmad, A. Effect of 28-homobrassinolide on the drought stress-induced changes in photosynthesis and antioxidant system of *Brassica juncea* L. *Acta Physiol. Plant.* **2009**, *31*, 889–897. [CrossRef]
77. Pastenes, C.; Pimentel, P.; Lillo, J. Leaf movements and photoinhibition in relation to water stress in field-grown beans. *J. Exp. Bot.* **2004**, *56*, 425–433. [CrossRef]
78. Lawlor, D.W. Limitation to photosynthesis in water-stressed leaves: Stomata vs. metabolism and the role of ATP. *Ann. Bot.* **2002**, *89*, 871–885. [CrossRef] [PubMed]
79. Shahriari, R.; Karimi, L. Evaluation of cold tolerance in wheat germplasm by measuring chlorophyll content and leaf color. In *Abstracts Seventh Crop Science Congress of Iran*; Seventh Iranian Congress of Plant Breeding and Plants (Persian); Shahid Bahonar University of Kerman, Pazhuhesh Square, Emam Khomeini Blvd.: Kerman, Iran, 2001; p. 507.
80. Pandey, H.C.; Baig, M.J.; Bhatt, R.K. Effect of moisture stress on chlorophyll accumulation and nitrate reductase activity at vegetative and flowering stage in *Avena* species. *Agric. Sci. Res. J.* **2012**, *2*, 111–118.
81. Alberte, R.S.; Thornber, J.P.; Fiscus, E.L. Water stress effects on the content and organization of chlorophyll in mesophyll and bundle sheath chloroplasts of maize. *Plant Physiol.* **1977**, *59*, 351–353. [CrossRef]
82. Patro, L.; Mohapatra, P.K.; Biswal, U.C.; Biswal, B. Dehydration induced loss of photosynthesis in Arabidopsis leaves during senescence is accompanied by the reversible enhancement in the activity of cell wall β-glucosidase. *J. Photochem. Photobiol. B Biol.* **2014**, *137*, 49–54. [CrossRef] [PubMed]
83. Ghotbi-Ravandi, A.A.; Shahbazi, M.; Shariati, M.; Mulo, P. Effects of mild and severe drought stress on photosynthetic efficiency in tolerant and susceptible barley (*Hordeum vulgare* L.) genotypes. *J. Agron. Crop Sci.* **2014**, *200*, 403–415. [CrossRef]
84. Karataş, İ.; Öztürk, L.; Demir, Y.; Ünlükara, A.; Kurunç, A.; Düzdemir, O. Alterations in antioxidant enzyme activities and proline content in pea leaves under long-term drought stress. *Toxicol. Ind. Health* **2014**, *30*, 693–700. [CrossRef]
85. Nikolaeva, M.K.; Maevskaya, S.N.; Shugaev, A.G.; Bukhov, N.G. Effect of drought on chlorophyll content and antioxidant enzyme activities in leaves of three wheat cultivars varying in productivity. *Russ. J. Plant Physiol.* **2010**, *57*, 87–95. [CrossRef]
86. Bown, A.W.; Shelp, B.J. The metabolism and functions of [gamma]-aminobutyric acid. *Plant Physiol.* **1997**, *115*, 1. [CrossRef]
87. Li, W.; Liu, J.; Ashraf, U.; Li, G.; Li, Y.; Lu, W.; Gao, L.; Han, F.; Hu, J. Exogenous γ-aminobutyric acid (GABA) application improved early growth, net photosynthesis, and associated physio-biochemical events in maize. *Front. Plant Sci.* **2016**, *7*, 919. [CrossRef]
88. Zhu, Z.; Shi, Z.; Xie, C.; Gong, W.; Hu, Z.; Peng, Y. A novel mechanism of Gamma-aminobutyric acid (GABA) protecting human umbilical vein endothelial cells (HUVECs) against H_2O_2-induced oxidative injury. *Comp. Biochem. Physiol. Part C Toxicol. Pharmacol.* **2019**, *217*, 68–75. [CrossRef]
89. Kaplan, F.; Kopka, J.; Haskell, D.W.; Zhao, W.; Schiller, K.C.; Gatzke, N.; Sung, D.Y.; Guy, C.L. Exploring the temperature-stress metabolome of Arabidopsis. *Plant Physiol.* **2004**, *136*, 4159–4168. [CrossRef]
90. Fait, A.; Fromm, H.; Walter, D.; Galili, G.; Fernie, A.R. Highway or byway: The metabolic role of the GABA shunt in plants. *Trends Plant Sci.* **2008**, *13*, 14–19. [CrossRef]
91. Bashir, R.; Riaz, H.N.; Shafiq, M.; Parveen, N.; Alenazi, M.M.; Anwar, S.; Alebidi, A.I. Foliar application of Y-aminobutyric acid (gaba) improves vegetative growth, and the physiological and antioxidative potential of *Daucus Carota* L. under water deficit conditions. *Preprints* **2019**, 2019030227. [CrossRef]
92. Kinnersley, A.M.; Turano, F.J. Gamma aminobutyric acid (GABA) and plant responses to stress. *Crit. Rev. Plant Sci.* **2000**, *19*, 479–509. [CrossRef]
93. Krishnan, S.; Laskowski, K.; Shukla, V.; Merewitz, E.B. Mitigation of drought stress damage by exogenous application of a non-protein amino acid γ–aminobutyric acid on perennial ryegrass. *J. Am. Soc. Hortic. Sci.* **2013**, *138*, 358–366. [CrossRef]
94. Vijayakumari, K.; Puthur, J.T. γ-Aminobutyric acid (GABA) priming enhances the osmotic stress tolerance in Piper nigrum Linn. plants subjected to PEG-induced stress. *Plant Growth Regul.* **2016**, *78*, 57–67. [CrossRef]
95. Wang, C.; Fan, L.; Gao, H.; Wu, X.; Li, J.; Lv, G.; Gong, B. Polyamine biosynthesis and degradation are modulated by exogenous gamma-aminobutyric acid in root-zone hypoxia-stressed melon roots. *Plant Physiol. Biochem.* **2014**, *82*, 17–26. [CrossRef] [PubMed]
96. Mibei, E.K.; Owino, W.O.; Ambuko, J.; Giovannoni, J.J.; Onyango, A.N. Metabolomic analyses to evaluate the effect of drought stress on selected African Eggplant accessions. *J. Sci. Food Agric.* **2018**, *98*, 205–216. [CrossRef] [PubMed]
97. Merewitz, E.B.; Du, H.; Yu, W.; Liu, Y.; Gianfagna, T.; Huang, B. Elevated cytokinin content in ipt transgenic creeping bentgrass promotes drought tolerance through regulating metabolite accumulation. *J. Exp. Bot.* **2011**, *63*, 1315–1328. [CrossRef] [PubMed]
98. Marček, T.; Hamow, K.Á.; Végh, B.; Janda, T.; Darko, E. Metabolic response to drought in six winter wheat genotypes. *PLoS ONE* **2019**, *14*, e0212411.

99. Michaletti, A.; Naghavi, M.R.; Toorchi, M.; Zolla, L.; Rinalducci, S. Metabolomics and proteomics reveal drought-stress responses of leaf tissues from spring-wheat. *Sci. Rep.* **2018**, *8*, 5710. [CrossRef]
100. Salama, H.M.; Al Watban, A.A.; Al-Fughom, A.T. Effect of ultraviolet radiation on chlorophyll, carotenoid, protein and proline contents of some annual desert plants. *Saudi J. Biol. Sci.* **2011**, *18*, 79–86. [CrossRef]
101. Guttieri, M.J.; McLean, R.; Stark, J.C.; Souza, E. Managing irrigation and nitrogen fertility of hard spring wheats for optimum bread and noodle quality. *Crop Sci.* **2005**, *45*, 2049–2059. [CrossRef]
102. Ozturk, A.; Aydin, F. Effect of water stress at various growth stages on some quality characteristics of winter wheat. *J. Agron. Crop Sci.* **2004**, *190*, 93–99. [CrossRef]
103. Rizhsky, L.; Liang, H.; Mittler, R. The combined effect of drought stress and heat shock on gene expression in tobacco. *Plant Physiol.* **2002**, *130*, 1143–1151. [CrossRef]
104. Jin, R.; Wang, Y.; Liu, R.; Gou, J.; Chan, Z. Physiological and metabolic changes of purslane (*Portulaca oleracea* L.) in response to drought, heat, and combined stresses. *Front. Plant Sci.* **2016**, *6*, 1123. [CrossRef] [PubMed]
105. Ashoub, A.; Beckhaus, T.; Berberich, T.; Karas, M.; Brüggemann, W. Comparative analysis of barley leaf proteome as affected by drought stress. *Planta* **2013**, *237*, 771–781. [CrossRef]
106. Alvarez, S.; Roy Choudhury, S.; Pandey, S. Comparative quantitative proteomics analysis of the ABA response of roots of drought-sensitive and drought-tolerant wheat varieties identifies proteomic signatures of drought adaptability. *J. Proteome Res.* **2014**, *13*, 1688–1701. [CrossRef] [PubMed]
107. Suguiyama, V.F.; da Silva, E.A.; Meirelles, S.T.; Centeno, D.D.C.; Braga, M.R. Leaf metabolite profile of the Brazilian resurrection plant Barbacenia purpurea Hook. (*Velloziaceae*) shows two time-dependent responses during desiccation and recovering. *Front. Plant Sci.* **2014**, *5*, 96. [CrossRef] [PubMed]
108. Ford, K.L.; Cassin, A.; Bacic, A.F. Quantitative proteomic analysis of wheat cultivars with differing drought stress tolerance. *Front. Plant Sci.* **2011**, *2*, 44. [CrossRef]
109. Gołębiowska-Pikania, G.; Kopeć, P.; Surówka, E.; Janowiak, F.; Krzewska, M.; Dubas, E.; Nowicka, A.; Kasprzyk, J.; Ostrowska, A.; Malaga, S.; et al. Changes in protein abundance and activity induced by drought during generative development of winter barley (*Hordeum vulgare* L.). *J. Proteom.* **2017**, *169*, 73–86. [CrossRef] [PubMed]
110. Mohammadkhani, N.; Heidari, R. Effects of drought stress on soluble proteins in two maize varieties. *Turk. J. Biol.* **2008**, *32*, 23–30.
111. Ashraf, M.; Foolad, M.R. Roles of glycine betaine and proline in improving plant abiotic stress resistance. *Environ. Exp. Bot.* **2007**, *59*, 206–216. [CrossRef]
112. Bhosale, K.S.; Shinde, B. Influence of arbuscular mycorrhizal fungi on proline and chlorophyll content in Zingiber officinale Rosc grown under water stress. *Indian J. Fundam. Appl. Life Sci.* **2011**, *1*, 172–176.
113. Nayyar, H. Accumulation of osmolytes and osmotic adjustment in water-stressed wheat (*Triticum aestivum* L.) and maize (Zea mays) affected by calcium and its antagonists. *Environ. Exp. Bot.* **2003**, *50*, 253–264. [CrossRef]
114. Zhu, X.; Gong, H.; Chen, G.; Wang, S.; Zhang, C. Different solute levels in two spring wheat cultivars induced by progressive field water stress at different developmental stages. *J. Arid Environ.* **2005**, *62*, 1–14. [CrossRef]
115. Singh, M.; Kumar, J.; Singh, S.; Singh, V.P.; Prasad, S.M. Roles of osmoprotectants in improving salinity and drought tolerance in plants: A review. *Rev. Environ. Sci. Bio Technol.* **2015**, *14*, 407–426. [CrossRef]
116. Shahryar, N.; Maali-Amiri, R. Metabolic acclimation of tetraploid and hexaploid wheats by cold stress-induced carbohydrate accumulation. *J. Plant Physiol.* **2016**, *204*, 44–53. [CrossRef]
117. Arabzadeh, N. H aloxylon persicum and H aloxylon aphyllum. *Asian J. Plant Sci.* **2012**, *11*, 44–51. [CrossRef]
118. Kameli, A.; Lösel, D.M. Growth and sugar accumulation in durum wheat plants under water stress. *New Phytol.* **1996**, *132*, 57–62. [CrossRef]
119. Fazeli, F.; Ghorbanli, M.; Niknam, V. Effect of drought on water relations, growth and solute accumulation in two sesame cultivars. *Pak. J. Biol. Sci.* **2006**, *9*, 1829–1835. [CrossRef]
120. Akinci, S.; Losel, D.M. Effects of water stress and recovery periods on soluble sugars and starch content in cucumber cultivars. *Fresenius Environ. Bull.* **2010**, *19*, 164–171.
121. Lemoine, R.; La Camera, S.; Atanassova, R.; Dédaldéchamp, F.; Allario, T.; Pourtau, N.; Bonnemain, J.L.; Laloi, M.; Coutos-Thévenot, P.; Maurousset, L.; et al. Source-to-sink transport of sugar and regulation by environmental factors. *Front. Plant Sci.* **2013**, *4*, 272. [CrossRef] [PubMed]
122. Jain, N.K.; Roy, I. Effect of trehalose on protein structure. *Protein Sci.* **2009**, *18*, 24–36. [CrossRef]
123. Buitink, J.; Leprince, O. Glass formation in plant anhydrobiotes: Survival in the dry state. *Cryobiology* **2004**, *48*, 215–228. [CrossRef]
124. Shvaleva, A.L.; Silva, F.C.E.; Breia, E.; Jouve, J.; Hausman, J.F.; Almeida, M.H.; Maroco, J.P.; Rodrigues, M.L.; Pereira, J.S.; Chaves, M.M. Metabolic responses to water deficit in two Eucalyptus globulus clones with contrasting drought sensitivity. *Tree Physiol.* **2006**, *26*, 239–248. [CrossRef]
125. Gibson, S.I. Control of plant development and gene expression by sugar signaling. *Curr. Opin. Plant Biol.* **2005**, *8*, 93–102. [CrossRef] [PubMed]
126. Van den Ende, W.; Peshev, D. Sugars as antioxidants in plants. In *Crop Improvement under Adverse Conditions*; Springer: New York, NY, USA, 2013; pp. 285–307.
127. Li, T.H.; Li, S.H. Leaf responses of micropropagated apple plants to water stress: Nonstructural carbohydrate composition and regulatory role of metabolic enzymes. *Tree Physiol.* **2005**, *25*, 495–504. [CrossRef] [PubMed]

128. Moore, K.; Roberts, L.J. Measurement of lipid peroxidation. *Free Radic. Res.* **1998**, *28*, 659–671. [CrossRef]
129. Apel, K.; Hirt, H. Reactive oxygen species: Metabolism, oxidative stress, and signal transduction. *Annu. Rev. Plant Biol.* **2004**, *55*, 373–399. [CrossRef] [PubMed]
130. Berjak, P. The challenge of recalcitrant germplasm cryopreservation. *J. Hortic. Sci. Biotechnol.* **2006**, *81*, 781–782. [CrossRef]
131. Gill, S.S.; Tuteja, N. Reactive oxygen species and antioxidant machinery in abiotic stress tolerance in crop plants. *Plant Physiol. Biochem.* **2010**, *48*, 909–930. [CrossRef]
132. Rahal, A.; Kumar, A.; Singh, V.; Yadav, B.; Tiwari, R.; Chakraborty, S.; Dhama, K. Oxidative stress, prooxidants, and antioxidants: The interplay. *Biomed Res. Int.* **2014**, *2014*, 761264. [CrossRef]
133. Pandey, H.C.; Baig, M.J.; Chandra, A.; Bhatt, R.K. Drought stress induced changes in lipid peroxidation and antioxidant system in genus Avena. *J. Environ. Biol.* **2010**, *31*, 435–440.
134. Zlatev, Z.S.; Lidon, F.C.; Ramalho, J.C.; Yordanov, I.T. Comparison of resistance to drought of three bean cultivars. *Biol. Plant.* **2006**, *50*, 389–394. [CrossRef]
135. Tatar, O.; Gevrek, I. Lipid peroxidation and water content of wheat. *Asian J. Plant Sci.* **2008**, *7*, 409–412. [CrossRef]
136. Morabito, D.; Guerrier, G. The free oxygen radical scavenging enzymes and redox status in roots and leaves of Populus x Euramericana in response to osmotic stress, desiccation and rehydration. *J. Plant Physiol.* **2000**, *157*, 74–80. [CrossRef]
137. Abid, M.; Ali, S.; Qi, L.K.; Zahoor, R.; Tian, Z.; Jiang, D.; Snider, J.L.; Dai, T. Physiological and biochemical changes during drought and recovery periods at tillering and jointing stages in wheat (*Triticum aestivum* L.). *Sci. Rep.* **2018**, *8*, 4615. [CrossRef] [PubMed]
138. Hernandez, J.A.; Jimenez, A.; Mullineaux, P.; Sevilia, F. Tolerance of pea (*Pisum sativum* L.) to long-term salt stress is associated with induction of antioxidant defences. *Plant Cell Environ.* **2000**, *23*, 853–862. [CrossRef]

horticulturae

MDPI

Article

Impact of Drought and Flooding on Alkaloid Production in *Annona crassiflora* Mart

Ana Beatriz Marques Honório [1], Iván De-la-Cruz-Chacón [2], Mariano Martínez-Vázquez [3], Magali Ribeiro da Silva [4], Felipe Girotto Campos [1], Bruna Cavinatti Martin [1], Gustavo Cabral da Silva [1], Carmen Sílvia Fernandes Boaro [1,*] and Gisela Ferreira [1]

1 Plant Biology Sector, Biostatistics, Plant Biology, Parasitology and Zoology Department, Institute of Biosciences, São Paulo State University, UNESP, Street Prof. Dr. Antônio Celso Wagner Zanin 250, Botucatu 18618-689, SP, Brazil; beatriz.honorio@unesp.br (A.B.M.H.); felipe.girotto@unesp.br (F.G.C.); bruna.cavinatti@unesp.br (B.C.M.); gustavo.cabral-silva@unesp.br (G.C.d.S.); gisela.ferreira@unesp.br (G.F.)
2 Laboratorio de Fisiología y Química Vegetal, Instituto de Ciencias Biológicas, Universidad de Ciencias y Artes de Chiapas, UNICACH, Libramiento Norte Poniente 1150, Tuxtla Gutiérrez 29039, CHIS, Mexico; ivan.cruz@unicach.mx
3 Chemistry Institute, Universidad Nacional Autónoma de México, UNAM, Circuito Exterior S/N, C.U., Coyoacán 04510, CDMX, Mexico; marvaz@unam.mx
4 Forest, Soil and Environmental Sciences Department, Faculty of Agronomic Science, São Paulo State University, UNESP, Avenida Universitária 3780, Botucatu 18610-034, SP, Brazil; magali.ribeiro@unesp.br
* Correspondence: carmen.boaro@unesp.br; Tel.: +55-(14)-3880-0124

Citation: Honório, A.B.M.; De-la-Cruz-Chacón, I.; Martínez-Vázquez, M.; da Silva, M.R.; Campos, F.G.; Martin, B.C.; da Silva, G.C.; Fernandes Boaro, C.S.; Ferreira, G. Impact of Drought and Flooding on Alkaloid Production in *Annona crassiflora* Mart. *Horticulturae* **2021**, *7*, 414. https://doi.org/10.3390/horticulturae7100414

Academic Editor: Yanyou Wu

Received: 26 September 2021
Accepted: 14 October 2021
Published: 19 October 2021

Abstract: The Brazilian Cerrado is the second largest Brazilian biome. In recent decades, a reduction in rainfall has indicated an extension of the dry season. Among the many native species of the Cerrado of the Annonaceae family and used in folk medicine, *Annona crassiflora* Mart. has fruits of high nutritional value and its by-products are sources of bioactive compounds, such as alkaloids. The aim of the study was to investigate how water stress impacts the production of alkaloids. The study was carried out in a nursery, and the knowledge was flood, field capacity and drought. Gas exchange, chlorophyll *a* fluorescence, antioxidant enzymes, total soluble sugars, starch, reducing sugars, sucrose, total alkaloids and liriodenine were analyzed. We observed that plants subjected to drought had an increase in the production of total alkaloids and liriodenine, without a reduction in photosynthetic metabolism. Plants kept under drought and flood conditions dissipated higher peroxidase activity, while catalase was higher in flooded plants. Starch showed the highest concentration in flooding plants without differing from drought plants; the lowest trehalose concentrations were found in both drought and flooding plants. The drought stimulated the synthesis of total alkaloids and liriodenine without reducing the primary metabolism, which suggests adaptation to Cerrado conditions.

Keywords: Annonaceae; antioxidant enzymes; carbohydrates; liriodenine; photosynthesis

1. Introduction

The Brazilian Cerrado is the second largest Brazilian biome, considered one of the 25 global biodiversity hotspots, present in more than twelve states and occupying approximately 25% of the national territory, with native flora characterized by small and twisted trees [1].

This biome covers important aquifers and rivers [2] and is located in the central area of the country, being the origin of large hydrographic regions in Brazil and in the South American continent [3,4]. However, Lee et al. [5], Debortoli et al. [6] and Penereiro et al. [7] reported reduction of approximately 70 mm in annual precipitation in the Cerrado region between 1979 and 2006, indicating an extension of the dry season. Furthermore, although with great biological diversity, the biome is under continuous threat of extinction due to the expansion of agriculture and pastures, as occurs with other biomes [2,8].

The Cerrado vegetation is exposed to high irradiances (1500 to 2500 μmol. m^{-2} s^{-1}), high temperatures (25–40 °C at midday) and in the dry season, low relative humidity (10 to 20%) [9]. Although there is abundance of light, the seasonality of the rainfall regime is one of the factors that limits vegetation growth, leading to greater investment in root formation to explore deeper soil layers [3]. As a consequence, species present in the Cerrado biome tend to have smaller specific leaf area; on the other hand, they invest more in the bark in order to guarantee their survival in situations of water and temperature stress [10], and present lower growth rates and greater hydraulic conductance per leaf area unit when compared to species from other biomes [11,12].

Many native Cerrado species belong to the Annonaceae family and are widely used in folk medicine for the treatment of different diseases [13,14]. Native *Annona crassiflora* Mart., widely spread throughout the Cerrado biome, popularly known as araticum do cerrado, marolo, araticum cortiça or bruto, is among the species with the most consumed fruits in this biome, with pleasant sensory characteristics and high nutritional value, rich in phenolic and oligosaccharide compounds [15], carotenoids and vitamins [16], in addition to alkaloids found in by-products such as leaves and stem bark [17], representing a natural source of bioactive compounds due to their antioxidant properties [14] and seeds with high lipid yield [18].

Alkaloids make up the most diverse group among nitrogenous compounds. Multiple biological activities have alkaloid origins, and there are several drugs available on the market produced from natural plant alkaloids [19]. Several alkaloids are found in *Annona crassiflora* Mart. leaves, peels and stems. Gonçalves et al. [20] isolated two alkaloids, namely atherospermidine and liriodenine, from the stem; Pereira et al. [21] isolated and characterized alkaloid stephelagin from the fruit peel and Egydio et al. [22] and Ferraz et al. [23] identified dimethoxy-dihydroxy-tetrahydroprotoberberine, isolaurelin, xylopine, anonaine, anoretin and romucosine in leaves.

Liriodenine, an alkaloid found in abundance in the Annonaceae family [20,24–27], has several potent biological activities [28,29], including potential antibacterial [24], antiprotozoal [30,31], cytotoxic [32,33] and antifungal activities [34]. In particular, it has activity against more than 20 phytopathogens, including *Rhizopus stolonifer* and *Aspergillus glaucus*, fungi that impair seed germination [25].

Although there are several reports of alkaloids in Annonaceae, so far, there are no reports on how drought conditions, similar to those periodically found in the Cerrado, impact alkaloid production in the species, or how the species tolerate flooding conditions and how these conditions would reflect on the synthesis of specialized metabolites such as alkaloids. Thus, the aim of this study was to investigate how water stress impacts the production of total alkaloids and liriodenine in *Annona crassiflora* Mart.

2. Materials and Methods

2.1. Plant Material

Annona crassiflora Mart. seedlings were obtained in the municipality of Paraguaçu—Minas Gerais (April 2019) and transported to the nursery of the Department of Forest Science of the Faculty of Agronomic Sciences, Unesp—Botucatu (coordinates 22°51′ latitude S and 48°26′ longitude W), where they were submitted to a 6-month acclimation period. Transplantation to 5-L polyethylene pots was carried out in November 2019. To fill the pots, medium-texture Dystrophic Red Latosol was used [35,36], collected from the surface layer (0–20 cm in depth).

During the period in which the experiment was conducted, plants were submitted to humidity and temperature conditions shown in Figure 1.

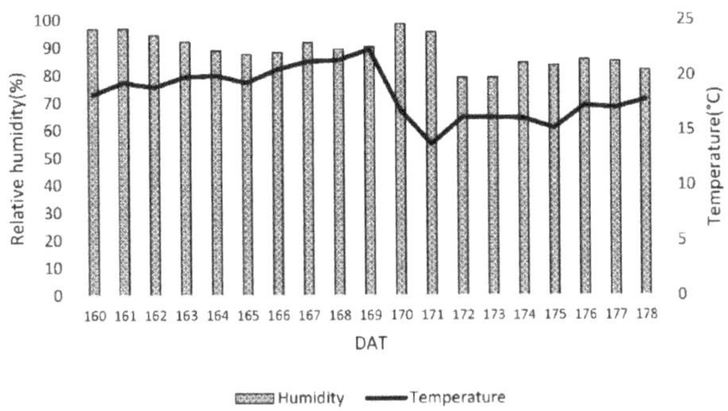

Figure 1. Average humidity and temperature between the 13th and 31st of May 2020 in the seedling nursery of the Department of Forestry Science of the Faculty of Agronomic Sciences (FCA) (DAT: days after transplanting).

2.2. *Experimental Design*

The experimental design was completely randomized, with three water stress levels (Flooding (>0.01 MPa); Field Capacity (−0.01 MPa) (Control) and Drought (−1.5 MPa)), with six replicates of two plants per plot. Plants remained in treatment (13–31 May 2020) until the permanent wilting point was reached. Two days after this stress condition was reached, the entire experiment was collected.

From the water retention curve, the percentage of water needed for the soil to reach −0.01 MPa (Field capacity) (Control) and −1.5 MPa (Drought) was calculated, which corresponded to 16% and 8% of water, respectively. After placing pots under the established conditions, they were weighed daily and the evapotranspiration difference was replaced to maintain previous conditions (−0.01 MPa and −1.5 MPa). To maintain plants under flooding, pots were kept in flooded trays throughout the experiment.

The moisture values corresponding to the water retention tension obtained through the soil water retention curve are shown below (Table 1).

Table 1. Water retention tension.

Tension	Humidity (%)
Flooding (>0.01 MPa)	37
Field capacity (−0.01 MPa)	16
Drought (−1.5 MPa)	8

2.3. *Gas Exchanges*

Gas exchanges were monitored weekly in all treatments from 9:00 a.m. to 11:00 a.m., with the aid of a CO_2 gas and infrared gas analyzer ("InfraRed Gas Analyzer—IRGA", model GSF 3000 Fl WALZ, Germany) with saturating light of 450 m^{-2} s^{-1} determined by means of a light curve. For monitoring, six replicates (1 plant per replicate) of each treatment were evaluated, taking measurements on the 2nd and 3rd fully expanded leaves.

CO_2 assimilation rate (*Anet*, μmol CO_2 m^{-2} s^{-1}), transpiration rate (*E*, mmol water vapor m^{-2} s^{-1}) and stomatal conductance (*gs*, mol m^{-2} s^{-1}) were determined. Water use efficiency (WUE, μmol CO_2 (mmol H_2O^{-1})) was calculated using the relationship between assimilated CO_2 and the transpiration rate (*Anet/E*). The apparent carboxylation efficiency was calculated according to the relationship between CO_2 assimilation rate and leaf intercellular CO_2 concentration (*Anet/Ci*, mol m^{-2} s^{-1} Pa^{-1}).

2.4. Chlorophyll a Fluorescence

Chlorophyll *a* fluorescence was performed from 9:00 a.m. to 11:00 a.m. using a fluorometer (LED-Array/PAM-Module3055-FL) on 18 plants (six replicates of 1 plant each) and leaves were acclimated to a period of 30 min in the dark by covering them with aluminum foil; then, an actinic light pulse of 4500 µmol m^{-2} s^{-1} was applied to obtain Fm (maximum dark-adapted fluorescence) and Fm′ (maximum light-adapted fluorescence). In addition to the maximum leaf light-adapted and dark-adapted fluorescence, Fo (minimum dark-adapted fluorescence) and Fo′ values (minimum light-adapted fluorescence) were also obtained.

The maximum quantum yield (Fv/Fm) [37], effective quantum yield (ϕPSII) [38], photochemical quenching (qP) [39], non-photochemical quenching (NPQ) [40] and electron transport rate (ETR) were calculated through Fm, Fo, Fm′ and Fo′, considering that 84% of light is absorbed by chlorophyll, with 50% of photons activating photosystem II chlorophyll and 50% photosystem I and photosystem II energy that cannot be dissipated (*Ex*), quantum yield of unregulated non-photochemical energy loss in photosystem II (ϕNO) and quantum yield of regulated non-photochemical energy loss in photosystem II (ϕNPQ) [41].

2.5. Carbohydrate Concentration

Total soluble sugars were extracted from the leaf material obtained from a pool of samples of two plants per replicate (six replicates per treatment), according to Garcia et al. [42], with minor modifications, and starch was extracted according to Clegg [43]. The procedure to determine the concentration of total soluble sugars was performed according to Morris [44]; for starch, it was described by Yemm and Folkes [45]; for reducing sugars, it was determined by Miller [46]; and for sucrose, it was established by Passos [47], with minor modifications.

2.6. Activity of Antioxidant Enzymes, Hydrogen Peroxide and Lipid Peroxidation

The extraction of antioxidant enzymes was performed as described by Kar and Mishra [48] from leaf material obtained from a pool of samples of two plants per replicate (six replicates per treatment). The activities of superoxide dismutase, EC 1.15.1.1 and catalase EC 1.11.1.6 enzymes were determined by the method of Peixoto et al. [49]; the activity of the peroxidase EC 1.11.1.7 enzyme was established according to Teisseire and Guy [50]; and soluble proteins were quantified as described by Bradford [51].

The hydrogen peroxide content was determined by the method of Alexieva et al. [52] and lipid peroxidation was determined according to methodology proposed by Heauth and Packer [53], and both analyses were obtained using a pool of leaf material from two plants per replicate (six replicates per treatment).

2.7. Extraction of Total Alkaloids

Total alkaloids were extracted from the root material of 18 *A. crassiflora* plants (six replicates of 2 plants each); the material was stored in a greenhouse with forced air circulation at 30 °C for ten days, and subsequently ground to obtain 1 g of dry mass for each replicate. Alkaloids were extracted from roots previously dried using the acid–base method. After thorough grinding, the plant material was moistened with a saturated sodium carbonate (Na_2CO_3) solution and left to dry for 48 h at room temperature. Alkaloids were extracted with chloroform ($CHCl_3$) by constant stirring for 1 h and then filtered and washed with distilled water. The $CHCl_3$ phases were extracted into a 1 M hydrochlo-ric acid (HCl) solution before being alkalinized to pH 9.5 with a saturated solution of Na_2CO_3. The alkaline solution was then re-extracted with $CHCl_3$, dried with anhydrous sodium sulfate (Na_2SO_4), filtered and evaporated at approximately 25 °C to obtain total alkaloids [25].

2.8. Quantification of Total Alkaloids and Liriodenine

To determine the total alkaloid content, the 18 samples were stored at room temperature, re-solubilized with $CHCl_3$ and transferred to quartz cuvettes. The absorbance of each solution was obtained by spectrophotometer at 254 nm wavelength using liriodenine as the standard for the elaboration of the standard curve ($y = 0.0881x - 0.0112$, $R^2 = 0.9949$).

After obtaining the extract, liriodenine was quantified using ultra-high-performance liquid chromatograph (UHPLC—Thermo Fisher-Scientific®, Waltham, MA, USA) with a gradient pump and UV-Vis detector using C 18 reverse phase column (150 × 4, 6 mm and 5 μm in particle diameter). The mobile phase was 30:70 water (pH 3.5 with acetic acid) and cratic isomethanol, with a flow rate of 1 mL/min, keeping the column temperature at 30 °C. Detection was carried out in UV at 254 nm. For liriodenine quantification, calibration curves were performed by analyzing the stock solution series ($y = 0.3595x - 0.0011$; $R^2 = 0.9989$ for samples with up to 10 μg of liriodenine in the extract and $y = 0.3658x + 1.142$; $R^2 = 0.9992$ for samples with more than 10 μg [25].

2.9. Statistical Analysis

Data were submitted to analysis of variance (ANOVA) using the SigmaPlot software Version 12 and means were compared by the Tukey test at 5% ($p < 0.05$) [54]. To present the biochemical variables, a radar chart was used. Input variables were initially standardized as a result of the different units, using the scale command from the basic package of the R computing environment, which centers the mean at zero and changes the scale to standard deviation [55].

3. Results

Annona crassiflora plants showed, in general, that the ability to adapt to water restriction conditions (drought: −1.5 MPa) reflected in the increase in specialized metabolism, unlike what occurred under flooding conditions (Figures 2 and 3, Table 2). In this experiment, an increase in the production of total alkaloids without the occurrence of reductions in the photosynthetic metabolism of plants (gas exchange and chlorophyll *a* fluorescence) was observed when *A. crassiflora* plants were kept under drought stress conditions (Figures 4 and 5).

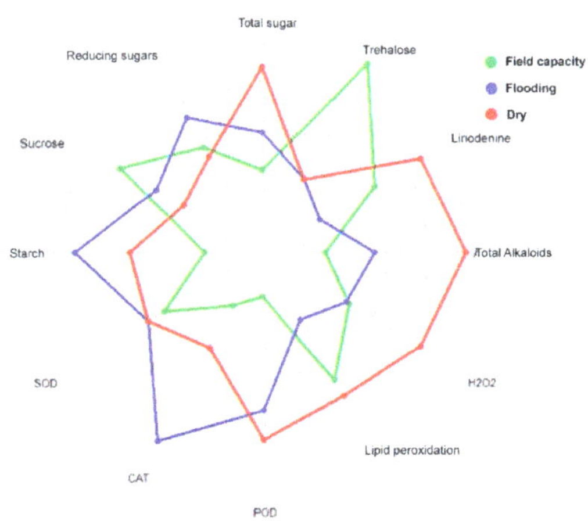

Figure 2. Biochemical variables: (liriodenine; total alkaloids; hydrogen peroxide (H_2O_2); lipoperoxide, peroxidase (POD); catalase (CAT); superoxide dismutase (SOD); starch; sucrose; reducing sugar; total sugars; trehalose) obtained from young *A. crassiflora* plants submitted to three water condition levels (Field Capacity (−0.01 MPa); Flooding; Drought (−1.5 MPa)) at 18 days after the beginning of treatments. Variables represented in the graph and that showed significant differences in statistical analysis by Tukey tests at 5% are shown in Table 2.

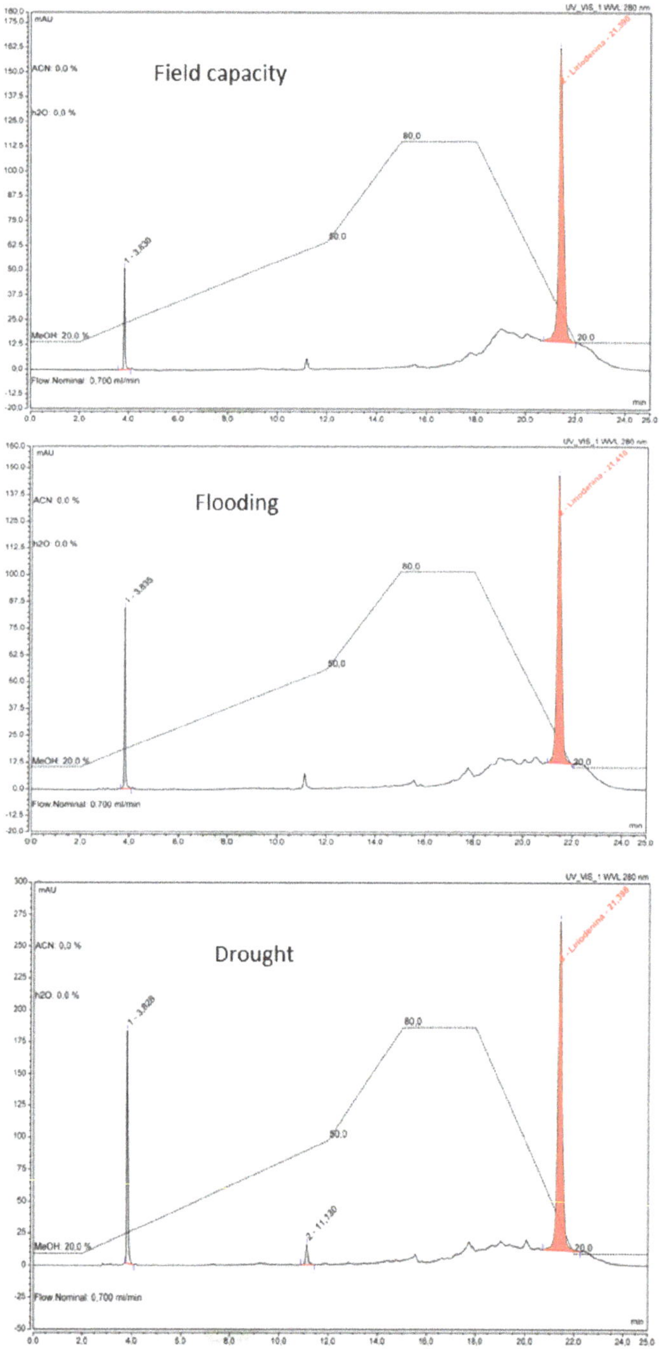

Figure 3. Chromatogram indicating liriodenine obtained from young *A. crassiflora* plants submitted to three water condition levels [Field Capacity (−0.01 MPa); Flooding and Drought (−1.5 MPa)] at 18 days after the beginning of treatments.

Table 2. Biochemical variables: liriodenine (μg.g^{-1}), total alkaloids (μg.g^{-1}), peroxidase (POD, μmol prupurogallin min^{-1} mg prot^{-1}), catalase (CAT, μKat μg^{-1} protein), starch (μg.g^{-1} FW) and trehalose (μg.g^{-1} FW) obtained from young *A. crassiflora* plants submitted to three water condition levels (Field Capacity (−0.01 MPa), Flooding and Drought (−1.5 MPa) at 18 days after the beginning of treatments.

Water Condition	Total Alk	Liriodenine	CAT	POD	Trehalose	Starch
Field capacity	54.26 B	10.8770 AB	0.0101 B	0.2681 B	83.90 A	69.713 B
Flooding	63.49 AB	8.4098 B	0.1034 A	0.6655 A	12.46 B	104.089 A
Drought	80.87 A	13.4374 A	0.0391 B	0.7656 A	13.16 B	89.381 AB

Averages followed by the same letter not differ based on the Tukey 5% significance test. Mean ± standard deviation (n = 4).

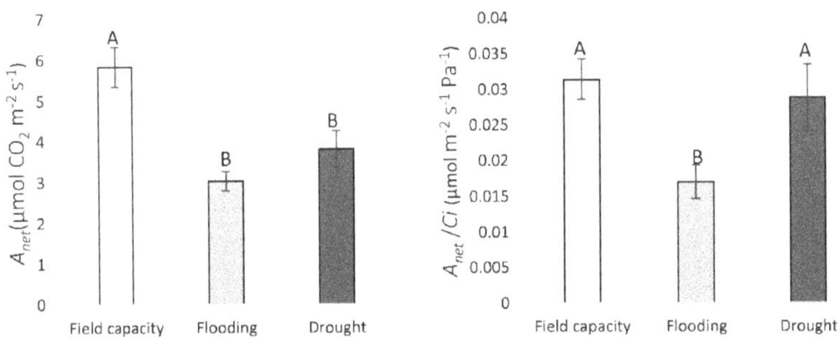

Figure 4. Carbon assimilation rate (A_{net}) and Rubisco carboxylation efficiency (A_{net}/Ci) in *Annona crassiflora* plants kept under Field capacity (−0.01 MPa), Flooding and Drought (−1.5 MPa) conditions at 5, 10 and 18 days after the application of treatments, respectively, and 168, 173 and 178 days after transplanting. Capital letters indicate significant differences among treatments ($p < 0.05$).

At the same time that plants showed adaptation to water restriction conditions characterized by responses observed in primary metabolism, the production of total alkaloids and liriodenine was also increased. Plants kept under drought conditions produced higher concentrations of total alkaloids in relation to those kept in soil with maximum water availability (field capacity), while saturated soil did not cause significant variations in total alkaloids but reduced the liriodenine concentration in relation to drought soils.

In this context, plants kept under water restriction showed greater carboxylation efficiency of the Rubisco enzyme (A_{net}/Ci) compared to plants kept under flooding (Figure 4). However, in both conditions, the carbon assimilation rate (A_{net}) was lower in relation to plants without water restriction (Field Capacity) and without differences in relation to Ci (data not shown). The other gas exchange variables did not show significant differences (stomatic conductance (gs), transpiration (E), vapor pressure deficit (VPD) and water use efficiency (WUE).

The chlorophyll *a* fluorescence was also impacted by treatments, and plants kept under drought conditions had higher maximum quantum yield (Fv/Fm), effective quantum yield (ϕPSII), potential quantum efficiency values (Fv′/Fm′) and lower fraction of energy dissipated in the form of heat (D) compared to plants kept under flooding conditions (Figure 5). Photosystem II energy that cannot be dissipated and used in the photochemical phase (Ex), photochemical quenching (qP), non-photochemical quenching (NPQ), electron transport rate (ETR), quantum yield of unregulated non-photochemical energy loss in photosystem II (ϕNO) and quantum yield of regulated non-photochemical energy loss in photosystem II (ϕNPQ) did not show significant differences.

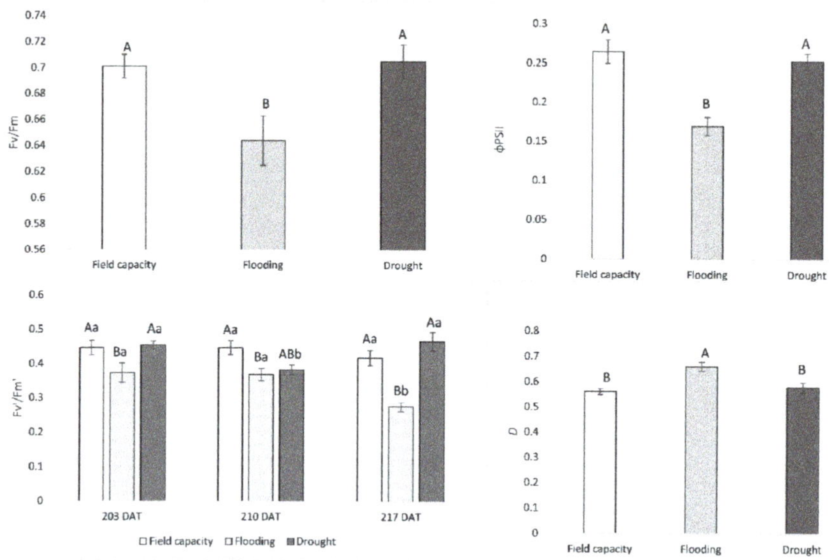

Figure 5. Maximum quantum yield (Fv/Fm), potential quantum efficiency (Fv′/Fm′), effective quantum yield (ϕPSII) and energy dissipated in the form of heat (*D*) in *Annona crassiflora* plants kept under Field capacity (−0.01 MPa), Flooding and Drought (−1.5 MPa) conditions at 5, 10 and 18 days after the application of treatments, respectively, at 168, 173 and 178 days after transplanting. Uppercase letters indicate significant differences among treatments; lowercase letters indicate differences among times ($p < 0.05$).

Starch and trehalose were affected depending on the conditions in which plants were kept, while total sugars, reducing sugars and sucrose did not show significant differences. Starch was found in higher concentrations in plants kept under flooding but without differing from plants kept under drought, and the lowest trehalose concentrations were found both in leaves of plants kept under both drought and flooding, indicating that this sugar may have been translocated to roots and used in order to neutralize the damage caused by stress (Figure 2, Table 2).

In general, the enzymatic system acted satisfactorily, preventing membrane damage, since there was no difference in lipid peroxidation and hydrogen peroxide among treatments, possibly indicating that the antioxidant enzymes inhibited the activity of reactive oxygen species. In plants kept under drought soil and flooding, higher peroxidase activity (POD) was observed, while catalase activity (CAT) was higher only in plants kept under flooding. Thus, flooded plants required greater enzymatic activity (Figure 2, Table 2).

4. Discussion

The increase in alkaloid production in *A. crassiflora* plants kept under drought stress conditions seems to be related to their ability to adapt to the Cerrado conditions, which has well-defined drought periods [56], since the photosynthetic process was preserved, ensuring both primary and specialized metabolism. This ability to adapt to the Cerrado conditions seems to be specific, since under flooding, reductions in primary metabolism were evident, affecting the specialized metabolism, especially the synthesis of alkaloid liriodenine.

The fact that drought stress causes increases in specialized metabolism substances such as alkaloids has been shown by several authors, such as Ghorbanpour and Hatami [57] in work with *Hyoscyamus niger*; Kleinwächter et al. [58] with thyme (*Thymus vulgaris*); Kleinwächter and Selmar [59] with spices and medicinal plants; and Liu et al. [29] with *Catharanthus roseus*. Specifically with the genus *Annona*, Castro-Moreno et al. [60] found the highest liriodenine concentration in *Annona lutescens* roots at the end of the dry season (about 377 μmol/g), which is the first report of the presence of liriodenine in *Annona crassiflora* roots under water stress (about 80.9 μg/g of total alkaloids and 13.47 μg/g of liriodenine). Some periodic collections of *Annona* species tissues in an annual cycle allow us to point out that although the biosynthesis of alkaloids is distributed throughout the plant, the roots generally accumulate the greatest number of alkaloids and produce a higher yield regardless of the phenological stage of plants [61].

In general, water restriction conditions lead to stomatal closure and reduced CO_2 absorption and, as a consequence, there is a considerable decrease in the consumption of NADPH + H^+ for CO_2 fixation via the Calvin cycle, which generates an excess supply and accumulation of this equivalent reducer. Thus, metabolic processes are directed towards the synthesis of highly reduced compounds such as isoprenoids, phenols and alkaloids with the use of these accumulated reducing agents [62]. In this context, although there were no significant differences in stomatal conductance in *A. crassiflora*, reductions in CO_2 assimilation rates (A_{net}) were observed due to water restriction (drought) and flooding, which corroborates the results of Simonneau et al. [63] and Oliveira and Gualtieri [64], respectively, and may have led to lower CO_2 fixation in the Calvin cycle [65]. However, only *A. crassiflora* plants kept under drought conditions showed greater synthesis of total alkaloids in relation to those kept under field capacity, in addition to having higher carboxylation efficiency (A_{net}/Ci) and higher liriodenine concentration in relation to flooded plants, which indicates that these responses were evident when plants were under water restriction.

The high carboxylation efficiency in *A. crassiflora* indicates adaptation for survival in environments with periods of low water availability, which is justified by the fact that the species is native to the Cerrado, unlike the results obtained by Mantoan et al. [66] with *A. emarginata*, showing a decrease in the carboxylation efficiency under irrigation suspension conditions, which may be related to the fact that *A. emarginata* is a species present in the Atlantic Forest, an environment with greater water availability. In addition, in plants submitted to water stress events, increased damage to the photosynthetic apparatus is observed, causing changes in chlorophyll *a* fluorescence patterns, which changes the light energy dissipation pathways and increases plant stress [67]. However, in this experiment, plants kept under drought conditions did not show reductions in the chlorophyll *a* fluorescence pattern, indicating that there was no significant damage to the photosynthetic apparatus. Under flooding, damage is evidenced by low chlorophyll *a* fluorescence values (ϕPSII, Fv/Fm, Fv′/Fm′) and high energy dissipation in the form of heat (*D*) (Figure 5), which resulted in lower carboxylation efficiency of the enzyme ribulose 1,5-bisphosphate carboxylase (rubisco) (Figure 4). Thus, flooding directly affected the photosynthetic apparatus in *A. crassiflora*, reducing its efficiency regardless of stomatal conductance, as proposed by Parolin and Wittmann [68] and Oliveira and Gualtieri [64]. In studies with *Annona glabra*, this reduction in photosynthetic efficiency did not occur, which reinforces its characteristics of adaptation to restinga, a highly flooded environment [69].

The highest effective quantum yield (ϕPSII), maximum quantum yield in the dark (Fv/Fm) and potential quantum efficiency (Fv′/Fm′) values observed in *A. crassiflora* plants kept under drought conditions (similar to plants kept under field capacity) indicate that the energy generated may have been destined both for the production of carbon skeletons used in primary metabolism and for specialized metabolism [58,70]. On the other hand, in plants submitted to flooding conditions, the low quantum yield in the dark (Fv/Fm), low potential quantum efficiency (Fv′/Fm′) and greater heat dissipation of the antenna (*D*) indicate a photoprotection mechanism to minimize damage to the photosystem [65].

Thus, in addition to the increase in NADPH + H that can be used for the synthesis of specialized metabolites such as alkaloids, an increase in the production of free radicals from the energy generated in the system is observed [58], and therefore, in stress situations, increases in the production of hydroxyl radicals (OH), superoxides (O_2^-), hydrogen peroxide molecules (H_2O_2) and singlet oxygen ($^1O_2$) are observed, originating from redox reactions that can be in the form of free radicals or in the molecular form of a non-radical [71,72]. Lipoperoxides are the result of the interaction between free radicals and fatty acids in cell membranes, and when this process occurs, the cell membrane integrity is compromised, resulting in the production of carboxylic compounds such as monoaldehyde [73]. To protect itself and try redox homeostasis, the plant has antioxidant enzymes, such as superoxide dismutase (SOD), catalase (CAT) and peroxidases (POD) [74].

The enzyme that acts first is SOD, catalyzing the dismutation of two $O_2^{\bullet -}$ radicals, generating H_2O_2 and O_2. Then, CAT, which is one of the main enzymes acting in the elimination of H_2O_2 generated during photorespiration and β-oxidation of fatty acids, converts two H_2O_2 molecules into water and molecular oxygen. Subsequently, POD, located in the cytosol and vacuoles, catalyzes reactions that use H_2O_2 as oxidants, so this reactive oxygen species (ROS) is also eliminated, even when SOD activity is low [74,75].

Within this complex enzyme system, no significant superoxide dismutase (SOD) activity was observed in *A. crassiflora*; however, CAT activity was higher in flooded plants, which seems to be a specific characteristic of saturated soils [76] and indicates an attempt to reduce H_2O_2 accumulated by stress in order to avoid damage to lipids, proteins and nucleic acids and ensure flooding tolerance [77]. POD activity was also observed, both in plants kept in flooding and in drought, which indicates the continued elimination of reactive oxygen species (ROS) from the system to avoid damage to cells. In this context, POD seems to have been efficient, since no differences were observed in the hydrogen peroxide or lipoperoxide concentration between treatments and control (Figure 2 heat map), which could mean higher malondialdehyde concentrations, Which, in turn, would represent damage to cell membranes [74].

It is noteworthy that to avoid the deleterious effect caused by reactive oxygen species (ROS) in plant tissues, especially in root regions, plants can show greater activity of antioxidant enzymes, especially POD, and accumulate amino acids in roots. Thus, with the greater allocation of amino acids to roots, plants can increase their nitrogen reserve by synthesizing alkaloids [57], which would explain their higher production in *A. crassiflora* roots, especially when submitted to drought stress.

Roots are the main alkaloid production organ in plants of the genus *Annona* [25] and alkaloid accumulation in certain situations, as observed in *A. crassiflora* roots, may indicate osmotic adjustment due to the accumulation of precursor osmolytes, such as amino acids, carbohydrates and sugars such as starch and trehalose [78]. The accumulation of osmolytes can alter the water potential and favor water absorption even in soils with water restriction and generate greater alkaloid [79], which may have occurred with *A. crassiflora*. Thus, even under reduced soil water conditions, gas exchange and fluorescence were not negatively affected, and alkaloid production still occurred.

Regarding the presence of osmolytes, starch interconversion into other sugars that act as osmoprotectors can be considered, with the ability to influence the carbon allocation for the entire plant, mitigating the negative effect of stress caused by water restriction. Furthermore, in photosynthetic cells, starch can be synthesized and temporarily stored in chloroplasts, and this "transient" starch is synthesized and degraded within a 24-hour period [80–82]. Flooding caused an increase in starch concentration in *A. crassiflora* leaves, as reported in wheat (*Triticum* spp.) [83,84], which may be a result of the rapid inhibition of plant growth at the beginning of flooding, leading to lower consumption of sugars produced by photoassimilation, in addition to reduction in photosynthesis [85–87].

Another aspect to be observed regarding the capacity of *A. crassiflora* plants to tolerate abiotic stress was the presence of trehalose both in flooded plants and in those under drought stress. As this non-transport-reducing disaccharide acts as a biostimulant in stress tolerance, its lower concentrations in leaves may indicate its translocation to roots and use in order to neutralize the damage that could be caused by stress, since two glucose molecules are generated from the hydrolysis of trehalose [88]. In addition, trehalose synthesis is induced from some stress condition in order to protect enzymes, proteins and lipid membranes against denaturation under stress situations, also playing an osmoprotective role [78,89].

In summary, when *A. crassiflora* plants are under stress situations, e.g., due to lack of water, CO_2 assimilation tends to decrease and as a consequence, a smaller amount of $NADPH_2$ is consumed within the Calvin cycle. Thus, much of the energy produced should be dissipated, and, despite the action of non-photochemical mechanisms, such as photorespiration and the xanthophyll cycle, these are potentiated in this situation, and numerous electrons are still transferred to molecular oxygen, generating ROS. Under this situation, plants activate their antioxidant system (SOD, POD and CAT), thus blocking the harmful effect of ROS, leading to a strong increase in the reduction potential of the reducing equivalent ($NADPH_2$), which can be directed to the synthesis of specialized metabolites [62], which seems to have occurred with *A. crassiflora* plants. Furthermore, starch and trehalose play a role in mitigating the effects of reduced soil water availability to ensure root water absorption (osmoregulation). Thus, *A. crassiflora* plants under drought stress showed increases in the content of total alkaloids, specifically liriodenine.

5. Conclusions

A. crassiflora plants are affected by flooding and drought conditions. Drought generates a stimulus signal for the synthesis of total alkaloids and liriodenine without reducing primary productivity, which denotes rusticity and adaptation of the species to the Cerrado conditions. On the other hand, flooding stress is harmful to the photosynthetic apparatus, which does not result in increased alkaloid production and reduces liriodenine production.

Author Contributions: A.B.M.H.: Conceptualization, data collection, data curation, formal analysis, investigation, methodology, writing—original draft, writing—review and editing. I.D.-l.-C.-C.: Investigation, methodology, resources, writing—review. M.M.-V.: Investigation, methodology. M.R.d.S.: investigation, methodology, resources, writing—review. F.G.C.: Data collection, data curation, formal analysis, investigation, methodology, writing—original draft, writing—review and editing. B.C.M.: Data collection, investigation, methodology. G.C.d.S.: Data collection, investigation, methodology. C.S.F.B.: Formal analysis, investigation, methodology, resources, writing—original draft, writing—review and editing, supervision. G.F.: Conceptualization, data curation, validation, investigation, methodology, resources, writing—original draft, writing—review and editing, supervision. All authors have read and agreed to the published version of the manuscript.

Funding: This study was financed in part by the Coordenação de Aperfeiçoamento de Pessoal de Nível Superior—Brasil (CAPES)—Finance Code 001.

Acknowledgments: To Coordenação de Aperfeiçoamento de Pessoal de Nível Superior—Brasil—CAPES, for the doctoral scholarship granted to the first author.

Conflicts of Interest: The authors declare no conflict of interest.

References

1. Prado, L.G.; Arruda, H.S.; Araujo, N.M.P.; Braga, L.E.D.O.; Banzato, T.P.; Pereira, G.A.; Figueiredo, M.C.; Ruiz, A.L.T.G.; Eberlin, M.N.; de Carvalho, J.E.; et al. Antioxidant, antiproliferative and healing properties of araticum (*Annona crassiflora* Mart.) peel and seed. *Food Res. Int.* **2020**, *133*, 109168. [CrossRef]
2. Rodrigues, J.A.M.; de Oliveira Andrade, A.C.; Viola, M.R.; Ferreira, D.D.; de Mello, C.R.; Thebaldi, M.S. Hydrological modeling in a basin of the Brazilian Cerrado biome. *Rev. Ambient. Agua* **2021**, *16*, 2–18. [CrossRef]
3. Lima, J.E.F.W. Situação e perspectivass sobre as águas do Cerrado. *Cienc. Cult.* **2011**, *63*, 27–29. [CrossRef]
4. Caldas, J.; Cerrados, E. Sustentabilidade Águas do Cerrado: Questão estratégica para o país. Manejo racional dos recursos hídricos é preocupação da pesquisa agropecuária uma vez que Cerrado é origem de grandes bacias hidrográficas. *Sustentabilidade* **2014**, 9–11. Available online: http://www.diadecampo.com.br/zpublisher/materias/Newsletter.asp?id=28947&secao=Agrotemas (accessed on 16 October 2021).
5. Lee, J.-E.; Lintner, B.R.; Boyce, C.K.; Lawrence, P.J. Land use change exacerbates tropical South American drought by sea surface temperature variability. *Geophys. Res. Lett.* **2011**, *38*. [CrossRef]
6. Debortoli, N.S.; Dubreuil, V.; Funatsu, B.; Delahaye, F.; De Oliveira, C.H.; Rodrigues-Filho, S.; Saito, C.; Fetter, R. Rainfall patterns in the Southern Amazon: A chronological perspective (1971–2010). *Clim. Chang.* **2015**, *132*, 251–264. [CrossRef]
7. Penereiro, J.C.; Badinger, A.; Maccheri, N.A.; Meschiatti, M.C. Distribuições de Tendências Sazonais de Temperatura Média e Precipitação nos Biomas Brasileiros. *Rev. Bras. Meteorol.* **2018**, *33*, 97–113. [CrossRef]
8. Aghajanlou, F.; Mirdavoudi, H.; Shojaee, M.; Mac Sweeney, E.; Mastinu, A.; Moradi, P. Rangeland Management and Ecological Adaptation Analysis Model for *Astragalus curvirostris* Boiss. *Horticulturae* **2021**, *7*, 67. [CrossRef]

9. Palhares, D.; Franco, A.C.; Zaidan, L.P. Respostas fotossintéticas de plantas do cerrado nas estações seca e chuvosa. *Rev. Bras. Biociências* **2010**, *8*, 213–220.
10. Hoffmann, W.A.; Adasme, R.; Haridasan, M.; De Carvalho, M.T.; Geiger, E.L.; Pereira, M.A.B.; Gotsch, S.G.; Franco, A.C. Tree topkill, not mortality, governs the dynamics of savanna-forest boundaries under frequent fire in central Brazil. *Ecology* **2009**, *90*, 1326–1337. [CrossRef]
11. Hoffmann, W.A.; Franco, A.C. The importance of evolutionary history in studies of plant physiological ecology: Examples from cerrados and forests of central Brazil. *Braz. J. Plant Physiol.* **2008**, *20*, 247–256. [CrossRef]
12. Yousefi, A.R.; Rashidi, S.; Moradi, P.; Mastinu, A. Germination and seedling growth responses of *Zygophyllum fabago*, *Salsola kali* L. and *Atriplex canescens* to peg-induced drought stress. *Environments* **2020**, *7*, 107. [CrossRef]
13. Justino, A.B.; Pereira, M.N.; Vilela, D.D.; Peixoto, L.G.; Martins, M.M.; Teixeira, R.R.; Miranda, N.C.; da Silva, N.M.; de Sousa, R.M.F.; de Oliveira, A.; et al. Peel of araticum fruit (*Annona crassiflora* Mart.) as a source of antioxidant compounds with α-amylase, α-glucosidase and glycation inhibitory activities. *Bioorg. Chem.* **2016**, *69*, 167–182. [CrossRef]
14. Roesler, R.; Malta, L.G.; Carrasco, L.C.; Pastore, G.C. Food Chemistry and Toxicology Evaluation of the Antioxidant Properties of the Brazilian Cerrado Fruit *Annona crassiflora* (Araticum). *Food Chem. Toxicol.* **2006**, *71*, 102–107.
15. Joca, T.A.C.; Oliveira, D.C.; de Zotz, G.; Winkler, U.; Moreira, A.S.F.P. The velamen of epiphytic orchids: Variation in structure and correlations with nutrient absorption. *Flora Morphol. Distrib. Funct. Ecol. Plants* **2017**, *230*, 66–74. [CrossRef]
16. Cardoso, L.D.M.; Oliveira, D.D.S.; Bedetti, S.D.F.; Martino, H.S.D.; Pinheiro-Sant'Ana, H.M. Araticum (*Annona crassiflora* Mart.) from the Brazilian Cerrado: Chemical composition and bioactive compounds. *Fruits* **2013**, *68*, 121–134. [CrossRef]
17. Modesto, E.C.; Santos, G.T.; dos Vilela, D.; Gonçalvez, G.D.; Matsushita, M. Efeitos nutricionais de dietas ricas em ácidos graxos poliinsaturados para os ruminantes e alguns benefícios para o homem. *Arq. Ciências Veterinárias E Zool. UNIPAR* **2002**, *5*, 119–134.
18. Luzia, D.M.M.; Jorge, N. Bioactive substance contents and antioxidant capacity of the lipid fraction of *Annona crassiflora* Mart. seeds. *Ind. Crops Prod.* **2013**, *42*, 231–235. [CrossRef]
19. Debnath, B.; Singh, W.S.; Das, M.; Goswami, S.; Singh, M.K.; Maiti, D.; Manna, K. Role of plant alkaloids on human health: A review of biological activities. *Mater. Today Chem.* **2018**, *9*, 56–72. [CrossRef]
20. Gonçalves, M.A.; Lara, T.A.; Pimenta, L.P.S. Oxaporphynic alkaloids of *Annona crassiflora* wood Mart. *Annu. Meet. Braz. Chem. Soc.* **2006**, *2*, 1–2.
21. Silva, M.A.; Da Silva, G.A.; da Marques, M.J.; Bastos, R.G.; da Silva, A.F.; Rosa, C.P.; Espuri, P.F. Triagem fitoquímica, atividade antioxidante e leishmanicida do extrato hidroetanólico 70% (v/v) e das frações obtidas de (*Annona crassiflora* Mart.). *Rev. Fitos* **2017**, *10*, 505–517. [CrossRef]
22. Egydio, A.P.M.; Valvassoura, T.A.; Santos, D.Y.A.C. Geographical variation of isoquinoline alkaloids of *Annona crassiflora* Mart. from cerrado, Brazil. *Biochem. Syst. Ecol.* **2013**, *46*, 145–151. [CrossRef]
23. Ferraz, C.R.; Silva, D.B.; Prado, L.C.D.S.; Canabrava, H.A.N.; Bispo-Da-Silva, L.B. Antidiarrhoeic effect and dereplication of the aqueous extract of *Annona crassiflora* (Annonaceae). *Nat. Prod. Res.* **2019**, *33*, 563–567. [CrossRef] [PubMed]
24. Costa, E.V.; Da Cruz, P.E.O.; De Lourenço, C.C.; De Souza Moraes, V.R.; De Lima Nogueira, P.C.; Salvador, M.J. Antioxidant and antimicrobial activities of aporphinoids and other alkaloids from the bark of *Annona salzmannii* A. DC. (Annonaceae). *Nat. Prod. Res.* **2013**, *27*, 1002–1006. [CrossRef] [PubMed]
25. De La Cruz Chacón, I.; González-Esquinca, A.R. Liriodenine alkaloid in *Annona diversifolia* during early development. *Nat. Prod. Res.* **2012**, *26*, 42–49. [CrossRef] [PubMed]
26. González-Esquinca, A.R.; De-La-Cruz-Chacón, I.; Castro-Moreno, M.; Orozco-Castillo, J.A.; Riley- Saldaña, C.A. Alkaloids and acetogenins in Annonaceae development: Biological considerations. *Rev. Bras. Frutic.* **2014**, *36*, 01–16. [CrossRef]
27. Sousa, M.C.; Bronzatto, A.C.; González-Esquinca, A.R.; Campos, F.G.; Dalanhol, S.J.; Boaro, C.S.F.; Martins, A.L.; da Silva Almeida, J.R.G.; Costa, E.V.; De-la-Cruz-Chacón, I.; et al. The production of alkaloids in *Annona cacans* seedlings is affected by the application of GA4+7 + 6-Benzyladenine. *Biochem. Syst. Ecol.* **2019**, *84*, 47–51. [CrossRef]
28. Chen, C.-Y.; Wu, H.-M.; Chao, W.-Y.; Lee, C.-H. Review on pharmacological activities of liriodenine. *Afr. J. Pharm. Pharmacol.* **2014**, *8*, 364–371. [CrossRef]
29. Liu, Y.; Meng, Q.; Duan, X.; Zhang, Z.; Li, D. Effects of PEG-induced drought stress on regulation of indole alkaloid biosynthesis in *Catharanthus roseus*. *J. Plant Interact.* **2017**, *12*, 87–91. [CrossRef]
30. Costa, E.V.; Pinheiro, M.L.B.; Xavier, C.M.; Silva, J.R.A.; Amaral, A.C.F.; Souza, A.D.L.; Barison, A.; Campos, F.R.; Ferreira, A.G.; Machado, G.M.C.; et al. A pyrimidine-β-carboline and other alkaloids from *Annona foetida* with antileishmanial activity. *J. Nat. Prod.* **2006**, *69*, 292–294. [CrossRef]
31. Cota, L.G.; Vieira, F.A.; Melo Júnior, A.F.; Brandão, M.M.; Santana, K.N.O.; Guedes, M.L.; Oliveira, D.A. Genetic diversity of *Annona crassiflora* (Annonaceae) in northern Minas Gerais State. *Genet. Mol. Res.* **2011**, *10*, 2172–2180. [CrossRef] [PubMed]
32. De Lima, J.P.S.; Pinheiro, M.L.B.; Antonio, A.M.; José, J.L.; Santos, D.M.F.; Barison, A.; Silva-Jardim, I.; Costa, E.V. In Vitro Atileishmanial and Cytotoxic Activities of *Annona mucosa* (Annonaceae). *Rev. Virtual Quim.* **2012**, *4*, 692–702. [CrossRef]
33. Khan, T.M.; Gul, N.S.; Lu, X.; Wei, J.H.; Liu, Y.C.; Sun, H.; Liang, H.; Orvig, C.; Chen, Z.F. In vitro and in vivo anti-tumor activity of two gold(III) complexes with isoquinoline derivatives as ligands. *Eur. J. Med. Chem.* **2019**, *163*, 333–343. [CrossRef] [PubMed]
34. De-la-Cruz-Chacón, I.; López-Fernández, N.Y.; Riley-Saldaña, C.A.; Castro-Moreno, M.; González-Esquinca, A.R. Antifungal activity in vitro of *Sapranthus microcarpus* (Annonaceae) against phytopathogens. *Acta Bot. Mex.* **2019**, *126*, 1420. [CrossRef]

35. Sousa, M.A.D.L.B.; Cavalheiro, F. Planejamento paisagístico do Campus Universitário da Faculdade de Ciências Agronômicas, UNESP, Botucatu, S.P. *Acta Bot. Bras.* **1987**, *1*, 155–163. [CrossRef]
36. Santos, H.G.; dos Jacomine, P.K.T.; Anjos, L.H.C.; dos Oliveira, V.Á.; de Lumbreras, J.F.; Coelho, M.R.; de Almeida, J.A.; de Araujo Filho, J.C.; de Oliveira, J.B.; Cunha, T.J.F. *Sistema Brasileiro de Classificação de Solos*, 5th ed.; EMBRAPA: Brasília, DF, Brazil, 2018; pp. 195–212.
37. Kitajima, M.; Butler, W.L. Quenching of chlorophyll fluorescence and primary photochemistry in chloroplasts by dibromothymoquinone. *BBA-Bioenerg.* **1975**, *376*, 105–115. [CrossRef]
38. Genty, B.; Briantais, J.M.; Baker, N.R. The relationship between the quantum yield of photosynthetic electron transport and quenching of chlorophyll fluorescence. *Biochim. Biophys. Acta-Gen. Subj.* **1989**, *990*, 87–92. [CrossRef]
39. Schreiber, U.; Schliwa, U.; Bilger, W. Continuous recording of photochemical and non-photochemical chlorophyll fluorescence quenching with a new type of modulation fluorometer. *Photosynth. Res.* **1986**, *10*, 51–62. [CrossRef]
40. Bilger, W.; Björkman, O. Role of the xanthophyll cycle in photoprotection elucidated by measurements of light-induced absorbance changes, fluorescence and photosynthesis in leaves of *Hedera canariensis*. *Photosynth. Res.* **1990**, *25*, 173–185. [CrossRef]
41. Klughammer, C.; Schreiber, U. Complementary PS II quantum yields calculated from simple fluorescence parameters measured by PAM fluorometry and the Saturation Pulse method. *PAM Appl. Notes* **2008**, *1*, 27–35.
42. Garcia, I.S.; Souza, A.; Barbedo, C.J.; Dietrich, S.M.C.; Figueiredo-Ribeiro, R.C.L. Changes in soluble carbohydrates during storage of *Caesalpinia echinata* LAM. (Brazilwood) seeds, an endangered leguminous tree from the Brazilian atlantic forest. *Braz. J. Biol.* **2006**, *66*, 739–745. [CrossRef] [PubMed]
43. Clegg, K.M. The application of the anthrone reagent to the estimation of starch in cereals. *J. Sci. Food Agric.* **1956**, *7*, 40–44. [CrossRef]
44. Morris, D.L. Quantitative Determination of Carbohydrates with *Dreywood's* Anthrone Reagent. American Association for the Advancement of Science Stable. *Am. Assoc. Adv. Sci.* **1948**, *107*, 254–255.
45. Yemm, E.W.; Folkes, B.F. The Estimation of Carbhydrates in Plant Extracts by Anthrone. *Int. J. Pharm. Pract.* **1954**, *57*, 508–514. [CrossRef]
46. Miller, G.L. Use of Dinitrosalicylic Acid Reagent for Determination of Reducing Sugar. *Anal. Chem.* **1959**, *31*, 426–428. [CrossRef]
47. Passos, L.P. *Métodos Analíticos e Laboratoriais em Fisiologia Vegetal*, 1st ed.; EMBRAPA: Coronel Pacheco, MG, Brazil, 1996; p. 223.
48. Kar, M.; Mishra, D. Catalase, Peroxidase and Polyphenoloxidase Activities during Rice Leaf Senescence. *Plant Physiol.* **1976**, *57*, 315–319. [CrossRef]
49. Peixoto, P.H.P.; Cambraia, J.; Sant'Anna, R.; Mosquim, P.R.; Moreira, M.A. Aluminum effects on lipid peroxidation and on the activities of enzymes of oxidative metabolism in sorghum. *Rev. Bras. Fisiol. Veg.* **1999**, *11*, 137–143.
50. Teisseire, H.; Guy, V. Copper-induced changes in antioxidant enzymes activities in fronds of duckweed (*Lemna minor*). *Plant Sci.* **2000**, *153*, 65–72. [CrossRef]
51. Bradford, M.M. A Rapid and Sensitive Method for the Quantitation of Microgram Quantities of Protein Utilizin the Principle of Protein-Dye Binding. *Anal. Biochem.* **1976**, 248–254. [CrossRef]
52. Alexieva, V.; Sergiev, I.; Mapelli, S.; Karanov, E. The effect of drought and ultraviolet radiation on growth and stress markers in pea and wheat V. *Trop. Ecol.* **2001**, *45*, 315–325. [CrossRef]
53. Heatuh, R.L.; Packer, L. Photoperoxidation in Isolated Chloroplasts. *Archives of Biochemistry and Biophysics* **1968**, *125*, 850–857. [CrossRef]
54. Costa, J.R. Técnicas Experimentais aplicadas às Ciências Agrárias. *Embrapa Agrobiol.* **2003**, 7–102.
55. Becker, R.A.; Chambers, J.M. Auditing of Data Analyses. *SIAM J. Sci. Stat. Comput.* **1988**, *9*, 747–760. [CrossRef]
56. Nascimento, D.T.F.; Novais, G.T. Clima do Cerrado: Dinâmica atmosférica e características, variabilidades e tipologias climáticas. *Élisée-Rev. De Geogr. Da UEG* **2020**, *9*, e922021.
57. Ghorbanpour, M.; Hatami, M. Role of plant growth promoting rhizobacteria on antioxidant enzyme activities and tropane alkaloid production of *Hyoscyamus niger* under water deficit stress Role of plant growth promoting rhizobacteria on antioxidant enzyme activities and tropane alkaloid. *Turk. J. Biol.* **2013**, *37*, 350–360. [CrossRef]
58. Kleinwächter, M.; Selmar, D. New insights explain that drought stress enhances the quality of spice and medicinal plants: Potential applications. *Agron. Sustain. Dev.* **2015**, *35*, 121–131. [CrossRef]
59. Kleinwächter, M.; Paulsen, J.; Bloem, E.; Schnug, E.; Selmar, D. Moderate drought and signal transducer induced biosynthesis of relevant secondary metabolites in thyme (*Thymus vulgaris*), greater celandine (*Chelidonium majus*) and parsley (*Petroselinum crispum*). *Ind. Crops Prod.* **2015**, *64*, 158–166. [CrossRef]
60. Castro-Moreno, M.; Tinoco-Ojangurén, C.L.; Cruz-Ortega, M.D.R.; González-Esquinca, A.R. Influence of seasonal variation on the phenology and liriodenine content of *Annona lutescens* (Annonaceae). *J. Plant Res.* **2013**, *126*, 529–537. [CrossRef]
61. De-la-Cruz-Chacón, I.; Riley-Saldaña, C.A.; Arrollo-Gómez, S.; Sancristóbal-Domínguez, T.J.; Castro-Moreno, M.; González-Esquinca, A.R. Spatio-Temporal Variation of Alkaloids in *Annona purpurea* and the Associated Influence on Their Antifungal Activity. *Chem. Biodivers.* **2019**, *16*, 1–14. [CrossRef]
62. Selmar, D.; Kleinwächter, M. Stress enhances the synthesis of secondary plant products: The impact of stress-related over-reduction on the accumulation of natural products. *Plant Cell Physiol.* **2013**, *54*, 817–826. [CrossRef]
63. Simonneau, T.; Lebon, E.; Coupel-Ledru, A.; Marguerit, E.; Rossdeutsch, L.; Ollat, N.; Soni, A.; Kumari, P.; Dhakar, S.; Kumar, N.; et al. Physiological behavior of *Annona muricata, Dipteryx odorata* and *Copaifera langsdorffii* in response to water and light stress. *Front. Plant Sci.* **2017**, *62*, 1–8. [CrossRef]

64. de Oliveira, A.K.M.; Gualtieri, S.C.J. Trocas gasosas e grau de tolerância ao estresse hídrico induzido em plantas jovens de *Tabebuia aurea* (Paratudo) submetidas a alagamento. *Ciência Florest.* **2017**, *27*, 181–191. [CrossRef]
65. Selmar, D.; Kleinwächter, M. Influencing the product quality by deliberately applying drought stress during the cultivation of medicinal plants. *Ind. Crops Prod.* **2013**, *42*, 558–566. [CrossRef]
66. Mantoan, L.P.B.; Rolim de Almeida, L.F.; Macedo, A.C.; Ferreira, G.; Boaro, C.S.F. Photosynthetic adjustment after rehydration in *Annona emarginata*. *Acta Physiol. Plant.* **2016**, *38*, 1–11. [CrossRef]
67. Kalaji, H.M.; Jajoo, A.; Oukarroum, A.; Brestic, M.; Zivcak, M.; Samborska, I.A.; Cetner, M.D.; Łukasik, I.; Goltsev, V.; Ladle, R.J. Chlorophyll a fluorescence as a tool to monitor physiological status of plants under abiotic stress conditions. *Acta Physiol. Plant.* **2016**, *38*, 102. [CrossRef]
68. Parolin, P.; Wittmann, F. Struggle in the flood: Tree responses to flooding stress in four tropical floodplain systems. *AoB Plants* **2010**, *2010*, 1–19. [CrossRef] [PubMed]
69. Mielke, M.S.; Matos, E.M.; Couto, V.B.; Almeida, A.-A.F.; de Gomes, F.P.; Mangabeira, P.A.O. Some photosynthetic and growth responses of *Annona glabra* L. seedlings to soil flooding. *Acta Bot. Bras.* **2005**, *19*, 905–911. [CrossRef]
70. Nowak, M.; Kleinwächter, M.; Manderscheid, R.; Weigel, H.J.; Selmar, D. Drought stress increases the accumulation of monoterpenes in sage (*Salvia officinalis*), an effect that is compensated by elevated carbon dioxide concentration. *J. Appl. Bot. Food Qual.* **2010**, *83*, 133–136.
71. Junior, U.O.B.; Lima, M.D.R.; Barbosa, M.A.M.; Batista, B.L.; Lobato, A. Biochemical Responses of Two Species of Eucalyptus Exposed to Aluminium Toxicity: Oxidative Stress and Antioxidant Metabolism. *Not. Bot. Horti Agrobot. Cluj-Napoca* **2016**, *44*, 107–115. [CrossRef]
72. Sataloff, R.T.; Johns, M.M.; Kost, K.M.; Arndt, S.K.; Irawan, A.; Sanders, G.J.; Boanares, D.; Kozovits, A.R.; Lemos-Filho, J.P.; Isaias, R.M.S.; et al. Responses of photosynthesis, lipid peroxidation and antioxidant system in leaves of *Quercus mongolica* to elevated O3. *J. Exp. Bot.* **2014**, *69*, 907–912. [CrossRef]
73. Silva, A.A.; da Gonçalves, R.C. Espécies reativas do oxigênio e as doenças respiratórias em grandes animais. *Ciência Rural* **2010**, *40*, 994–1002. [CrossRef]
74. Barbosa, M.R.; de Araújo Silva, M.M.; Willadino, L.; Ulisses, C.; Camara, T.R. Geração e desintoxicação enzimática de espécies reativas de oxigênio em plantas. *Cienc. Rural* **2014**, *44*, 453–460. [CrossRef]
75. Lignan, C.; Suberin, L.; Stilbenes, C.; Tannin, F. Phenylpropanoids comprise a multitude of plant secondary metabolites and cell wall components. In *Plant Biochemistry*, 4th ed.; Elsevier Academic Press: Amsterdam, The Netherlands, 2011; pp. 435–454.
76. Møller, I.M.; Jensen, P.E.; Hansson, A. Oxidative modifications to cellular components in plants. *Annu. Rev. Plant Biol.* **2007**, *58*, 459–481. [CrossRef]
77. Hossain, Z.; López-Climent, M.F.; Arbona, V.; Pérez-Clemente, R.M.; Gómez-Cadenas, A. Modulation of the antioxidant system in citrus under waterlogging and subsequent drainage. *J. Plant Physiol.* **2009**, *166*, 1391–1404. [CrossRef]
78. Cortina, C.; Culiáñez-Macià, F.A. Tomato abiotic stress enhanced tolerance by trehalose biosynthesis. *Plant Sci.* **2005**, *169*, 75–82. [CrossRef]
79. Jaleel, C.A.; Manivannan, P.; Kishorekumar, A.; Sankar, B.; Gopi, R.; Somasundaram, R.; Panneerselvam, R. Alterations in osmoregulation, antioxidant enzymes and indole alkaloid levels in *Catharanthus roseus* exposed to water deficit. *Colloids Surf. B Biointerfaces* **2007**, *59*, 150–157. [CrossRef] [PubMed]
80. Krasensky, J.; Jonak, C. Drought, salt, and temperature stress-induced metabolic rearrangements and regulatory networks. *J. Exp. Bot.* **2015**, *63*, 1593–1608. [CrossRef] [PubMed]
81. Liu, B.; Liang, J.; Tang, G.; Wang, X.; Liu, F.; Zhao, D. Drought stress affects on growth, water use efficiency, gas exchange and chlorophyll fluorescence of Juglans rootstocks. *Sci. Hortic.* **2019**, *250*, 230–235. [CrossRef]
82. Thalmann, M.; Santelia, D. Starch as a determinant of plant fitness under abiotic stress. *New Phytol.* **2017**, *214*, 943–951. [CrossRef] [PubMed]
83. Ahmed, F.; Ra, M.; Ismail, M.R.; Juraimi, A.S.; Rahim, H.A.; Asfaliza, R.; Latif, M.A. Waterlogging Tolerance of Crops: Breeding Mechanism of Tolerance Molecular Approaches and Future Prospects. *BioMed Res. Int.* **2013**, *2013*. [CrossRef]
84. Malik, A.I.; Colmer, T.D.; Lambers, H.; Setter, T.L.; Schortemeyer, M. Short-term waterlogging has long-term effects on the growth and physiology of wheat. *New Phytol.* **2002**, *4*, 225–236. [CrossRef]
85. Huang, B.; Johnson, J.W. Root respiration and carbohydrate status of two wheat genotypes in response to hypoxia. *Annals of Botany* **1995**, *75*, 427–432. [CrossRef]
86. Malik, A.I.; Colmer, T.D. Changes in physiological and morphological traits of roots and shoots of wheat in response to different depths of waterlogging. *Funct. Plant Biol.* **2001**, *28*, 1121–1131. [CrossRef]
87. Setter, T.L.; Waters, I.; Sharma, S.K.; Singh, K.N.; Kulshreshtha, N.; Yaduvanshi, N.P.S.; Ram, P.C. Review of wheat improvement for waterlogging tolerance in Australia and India: The importance of anaerobiosis and element toxicities associated with different soils. *Ann. Bot.* **2008**, *103*, 221–235. [CrossRef]
88. Fernandez, O.; Béthencourt, L.; Quero, A.; Sangwan, R.S.; Clément Christophe, C. Trehalose and plant stress responses: Friend or foe? *Trends Plant Sci.* **2010**, *15*, 409–417. [CrossRef]
89. Paul, M.J.; Primavesi, L.F.; Jhurreea, D.; Zhang, Y. Trehalose metabolism and signaling. *Annu. Rev. Plant Biol.* **2008**, *59*, 417–441. [CrossRef] [PubMed]

horticulturae

MDPI

Article

The Differential Response of Intracellular Water Metabolism Derived from Intrinsic Electrophysiological Information in *Morus alba* L. and *Broussonetia papyrifera* (L.) Vent. Subjected to Water Shortage

Rui Yu [1], Yanyou Wu [2,*] and Deke Xing [1]

[1] Key Laboratory of Modern Agricultural Equipment and Technology, Ministry of Education, Institute of Agricultural Engineering, Jiangsu University, Zhenjiang 212013, China; 2111316005@stmail.ujs.edu.cn (R.Y.); xingdeke@ujs.edu.cn (D.X.)

[2] Research Center for Environmental Bio-Science and Technology, State Key Laboratory of Environmental Geochemistry, Institute of Geochemistry, Chinese Academy of Sciences, Guiyang 550081, China

* Correspondence: wuyanyou@mail.gyig.ac.cn; Tel.: +86-851-84391746

Abstract: Plant electrical signals can quickly respond to the shifting environment. Almost all life activities of plants are dependent on water. The measurement of plant electrophysiological indices provides a more convenient method for studying the intracellular water utilization. In this study, *Morus alba* L. (*Morus alba* or *M. alba*) and *Broussonetia papyrifera* (L.) Vent. (*Broussonetia papyrifera* or *B. papyrifera*) were experimental materials, and the parameters were measured in two habitats (waterfront, well-water and arid slopes, deficient-water). The physiological and electrophysiological responses of leaves to different habitats were analyzed. The theoretically intrinsic relationships between the clamping force and leaf impedance (Z), capacitive reactance (Xc), resistance (R), and inductive reactance (Xl) were revealed as 3-parameter exponential decay and linear models based on bioenergetics, respectively. Leaf intrinsic electrophysiological parameters were successfully obtained by using the above-mentioned relationships and were used to manifest metabolic activity in plants. The intracellular water-holding capacity (IWHC), water use efficiency (IWUE), water-holding time (IWHT), and water transfer rate (WTR) of plant leaves were defined based on the intrinsic electrophysiological parameters and were used to reflect the intracellular water metabolism. The correlation between the physiological and electrophysiological parameters of the two plant species in the two habitats was also analyzed. The results showed that *Morus alba* continuously adapted to the shifting environment, the intracellular water metabolism was insensitive to soil water shortage and was independent from the external physiological state. The intracellular water metabolism in *Broussonetia papyrifera* was very sensitive to soil water shortage, and both intracellular water metabolism and immediate physiological parameters could characterize the response of *Broussonetia papyrifera* growth and development to soil water.

Keywords: electrophysiological signals; intracellular water metabolism; bioenergetics; plant physiological information; water shortage response

Citation: Yu, R.; Wu, Y.; Xing, D. The Differential Response of Intracellular Water Metabolism Derived from Intrinsic Electrophysiological Information in *Morus alba* L. and *Broussonetia papyrifera* (L.) Vent. Subjected to Water Shortage. *Horticulturae* **2022**, *8*, 182. https://doi.org/10.3390/horticulturae8020182

Academic Editor: Riccardo Lo Bianco

Received: 26 December 2021
Accepted: 18 February 2022
Published: 21 February 2022

1. Introduction

Morus alba L. (*Morus alba* or *M. alba*) and *Broussonetia papyrifera* (L.) Vent. (*Broussonetia papyrifera* or *B. papyrifera*), belonging to the family Moraceae, grow fast, and adapt to adverse environments [1]. The species *M. alba* is the sole food source of the domesticated silkworm and is also an economically important perennial tree [2]. In addition, *M. alba* has edible, medicinal, animal feed, biological materials, ecological protection, and other uses with the developed society and scientific progress [3]. *M. alba* and *B. papyrifera* have great differences in water use efficiency mechanisms and drought resistance. They can be used as comparison materials to further study the differential response of intracellular water metabolism to water shortage based on electrophysiology.

Photosynthesis in green plants is of great significance to plant growth and development [4]. The photosynthetic rate will decrease under drought stress [5]. Plant leaf water potential is one of the main physiological indicators reflecting plant soil water status. When combined with stomatal conductance and other information, it can determine the plant drought resistance [6,7]. Researchers have reported that the increases of photosynthesis were accompanied by the increased leaf water potential [8,9]. Both photosynthetic indexes and leaf water potential represent the ability of plants to resist drought [10,11]. At the same time, almost all life activities in plants involve charge separation, electron movement, proton and dielectric transport, etc. [12]. The change in water content inevitably leads to changes in membrane permeability and ion concentration inside and outside the cell. Once a plant suffers from environmental stress, the moisture status, ion concentration and membrane permeability of its cells will change immediately, and thereafter change the electrical signal of plant. Therefore, electrophysiological indicators can be used as a theoretical basis for reflecting the water status of plants [13–15]. When the environment changes, plant electrical signals will change correspondingly [16–19]. Therefore, it is of great significance to study the intrinsic electrical parameters of plants and the related environmental factors and physiological responses, in order to evaluate the life phenomena of plant [20,21].

Drought can rapidly trigger plant electrical activity [22]. The ions and ionic groups in plant mesophyll cells are electrolytes, which have been considered concentric spherical capacitors with dual functions of inductance and resistance [23]. The mesophyll cells of plants can record electrical activity caused by external stimuli. Due to the external stimulation, changes in the structure and activity of the plant cell membrane directly affect the physiological process of plants, mainly reflected in changes in the stomatal state, photosynthetic rate and plant electrical signal [24]. In addition to the influence of soil water content on the electrical signals of plant leaves, the clamping force of the two electrodes of a self-made parallel-plate capacitor is also the main factor affecting the electrophysiological parameters [25]. This phenomenon is mainly due to the change in electrolyte concentration of mesophyll cells under different clamping forces. Therefore, the internal relationship between clamping force and electrophysiological parameters can be revealed under different water conditions, providing a fast and real-time method for monitoring the physiological state of plant leaves. At the same time, exploring the correlation between leaf water potential, water content, photosynthesis and electrophysiological information has important practical significance for studying the mechanism of intracellular water utilization in plants.

The water absorbed by roots is transported to the aboveground parts of the plant through vessels in the roots, stems and leaves; only 1–3% of the water is retained in the plant for photosynthesis and other life activities, and almost all of the rest is lost through transpiration [26,27]. To explore the relationship between plants and water, the physiological and molecular mechanisms of the plant response to drought stress are crucial for improving plant productivity and environmental efficiency [28–30]. Therefore, it is necessary to directly and quantitatively monitor the intracellular water status of plant leaves.

In our previous study, we found that the growth and development of *Morus alba* and *Broussonetia papyrifera* respond differently to the environment under short-term drought stress [31]. Can immediate extrinsic parameters and intrinsic electrophysiological parameters characterize the response mechanism of *M. alba* and *B. papyrifera* to water scarcity under prolonged drought conditions? What is the difference between the electrophysiological-based responses of *M. alba* and *B. papyrifera* in their intracellular water metabolism response patterns under water-deficient conditions? Water has a high dielectric constant and is a good condition for various electrolytes in cells to participate in chemical reactions [32]. Inevitably, the water metabolism of plants will not only lead to changes in intracellular and extracellular membrane permeability and ion concentration, but also changes in plant electrical signals [33]. The leaf resistance (R), capacitive reactance (Xc), impedance (Z), and inductive reactance (Xl) are related to the concentration of ions, ion groups and electric dipoles, and variations in the electrolyte concentration are caused by intracellular water metabolism. This study revealed for the first time the intrinsic mechanism of the clamping force and leaf Z, Xc, R, and Xl of two mulberry species in two habitats and established a physical model on this basis. Subsequently, the intrinsic electrophysiological parameters of the plant leaves were successfully obtained by using these equations. Then the intracellular water-holding capacity (IWHC), water use efficiency (IWUE), water-holding time (IWHT) and water transfer rate (WTR) of the intracellular water metabolism indexes in plant leaves were defined and applied according to the intrinsic electrophysiological parameters. This study aims to reveal the intrinsic mechanistic relationships between leaf Z, Xc, R, and Xl and exogenous stimuli, and provide a novel method for rapid monitoring plant physiological status. At the same time, this study was the first to discuss the correlation between leaf water potential, leaf water content, photosynthesis and electrophysiological information, which was of great significance for revealing plant leaf intracellular water metabolic efficiency.

2. Materials and Methods

2.1. Experimental Materials

M. alba and *B. papyrifera* were grown in two habitats, waterfront and arid slope soil, at Jiangsu University (N 32°11′ and E 119°27′). The average annual temperature, sunshine hours, and precipitation in this test area are 15.4 °C, 2051.7 h, and 1106 mm, respectively. The soil had a total organic matter content of 10.49 $g \cdot kg^{-1}$, a soil field water capacity of 25.5% and a pH value of 7.39. The tested leaves were sampled and measured at 9 a.m. to 11 a.m. on sunny days in June, and the measured temperature was 25.0 ± 2.0 °C.

2.2. Measurement of Electrophysiological Parameters of Plant Leaves under Different Clamping Forces

The electrophysiological parameters of plant leaves were measured by an LCR tester (Model 3532-50, Hioki, Nagano, Japan). Taken fully unfolded leaf from the fourth leaf position of each plant. The electrophysiological parameters of leaves were determined according to the method described by Xing et al. [34], the test parameters were further modified and described in our previous studies [25,35] (Figure S1).

2.3. Calculation of Intrinsically Electrophysiological Parameters of Plant Leaves

The calculation principle of leaf electrophysiological parameters has been described in our previous study [25,34,36] (Formula S1). According to bioenergetics, the model relationship between plant intrinsic electrophysiological parameters and clamping force was fitted according to the Nernst and Gibbs free energy equation.

2.4. Definition of the Intracellular Water Utilization Parameters

According to the model relationship between plant intrinsic electrophysiological parameters and clamping force, intracellular water utilization parameters of leaves were calculated by referring to the previous research method of Zhang et al. [25] (Formula S1).

2.5. Determination of Leaf Photosynthetic Parameters, Soil Moisture Content, Leaf Water Content

Net photosynthetic rate (P_N, μmol $(CO_2)m^{-2}$ s^{-1}) and transpiration rate (Tr, mmol m^{-2} s^{-1}) of *M. alba* and *B. papyrifera* were measured using li-6400 portable gas exchange measurement system (LI-COR, Lincoln, NE, USA) equipped with a blue/red light source from 9 a.m. to 11 a.m. in the two habitats [37,38]. According to the previous research method of Yu et al. [31], after the determination of the photosynthetic parameters, the soil moisture content and leaf water content of *M. alba* and *B. papyrifera* in the two habitats were measured using the drying method.

2.6. Determination of Leaf Water Potential

A water potential system (PSYPRO, Wescor, Inc., Logan, UT, USA) was used to determine the water potential values of *M. alba* and *B. papyrifera* in the two habitats. A hole punch with the corresponding diameter was selected to drill the hole, and it was quickly put into the sample chamber of the C-52 water potential probe. After balancing for 6 min, the measurement began. Three data points were measured each time, and the average value was used as the water potential measurement value of the leaf at this time.

2.7. Data Analyses

Data were analyzed using exploratory data analysis by SigmaPlot software (version 10.0, Systat Software Inc., San Jose, CA, USA) and SPSS software (version 21.0, SPSS Inc., Chicago, IL, USA). The statistical analysis included one-way analysis of variance (ANOVA), and significant differences between the means were tested using Tukey's test at 95% confidence. The data are shown as the means ± SE. Graphs were prepared using Origin Pro. 9.0 (Northampton, MA, USA).

3. Results

3.1. Soil Moisture Content of M. alba and B. papyrifera in Two Habitats

As illustrated in Table 1, the soil moisture content of *M. alba* and *B. papyrifera* was different in the two habitats, which is, the T_H level was significantly higher than that the T_L level.

Table 1. The soil moisture content of *M. alba* and *B. papyrifera* in two habitats.

Plants	Treatment	ξ_S (H_2O) (%)
M. alba	T_H	21.54 ± 0.89 a
	T_L	7.90 ± 0.54 b
B. papyrifera	T_H	21.26 ± 0.64 a
	T_L	7.77 ± 0.50 b

Note: values indicate the means ± SE, $n = 5$. Small letters indicate significant differences at 5% level ($p \leq 0.05$). T_H is waterfront soil. T_L is arid slopes soil.

3.2. Fitting Equation Parameters of M. alba and B. papyrifera in Two Habitats

Figure 1 randomly lists the fitting curves and equations of the relationship between leaf Z (Figure 1A), Xc (Figure 1B), R (Figure 1C), Xl (Figure 1D) and champing force (F) in Moraceae. The results show that the relationships of leaf Z, Xc, R, and Xl to the clamping force correlated well. Subsequently, the fitting equation coefficients of both the clamping force and leaf Z, Xc, R, and Xl in *M. alba* and *B. papyrifera* grown in two habitats were calculated separately (Table 2). The correlation coefficients (R^2) of the fitting equations of Z-F, Xc-F, R-F, and Xl-F for sixteen leaves of *B. papyrifera* in two habitats were 0.9709–0.9938, 0.9769–0.9949, 0.9620–0.9940, and 0.9770–0.9943, and those in *M. alba* in two habitats were 0.9733–0.9937, 0.9673–0.9931, 0.9764–0.9960, and 0.9676–0.9933, respectively. Moreover, all p values were less than 0.0001. This result shows that the fitting equations of Z-F, Xc-F, R-F, and Xl-F had good correlations, indicating the real existence of intrinsic mechanism relations between F and leaf Z, Xc, R, and Xl.

3.3. Electrophysiological Information of M. alba and B. papyrifera in Two Habitats

As shown in Table 3, the intrinsic electrophysiological parameters of *M. alba* and *B. papyrifera* in the two habitats are depicted, and the leaf intrinsic impedance (IZ), intrinsic resistance (IR), intrinsic inductive reactance (IXl), capacitive reactance (IXc), and intrinsic capacitance (IC) of *M. alba* at the T_H and T_L levels were not significantly different. Moreover, the leaf IZ, IR, IXl, and IXc of *B. papyrifera* at the T_H levels were significantly lower than those at the T_L levels, and the leaf IC of *B. papyrifera* at the T_H levels was significantly higher than that at the T_L levels. The IZ, IR, IXl, and IXc values of *B. papyrifera* at the T_L level were significantly increased by 325%, 335%, 331%, and 313%, respectively, compared to the T_H

levels. However, the IC values of *B. papyrifera* at T_L levels were significantly decreased by 77% compared to the T_H levels.

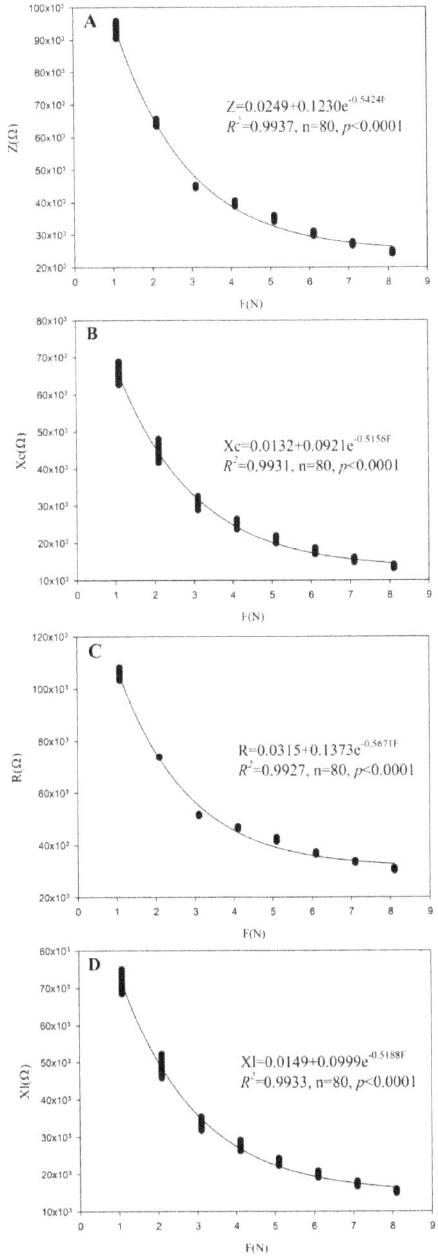

Figure 1. Fitting curves and equations of the relationships among leaf Z (**A**), Xc (**B**), R (**C**), Xl (**D**), and champing force (F) in Moraceae.

Table 2. Fitting equation parameters of *M. alba* and *B. papyrifera*.

Plants	Treatment	Z-F	Xc-F	R-F	Xl-F
		$y_0/k_1/b_1$ $R^2/p<$	$p_0/k_2/b_2$ $R^2/p<$	$g_0/k_3/b_3$ $R^2/p<$	$q_0/k_4/b_4$ $R^2/p<$
M. alba	T_H-1	0.0249/0.1230/0.5424 0.9937/0.0001	0.0132/0.0921/0.5156 0.9931/0.0001	0.0315/0.1373/0.5671 0.9927/0.0001	0.0149/0.0999/0.5188 0.9933/0.0001
	T_H-2	0.0288/0.2364/0.5862 0.9925/0.0001	0.0173/0.2489/0.7858 0.9674/0.0001	0.0365/0.2418/0.5473 0.9960/0.0001	0.0194/0.2611/0.7665 0.9700/0.0001
	T_H-3	0.0178/0.1831/0.5822 0.9779/0.0001	0.0072/0.0874/0.4052 0.9795/0.0001	0.0230/0.2495/0.6632 0.9764/0.0001	0.0088/0.0998/0.4278 0.9792/0.0001
	T_H-4	0.0255/0.2368/0.5748 0.9878/0.0001	0.0177/0.2925/0.6980 0.9845/0.0001	0.0296/0.2419/0.5606 0.9879/0.0001	0.0195/0.3059/0.6904 0.9847/0.0001
	T_L-1	0.0337/0.1395/0.5515 0.9922/0.0001	0.0236/0.1842/0.7192 0.9832/0.0001	0.0393/0.1409/0.5402 0.9914/0.0001	0.0258/0.1915/0.7078 0.9842/0.0001
	T_L-2	0.0371/0.0561/0.3261 0.9918/0.0001	0.0290/0.1020/0.3708 0.9908/0.0001	0.0413/0.0547/0.3380 0.9910/0.0001	0.0314/0.1052/0.3696 0.9909/0.0001
	T_L-3	0.0537/0.2467/0.9774 0.9733/0.0001	0.0337/0.1832/0.6666 0.9673/0.0001	0.0622/0.2995/1.1238 0.9783/0.0001	0.0373/0.1972/0.6847 0.9676/0.0001
	T_L-4	0.0277/0.1153/0.8068 0.9792/0.0001	0.0167/0.1170/0.7476 0.9797/0.0001	0.0332/0.1279/0.8939 0.9789/0.0001	0.0186/0.1240/0.7530 0.9797/0.0001
B. papyrifera	T_H-1	0.0298/0.2235/0.6178 0.9889/0.0001	0.0145/0.1875/0.6226 0.9844/0.0001	0.0403/0.2442/0.6525 0.9889/0.0001	0.0167/0.2014/0.6237 0.9850/0.0001
	T_H-2	0.0252/0.1182/0.4428 0.9933/0.0001	0.0129/0.1109/0.5128 0.9899/0.0001	0.0343/0.1210/0.4478 0.9910/0.0001	0.0148/0.1179/0.5084 0.9905/0.0001
	T_H-3	0.0277/0.2136/0.5703 0.9914/0.0001	0.0138/0.1335/0.5496 0.9853/0.0001	0.0369/0.2539/0.5912 0.9932/0.0001	0.0158/0.1477/0.5528 0.9863/0.0001
	T_H-4	0.0349/0.1283/0.3730 0.9938/0.0001	0.0194/0.0936/0.4646 0.9887/0.0001	0.0438/0.1399/0.3266 0.9940/0.0001	0.0219/0.1013/0.4534 0.9894/0.0001
	T_L-1	0.0980/0.5444/0.5290 0.9793/0.0001	0.0513/0.3909/0.4726 0.9949/0.0001	0.1225/0.6150/0.5545 0.9844/0.0001	0.0582/0.4256/0.4785 0.9943/0.0001
	T_L-2	0.1448/0.6665/0.6820 0.9840/0.0001	0.0880/0.5156/0.5623 0.9858/0.0001	0.1700/0.7525/0.7302 0.9804/0.0001	0.0975/0.5559/0.5698 0.9851/0.0001
	T_L-3	0.1104/0.7424/0.4926 0.9807/0.0001	0.0794/0.6497/0.5687 0.9861/0.0001	0.1461/0.9212/0.6337 0.9944/0.0001	0.0937/0.7578/0.6509 0.9844/0.0001
	T_L-4	0.1176/0.9636/0.8204 0.9709/0.0001	0.0635/0.6419/0.7601 0.9769/0.0001	0.1480/1.1231/0.8556 0.9620/0.0001	0.0717/0.7038/0.7656 0.9770/0.0001

Note: Z: impedance, Xc: capacitive reactance, R: resistance, Xl: inductive reactance, F: clamping force. T_H is waterfront soil. T_L is arid slopes soil.

Table 3. The electrophysiological parameters of *M. alba* and *B. papyrifera* in the two habitats.

Plants	Treatment	IZ (MΩ)	IR (MΩ)	IXl (MΩ)	IXc (MΩ)	IC (pF)
M. alba	T_H	0.22 ± 0.03 b	0.25 ± 0.03 b	0.21 ± 0.06 b	0.19 ± 0.06 b	358.55 ± 44.96 a
	T_L	0.18 ± 0.04 b	0.20 ± 0.06 b	0.18 ± 0.03 b	0.17 ± 0.02 b	325.33 ± 43.61 a
B. papyrifera	T_H	0.20 ± 0.03 b	0.23 ± 0.03 b	0.16 ± 0.02 b	0.15 ± 0.02 b	380.16 ± 45.24 a
	T_L	0.85 ± 0.09 a	1.00 ± 0.11 a	0.69 ± 0.08 a	0.62 ± 0.07 a	88.96 ± 10.86 b

Note: values indicate the means ± SE, $n = 5$. Small letters indicate significant differences at 5% level ($p \leq 0.05$). Different lowercase letters in the same column indicate significant differences in measurement factors. IZ: intrinsic impedance, IXc: intrinsic capacitive reactance, IR: intrinsic resistance, IXl: intrinsic inductive reactance, IC: intrinsic capacitance. T_H is waterfront soil. T_L is arid slopes soil.

3.4. Intracellular Water Utilization of *M. alba* and *B. papyrifera* in Two Habitats

The water use parameters of *M. alba* and *B. papyrifera* in the two habitats were obtained by using the corresponding parameters of the fitting equations. The results showed that there were significant differences in leaf specific effective thickness (d), IWUE, IWHC, IWHT, WTR, and water content of *M. alba* and *B. papyrifera* between the two habitats. As shown in Table 4, the leaf d of *M. alba* at the T_H levels was significantly higher than that at the T_L levels, the leaf d values of *M. alba* at the T_L levels were significantly decreased by 49% compared to the T_H levels, but the leaf IWUE, IWHT, and WTR of *M. alba* in the two habitats were not significantly different. However, the leaf d and WTR of *B. papyrifera* at the T_H levels were significantly higher than those at the T_L levels, and the d and WTR values of *B. papyrifera* at the T_L levels were significantly decreased by 78% and 89%, respectively, compared to the T_H levels, but the leaf IWUE and IWHT of *B. papyrifera* in the two habitats were not significantly different.

Table 4. Water use parameters of *M. alba* and *B. papyrifera* in the two habitats.

Plants	Treatment	d	IWUE	IWHT	WTR
M. alba	T_H	547.54 ± 58.45 a	0.12 ± 0.04 a	71.18 ± 5.14 a	93.18 ± 25.03 a
	T_L	279.29 ± 61.00 b	0.05 ± 0.01 a	53.03 ± 7.88 a	124.90 ± 35.21 a
B. papyrifera	T_H	468.86 ± 55.19 a	0.07 ± 0.02 a	72.88 ± 5.65 a	104.93 ± 19.79 a
	T_L	102.62 ± 12.17 c	0.13 ± 0.02 a	72.94 ± 4.17 a	11.66 ± 1.94 b

Note: values indicate the means ± SE, $n = 5$. Small letters indicate significant differences at 5% level ($p \leq 0.05$). Different lowercase letters in the same column indicate significant differences in measurement factors. d: Specific effective thickness, IWHT: intracellular water-holding time, IWUE: intracellular water use efficiency, WTR: dynamic water transfer rate. T_H is waterfront soil. T_L is arid slopes soil.

As shown in Figure 2, the IWHC and leaf water content of *M. alba* and *B. papyrifera* in the two habitats were displayed and basically consistent. The leaf IWHC and leaf water content of *M. alba* and *B. papyrifera* at the T_H levels were significantly higher than those at the T_L levels. The leaf IWHC values of *M. alba* and *B. papyrifera* at the T_L level were significantly decreased by 19% and 89%, respectively, compared to those at the T_H level. The leaf water content values of *M. alba* and *B. papyrifera* at the T_L level were significantly decreased by 18% and 25%, respectively, compared to those at the T_H level.

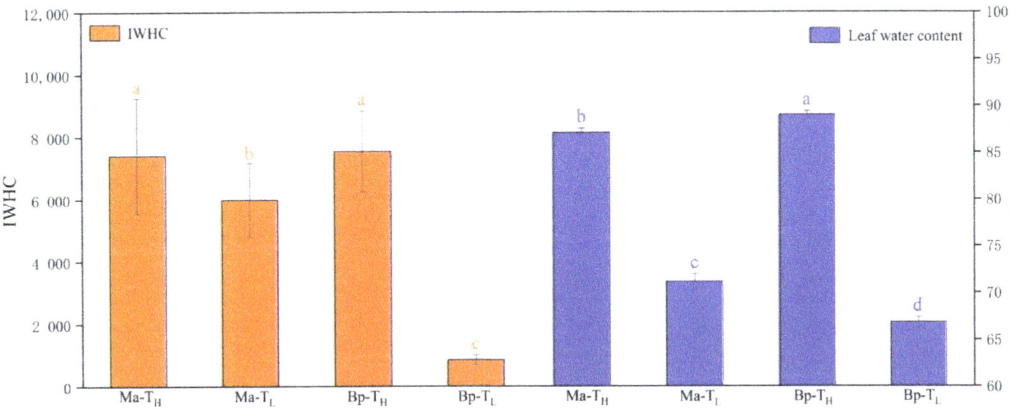

Figure 2. IWHC and leaf water content of *M. alba* and *B. papyrifera* in the two habitats. Note: values indicate the means ± SE, $n = 5$. Small letters indicate significant differences at 5% level ($p \leq 0.05$). IWHC: intracellular water-holding capacity, Bp-T_H: *B. papyrifera* at the T_H level, Bp-T_L: *B. papyrifera* at the T_L level, Ma-T_H: *M. alba* at the T_H level, Ma-T_L: *M. alba* at the T_L level.

3.5. Photosynthetic Parameters of *M. alba* and *B. papyrifera* in Two Habitats

The net photosynthetic rate (Figure 3A) and transpiration rate (Figure 3B) of the leaves of *M. alba* and *B. papyrifera* in the two habitats are shown in Figure 4. The net photosynthetic rate and transpiration rate of the leaves of *M. alba* and *B. papyrifera* at the T_H levels were significantly higher than those at the T_L levels. The leaf P_N values of *M. alba* and *B. papyrifera* at the T_L level were significantly decreased by 23% and 26%, respectively, compared to those at the T_H level. The leaf Tr values of *M. alba* and *B. papyrifera* at the T_L level were significantly decreased by 27% and 33%, respectively, compared to those at the T_H level.

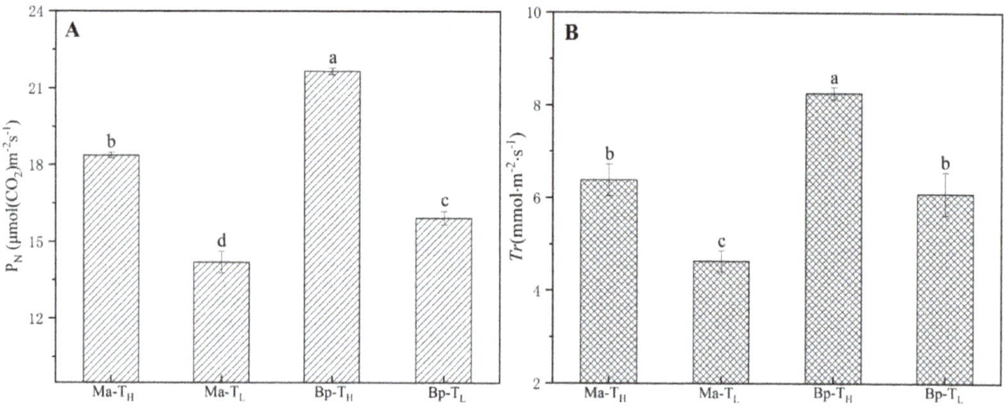

Figure 3. The net photosynthetic rate (P_N) (**A**) and transpiration rate (*Tr*) (**B**) of the leaves of *M. alba* and *B. papyrifera* in the two habitats. Note: Values indicate the means ± SE, n = 5. Small letters indicate significant differences at 5% level ($p \leq 0.05$). Bp-T_H: *B. papyrifera* at the T_H level, Bp-T_L: *B. papyrifera* at the T_L level, Ma-T_H: *M. alba* at the T_H level, Ma-T_L: *M. alba* at the T_L level.

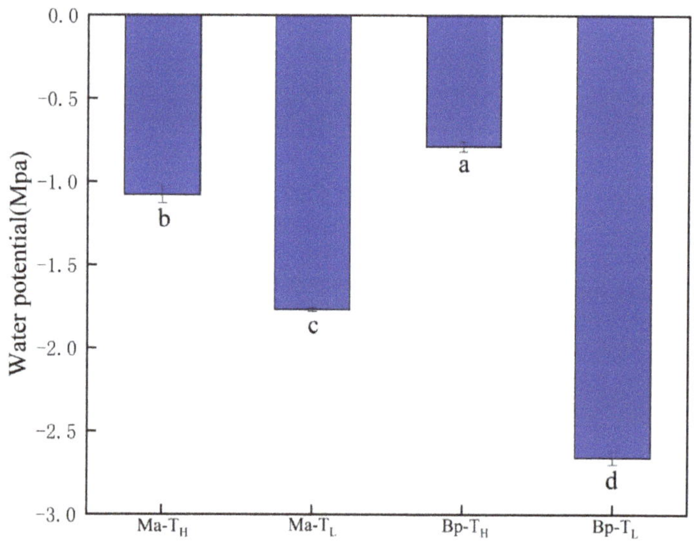

Figure 4. The leaf water potential of *M. alba* and *B. papyrifera* in the two habitats. Note: values indicate the means ± SE, n = 5. Small letters indicate significant differences at 5% level ($p \leq 0.05$). Bp-T_H: *B. papyrifera* at the T_H level, Bp-T_L: *B. papyrifera* at the T_L level, Ma-T_H: *M. alba* at the T_H level, Ma-T_L: *M. alba* at the T_L level.

3.6. Leaf Water Potential of M. alba and B. papyrifera in Two Habitats

The leaf water potential of *M. alba* and *B. papyrifera* in the two habitats is presented in Figure 4. The leaf water potential of the leaves of *M. alba* and *B. papyrifera* at the T_H levels was obviously higher than that at the T_L levels, with significant differences observed between the two habitats. The leaf water potential values of *M. alba* and *B. papyrifera* at the T_L level were significantly decreased by 39% and 70%, respectively, compared to those at the T_H level.

3.7. Correlation of Different Physiological Information and Intracellular Water Utilization Parameters

The Pearson correlation coefficients for the relationships between the different physiological information and intracellular water utilization parameters of *M. alba* and *B. papyrifera* are shown in Tables 5 and 6. In *M. alba* (Table 5), IZ showed a strong positive correlation with IR, strong negative correlation with WTR, and positive correlation with IXl and IXc. IR showed strong negative correlation with WTR. IXl was found to exhibit strong positive correlation with IXc and IWUE, strong negative correlation with IC and IWHC, and negative correlation with WTR. IXc showed strong positive correlation with IWUE, strong negative correlation with IC and IWHC, and negative correlation with WTR. IC was found to exhibit strong positive correlation with IWHC and negative correlation and IWUE. D showed positive correlation with Ψ and ξ_L. IWUE was negative correlation with WTR. Ψ showed strong positive correlation with ξ_L and P_N. ξ_L showed strong positive correlation with P_N. P_N exhibited strong positive correlation with Tr.

Table 5. Pearson correlation coefficients among different physiological parameters of *M. alba*.

	IR	IXl	IXc	IC	d	IWHC	IWUE	IWHT	WTR	Ψ	ξ_L	P_N	Tr
IZ	0.961 **	0.741 *	0.713 *	−0.622	0.131	−0.583	0.624	0.184	−0.940 **	0.289	0.260	0.261	0.123
IR		0.537	0.501	−0.404	0.215	−0.364	0.432	0.426	−0.857 **	0.277	0.245	0.247	0.069
IXl			0.999 **	−0.951 **	−0.151	−0.927 **	0.869 **	−0.477	−0.806 *	0.176	0.152	0.157	0.082
IXc				−0.957 **	−0.169	−0.935 **	0.868 **	−0.512	−0.785 *	0.163	0.140	0.146	0.080
IC					0.397	0.998 **	−0.714 *	0.642	0.705	0.119	0.131	0.073	0.026
d						0.449	0.295	0.649	−0.189	0.761 *	0.799 *	0.530	0.319
IWHC							−0.671	0.676	0.669	0.183	0.193	0.129	0.057
IWUE								−0.302	−0.720 *	0.549	0.550	0.406	0.259
IWHT									−0.061	0.373	0.372	0.307	0.115
WTR										−0.271	−0.252	−0.213	−0.122
Ψ											0.977 **	0.839 **	0.573
ξ_L												0.883 **	0.667
P_N													0.861 **

Note: ** correlation is significant at the 0.01 level (two-tailed). * Correlation is significant at the 0.05 level (two-tailed).

Table 6. Pearson correlation coefficients among different physiological parameters of *B. papyrifera*.

	IR	IXl	IXc	IC	d	IWHC	IWUE	IWHT	WTR	Ψ	ξ_L	P_N	Tr
IZ	0.998 **	0.970 **	0.979 **	−0.930 **	−0.858 **	−0.901 **	0.783 *	0.082	−0.893 **	−0.938 **	−0.931 **	−0.979 **	−0.885 **
IR		0.979 **	0.985 **	−0.927 **	−0.854 **	−0.897 **	0.792 *	0.066	−0.891 **	−0.935 **	−0.931 **	−0.977 **	−0.894 **
IXl			0.998 **	−0.931 **	−0.860 **	−0.899 **	0.756 *	−0.098	−0.880 **	−0.941 **	−0.937 **	−0.951 **	−0.917 **
IXc				−0.939 **	−0.866 **	−0.908 **	0.758 *	−0.077	−0.889 **	−0.948 **	−0.942 **	−0.960 **	−0.917 **
IC					0.786 *	0.996 **	−0.769 *	0.038	0.977 **	0.947 **	0.921 **	0.866 **	0.782 *
d						0.738 *	−0.400	−0.033	0.748 *	0.918 **	0.931 **	0.922 **	0.694
IWHC							−0.773 *	0.030	0.981 **	0.922 **	0.891 **	0.823 *	0.745 *
IWUE								0.004	−0.746 *	−0.617	−0.586	−0.657	−0.734 *
IWHT									−0.126	0.032	0.015	−0.111	0.169
WTR										0.901 **	0.875 **	0.826 *	0.710 *
Ψ											0.996 **	0.933 **	0.765 *
ξ_L												0.939 **	0.759 *
P_N													0.869 **

Note: ** correlation is significant at the 0.01 level (two-tailed). * Correlation is significant at the 0.05 level (two-tailed).

In *B. papyrifera* (Table 6), IZ showed strong positive correlation with IR, IXl and IXc, was highly negatively correlated with IC, d, IWHC, WTR, Ψ, ξ_L, P_N, and Tr, and positively correlated with IWUE. IR showed strong positive correlation with IXl and IXc, was strongly negatively correlated with IC, d, IWHC, WTR, Ψ, ξ_L, P_N, and Tr, and positively correlated with IWUE. IXl was strongly positively correlated with IXc, strongly negatively correlated with IC, d, IWHC, WTR, Ψ, ξ_L, P_N, and Tr, and positively correlated with IWUE. IXc was found to be strong negative correlation with IC, d, IWHC, WTR, Ψ, ξ_L, P_N, and Tr, and positively correlated with IWUE. IC was strongly positively correlated with IWHC, WTR, Ψ, ξ_L, and P_N, positive correlation with d and Tr, and negative correlation and IWUE. D was strongly positively correlated with Ψ, ξ_L, and P_N and positively correlated with IWHC and WTR. IWHC was found to be strongly positively correlated with WTR, Ψ, and ξ_L, positively correlated with P_N and Tr, and negatively correlated with IWUE. IWUE was negatively correlated with WTR. WTR was found to be strongly positively correlated with Ψ and ξ_L and showed positive correlation with P_N and Tr. Ψ was strongly positively correlated with ξ_L and P_N and positively correlated with Tr. Ξ_L was strongly positively correlated with P_N and showed positively correlation with Tr. P_N was strongly positively correlated with Tr.

4. Discussion

Water balance in plants is essential for plant growth. Leaf water potential can reflect the water status of plants and the influence degree of the soil-vegetation-atmosphere continuous system on the water in plants [39]. It can also reflect the ability of plants to absorb water from soil and maintain plant growth and development [6]. The results showed that leaf water content and water potential of *M. alba* and *B. papyrifera* were parallel to the soil water content. The leaf water content and water potential in waterfront soil were higher than those in arid slope soil, indicating that leaf water content and water potential could respond to soil water deficit immediately. Furthermore, lower leaf water content and water potential indicate soil water deficit, and plants can adapt to water shortage environments by reducing water demand [40].

Photosynthesis is the basis of plant growth and development and one of the indicators to observe the ability of plants adapting to the environmental stress. It is not only affected by the physiological characteristics of plants but also restricted by environmental factors. Water is an important factor involved in photosynthesis. With less soil moisture, photosynthesis will be reduced [41]. Therefore, studying the response of photosynthesis to stress can characterize the adaptability of plants to the environment. [42]. In this study, the P_N and Tr of *M. alba* and *B. papyrifera* in deficient-water soil were lower than those in well-water soil, indicating that soil water deficit would lead to stomatal closure on leaves, weaken transpiration, affect carbon dioxide absorption and reduce net photosynthetic rate, and plants could only maintain their own growth and development with limited water. The results also indicated that photosynthetic indexes could respond to soil water status quickly [43]. At the same time, chloroplasts are the main site of photosynthesis in plants, and low water potential results in damage to chloroplast thylakoid membrane structure, weakened electron transfer and phosphorylation, and decreased photosynthetic rate [44]. The response of photosynthesis to soil moisture was consistent with our previous research results [31].

Changes in electrical signals of plants are usually caused by changes in ion concentration, cell water status and membrane permeability of leaf cells immediately after plants are subjected to environmental stress [33]. In mesophyll cell, the cell and organelle are surrounded by cell membrane, so the cell can be regarded as a concentric circular capacitor with the dual functions of inductance and resistance. The ions and ionic groups in the mesophyll cell are electrolytes [23]. Almost all plant life activities are accompanied by the separation of electric charges, the movement of electrons, the transport of protons and media, etc. [36]. The clamping force stimulates the changes of plant leaves and immediately changes the membrane permeability of leaf cells, resulting in changes in electrolyte concentration and changes in leaf Z, R, Xc, and Xl. According to the Nernst equation, it revealed the Z (or R, Xc, Xl) = $y + ke^{-bF}$ of the theoretical internal relationship between Z, Xc, R, and Xl and the clamping force.

The Nernst equation is used to quantitatively describe the diffusion potential of an ion between A and B systems [25]. This equation relates chemical energy to galvanic electrode potential, making a significant contribution to electrochemistry. Meanwhile, the Nernst equation describes the relationship between the equilibrium potential of an ion and the ion concentration on both sides of the ion-permeable membrane [45]. The results showed that the fitting equations of Z-F, R-F, Xc-F, and Xl-F had good correlation (R^2 = 0.9620~0.9960, $p < 0.0001$), which highlighted the real existence of the above internal mechanism.

Part of the energy that reduces the internal energy of system converts into external work is defined as Gibbs free energy. In this study, the Gibbs free energy of leaf cell capacitor is theoretically equal to the work done by the clamping force [25]. The change in Gibbs free energy resulting from a reversible transition from one state to another at constant pressure and temperature is the amount of work available in that state change [46]. According to the Gibbs free energy, it also revealed the theoretically intrinsic relationships between leaves C and the clamping force, and the specific effective thickness (d) was defined. The results showed that the leaf d of *M. alba* and *B. papyrifera* grown in the waterfront soil was significantly higher than those of *M. alba* and *B. papyrifera* grown in the arid slope soil. These results revealed the life phenomenon in plants; that was, when plant water metabolism and growth were vigorous, the electrolyte concentration became low, water was abundant in leaf cell, and leaf d became high.

So far, spontaneous period and intrinsic electrical signals cannot be detected in plants. Most of the electrical signals in plants are induced by a variety of stimuli with low reproducibility [47]. In this study, the intrinsic electrophysiological indices of IZ, IR, IXc, IXl and IC of plant leaves were successfully determined by analyzing the intrinsic relationship between leaf Z, R, Xc and Xl and clamping force. The results showed that IC of leaf in waterfront soil was significantly higher than that in arid slope soil, with lower IZ, IR, IXl, and IXc, but the leaf IZ, IR, IXl, IXc, and IC of *M. alba* grown in waterfront soil and arid slope soil were not significantly different. Drought stress can affect plant water metabolism [5]. The results showed that the IWHC, WTR and water contents of *B. papyrifera* leaves grown under adequate water supply were higher, but there was no significant difference between IWHT and IWUE. In contrast to *B. papyrifera*, the d and IWHC of *M. alba* decreased with decreasing soil moisture, while the IWUE, IWHT and WTR did not change significantly. These results indicated that the intercellular water use characteristics of the two plant species were obviously different and had different response patterns. *M. alba* responded to water shortage by decreasing d. For *B. papyrifera* under water shortage, the effective thickness d of leaves was greatly reduced, the water utilization rate was increased, and the water transfer rate was decreased. These results were mainly due to the fast logarithmic growth period of *M. alba* leaves, short life cycle, low photosynthesis and slow overall growth (d), and water shortage had no effect on water metabolism of plants. However, the leaf life cycle of *B. papyrifera* was long and the growth was fast, which was affected by water deficit at any time. These revealed life phenomena peculiar to plants. In addition, soil water deficit and slowing down of the intracellular water transfer rate resulted in less water demand and therefore sufficient water for plants. As a result, leaf growth would be slow and there would be a larger sacrifice of leaf growth (d). This is consistent with the biological fact that plant biomass affects plant water use efficiency [48].

From Tables 5 and 6, correlation analysis between physiological indices and electrophysiological parameters of the leaves of *M. alba* and *B. papyrifera* was analyzed. As shown in Table 5, electrophysiological and intracellular water information of *M. alba* did not correlate with plant external physiological information (such as P_N, Tr, Ψ, and ξ_L). However, there were correlations between leaf water potential, water content and photosynthetic parameters, as well as between electrophysiological parameters and intracellular water use parameters, but they were not correlated with IWHT. The results indicated that the leaves of *M. alba* had a short life cycle and leaf function period, and fell off quickly; the leaves were thin, and the inner cells had little water, while the small cells were relatively stable in all aspects under soil water deficit [31]. Plant external physiological information did not affect the intracellular water status, indicating that the intracellular water metabolism of *M. alba* of leaves was independent of environmental water metabolism. Because *M. alba* was an acclimated species for a long time, it had gradually adapted to the shifting environment, and the rapid growth of new leaves was compatible with the activity of continuous leaf picking [3,49]. However, as shown in Table 6, contrary to *M. alba*, the electrophysiological indices of *B. papyrifera* were well correlated, except IWHT. These results showed that the leaves of *B. papyrifera* grew fast, had a long life cycle and leaf function period. The results indicated that both plant intrinsic electrophysiological information and external instant information in *B. papyrifera* were sensitive to soil moisture changes.

M. alba and *B. papyrifera* have different adaptive mechanisms to soil water deficit. The intrinsic water use efficiency and water transfer rate of *M. alba* leaves were not affected by the soil water deficit, thus maintaining the constant water retention time in leaves, and the growth of a single leaf was not affected. *M. alba* leaves had a short life cycle, so it responded to environmental changes by reducing the number of leaves. The leaf life cycle of *B. papyrifera* is long, the growth of an individual leaf was slow, and the water transport rate was reduced, in order to maintain the time of water supply to cope with the soil water shortage environment.

Previous studies have shown that the changes of plant electrophysiological parameters Z, R, and C can directly reflect the changes of plant water status [30,50–52]. However, Z, R, Xl, and Xc parameters alone cannot be used to determine water retention, transport and utilization in plant leaf cells [25]. In this study, IWHC, IWUE, IWHT, and WTR were used to accurately reveal the characteristics of intracellular water metabolism in leaves of *M. alba* and *B. papyrifera* in different habitats. The indices in this paper were defined by the intrinsic electrophysiological parameters of plant leaves and had the advantages of stability, accuracy, and representativeness. It is of great significance to accurately describe plant intracellular water metabolism.

5. Conclusions

Plant electrophysiological information can rapidly reflect soil water shortage conditions. In this study, the internal relationship between leaf Z, R, Xc, and Xl and the clamping force was revealed and established from the perspective of bioenergetics, and fitting equations were used to calculate the electrophysiological parameters IZ, IR, IXc, IXl, and IC. Then, the leaf intracellular water use parameters d, IWUC, IWUE, IWUT, and WTR of *M. alba* and *B. papyrifera* were defined based on the plant internal electrophysiological parameters. The mechanism of plant water metabolism was evaluated, and the correlation between plant physiological parameters (Ψ, ξ_L, P_N, Tr) and electrophysiological parameters was discussed. The results showed that the intracellular water metabolism based on electrophysiology in *M. alba* and *B. papyrifera* had different response patterns to soil water conditions. *M. alba* gradually adapted to the environment. These conclusions can be used to obtain plant intracellular water information and provide theoretical support for studying the adaptation mechanism of plants to shifting environment.

Supplementary Materials: The following are available online at https://www.mdpi.com/article/10.3390/horticulturae8020182/s1, Figure S1: The experimental setup used in the study and a schematic diagram of the parallel-plate capacitor; Formula S1: The deduction process of clamping force and electrophysiological parameter model.

Author Contributions: Conceptualization, methodology, funding acquisition, supervision, validation, Y.W.; data curation; formal analysis, writing—original draft, visualization, R.Y.; software, formal analysis, D.X. All authors have read and agreed to the published version of the manuscript.

Funding: This research was supported by the project of the National Key Research and Development Program of China (2016YFC0502602), the National Natural Science Foundation of China (No. U1612441-2), the Priority Academic Program Development (PAPD) of Jiangsu Higher Education Institutions, and the Graduate Innovative Projects of Jiangsu Province (2014 (KYLX_1061)).

Data Availability Statement: The datasets from (or analyzed during) the current study are available from the corresponding author upon reasonable request.

Conflicts of Interest: The authors declare no conflict of interest.

Abbreviations

C: capacitance, Z: impedance, Xc: capacitive reactance, R: resistance, Xl: inductive reactance, F: clamping force, IC: intrinsic capacitance, IZ: intrinsic impedance, IXc: intrinsic capacitive reactance, IR: intrinsic resistance, IXl: intrinsic inductive reactance, d: specific effective thickness, IWHC: intracellular water-holding capacity, IWUE: intracellular water use efficiency, IWHT: intracellular water-holding time, WTR: the dynamic water transfer rate. Ψ: water potential, ξ_L: leaf water content, P_N: the net photosynthetic rate. Tr: the transpiration rate.

References

1. Wu, Y.Y.; Liu, C.Q.; Li, P.P.; Wang, J.Z.; Xing, D.; Wang, B.L. Photosynthetic characteristics involved in adaptability to Karst soil and alien invasion of paper mulberry (*Broussonetia papyrifera* (L.) Vent.) in comparison with mulberry (*Morus alba* L.). *Photosynthetica* **2009**, *47*, 155–160. [CrossRef]
2. Chen, X.Y.; Zhang, T.; Wang, X.; Hamann, M.T.; Kang, J.; Yu, D.Q.; Chen, R.Y. A chemical investigation of the leaves of *Morus alba* L. *Molecules* **2018**, *23*, 1018. [CrossRef] [PubMed]
3. Vijayan, K.; Raju, P.J.; Tikader, A.; Saratchnadra, B.; Alkhayri, J.; Subramaniam, S. Biotechnology of mulberry (*Morus* L.)—A review. *Emir. J. Food Agric.* **2014**, *26*, 472–496. [CrossRef]
4. Ye, J.; Chen, W.F.; Feng, L.W.; Liu, G.Z.; Wang, Y.; Li, H.X.; Ye, Z.B.; Zhang, Y.Y. The chaperonin 60 protein SlCpn60a1 modulates photosynthesis and photorespiration in tomato. *J. Exp. Bot.* **2020**, *71*, 7224–7240. [CrossRef] [PubMed]
5. Samarah, N.H.; AL-Quraan, N.A.; Massad, R.S.; Welbaum, G.E. Treatment of bell pepper (*Capsicum annuum* L.) seeds with chitosan increases chitinase and glucanase activities and enhances emergence in a standard cold test. *Sci. Hortic.* **2020**, *269*, 109393. [CrossRef]
6. Tosin, R.; Pôças, I.; Novo, H.; Teixeira, J.; Fontes, N.; Graça, A.; Cunha, M. Assessing predawn leaf water potential based on hyperspectral data and pigment's concentration of *Vitis vinifera* L. in the Douro Wine Region. *Sci. Hortic.* **2021**, *278*, 109860. [CrossRef]
7. Bhusal, N.; Han, S.G.; Yoon, T.M. Impact of drought stress on photosynthetic response, leaf water potential, and stem sap flow in two cultivars of bi-leader apple trees (Malus × domestica Borkh.). *Sci. Hortic.* **2019**, *246*, 535–543. [CrossRef]
8. Arif, S.W.; Aqil, A.; Shamsul, H.; Qazi, F. Salt-induced modulation in growth, photosynthesis and antioxidant system in two varieties of *Brassica juncea*. *Saudi J. Biol. Sci.* **2013**, *20*, 183–193.
9. Vos, J.; Groenwold, J. Genetic differences in water-use efficiency, stomatal conductance and carbon isotope fractionation in potato. *Potato Res.* **1989**, *32*, 113–121. [CrossRef]
10. Horike, H.; Kinoshita, T.; Kume, A.; Hanba, Y. Responses of leaf photosynthetic traits, water use efficiency, and water relations in five urban shrub tree species under drought stress and recovery. *Trees* **2021**, *35*, 1–5. [CrossRef]
11. Hammoud, M.; Alturk, S.M.; Sebaaly, Z.E.; Sassine, Y.N. Drip vs. mini-sprinkler irrigation system on leaf water potential and various vegetative attributes of Annona squamosa under Lebanese conditions. *J. Agr. Sci.* **2021**, *43*, 338–346. [CrossRef]
12. Hopkins, W.G.; Huner, N.P.A. *Introduction to Plant Physiology*, 3rd ed.; John Wiley & Sons Inc.: New York, NY, USA, 2004; p. 27.
13. Choi, W.G.; Hilleary, R.; Swanson, S.J.; Kim, S.H.; Gilroy, S. Rapid, long-distance electrical and calcium signaling in plants. *Annu. Rev. Plant Biol.* **2016**, *67*, 287–307. [CrossRef] [PubMed]
14. Hedrich, R.; Salvador-Recatala, V.; Dreyer, I. Electrical wiring and long-distance plant communication. *Trends Plant Sci.* **2016**, *21*, 376–387. [CrossRef] [PubMed]
15. Nguyen, C.T.; Kurenda, A.; Stolz, S.; Chetelat, A.; Farmer, E.E. Identification of cell populations necessary for leaf-to-leaf electrical signaling in a wounded plant. *Proc. Natl. Acad. Sci. USA* **2018**, *115*, 10178–10183. [CrossRef] [PubMed]
16. Stahlberg, R.; Cleland, R.E.; Volkenburgh, E.V. Slow wave potentials—A propagating electrical signal unique to higher plants. In *Communication in Plants*; Springer: Berlin/Heidelberg, Germany, 2006; pp. 291–308.
17. Sorkin, M.L.; Nusinow, D.A. Time will tell: Intercellular communication in the plant clock. *Trends Plant Sci.* **2021**, *26*, 706–719. [CrossRef]
18. Fromm, J.; Lautner, S. Electrical signals and their physiological significance in plants. *Plant Cell Environ.* **2010**, *30*, 249–257. [CrossRef] [PubMed]
19. Lautner, S.; Grams, T.; Matyssek, R.; Fromm, J. Characteristics of electrical sifnals in poplar and responses in photosynthesis. *Plant Physiol.* **2005**, *138*, 2200–2209. [CrossRef]
20. Seo, P.J.; Mas, P. Stressing the role of the plant circadian clock. *Trends Plant Sci.* **2015**, *20*, 230–237. [CrossRef]
21. Asner, G.P.; Brodrick, P.G.; Anderson, C.B.; Vaughn, N.; Martin, R.E. Progressive forest canopy water loss during the 2012–2015 California drought. *Proc. Natl. Acad. Sci. USA* **2016**, *113*, E249. [CrossRef]
22. Gallé, A.; Lautner, S.; Flexas, J.; Fromm, J. Environmental stimuli and physiological responses: The current view on electrical signaling. *Environ. Exp. Bot.* **2015**, *114*, 15–21. [CrossRef]
23. Buckley, D.J.; Lefebvre, M.; Meijer, E.G.M.; Brown, D.C.W. A signal generator for electrofusion of plant protoplasts. *Comput. Electron. Agr.* **1990**, *5*, 179–185. [CrossRef]
24. Sanan-Mishra, N. Electric signaling and long-distance communication in plants. In *Sensory Biology of Plants*; Springer Nature: Singapore, 2019; pp. 509–535.
25. Zhang, C.; Wu, Y.Y.; Su, Y.; Xing, D.K.; Dai, Y.; Wu, Y.S.; Fang, L. A plant's electrical parameters indicate its physiological state: A study of intracellular water metabolism. *Plants* **2020**, *9*, 1256. [CrossRef] [PubMed]
26. Chaves, M.M.; Flexas, J.; Pinheiro, C. Photosynthesis under drought and salt stress: Regulation mechanisms from whole plant to cell. *Ann. Bot.-Lond.* **2009**, *103*, 551–560. [CrossRef] [PubMed]

27. David, T.S.; Henriques, M.O.; Kurz-Besson, C.; Nunes, J.; Valente, F.; Vaz, M.; Pereira, J.S.; Siegwolf, R.; Chaves, M.M.; Gazarini, L.C. Water-use strategies in two co-occurring Mediterranean evergreen oaks: Surviving the summer drought. *Tree Physiol.* **2007**, *27*, 793–803. [CrossRef] [PubMed]
28. Tranmotini, S.; Leeuwen, C.V.; Domec, J.C.; Destrac-Irvine, A.; Basteau, C.; Vitali, M.; Mosbach-Schulz, O.; Lovisolo, C. Impact of soil texture and water availability on the hydraulic control of plant and grape-berry development. *Plant Soil* **2013**, *368*, 215–230. [CrossRef]
29. Sun, Z.J.; Livingston, N.J.; Guy, R.D.; Ethier, G.J. Stable carbon isotope as indicators of increased water use efficiency and productivity in white spruce (*Picea glauca* (Moench) Voss) seedlings. *Plant Cell Environ.* **1996**, *19*, 887–894. [CrossRef]
30. González-Fernández, A.B.; Rodríguez-Pérez, J.R.; Marcelo, V.; Valenciano, J.B. Using field spectrometry and a plant probe accessory to determine leaf water content in commercial vineyards. *Agr. Water Manag.* **2015**, *156*, 43–50. [CrossRef]
31. Yu, R.; Wu, Y.Y.; Xing, D.K. Can Electrophysiological parameters substitute for growth, and photosynthetic parameters to characterize the response of Mulberry and paper Mulberry to drought? *Plants* **2021**, *10*, 1772. [CrossRef]
32. Martínez, J.P.; Silva, H.; Ledent, J.F.; Pinto, M. Effect of drought stress on the osmotic adjustment, cell wall elasticity and cell volume of six cultivars of common beans (*Phaseolus vulgaris* L.). *Eur. J. Agron.* **2007**, *26*, 30–38. [CrossRef]
33. Xing, D.K.; Chen, X.L.; Wu, Y.Y.; Chen, Q.; Li, L.; Fu, W.G.; Shu, Y. Leaf stiffness of two Moraceae species based on leaf tensity determined by compressing different external gripping forces under dehydration stress. *J. Plant Interact.* **2019**, *14*, 610–616. [CrossRef]
34. Xing, D.K.; Xu, X.J.; Wu, Y.Y.; Liu, Y.J.; Wu, Y.S.; Ni, J.H.; Azeem, A. Leaf tensity: A method for rapid determination of water requirement information in *Brassica napus* L. *J. Plant Interact.* **2018**, *13*, 380–387. [CrossRef]
35. Xing, D.K.; Chen, X.; Wu, Y.Y.; Zwiazek, J.J. Leaf physiological impedance and elasticity modulus in *Orychophragmus violaceus* seedlings subjected to repeated osmotic stress. *Sci. Hortic.* **2021**, *276*, 109763. [CrossRef]
36. Zhang, C.; Wu, Y.Y.; Su, Y.; Li, H.; Fang, L.; Xing, D.K. Plant's electrical information manifests the composition and nutrient transport characteristics of membrane proteins. *Plant Signal. Behav.* **2021**, *16*, 1918867. [CrossRef]
37. Chen, D.Q.; Wang, S.W.; Cao, B.B.; Cao, D.; Leng, G.H.; Li, H.B.; Yin, L.N.; Shan, L.; Deng, X.P. Genotypic variation in growth and physiological response to drought stress and re-watering reveals the critical role of recovery in drought adaptation in maize seedlings. *Front. Plant Sci.* **2016**, *6*, 1241. [CrossRef] [PubMed]
38. Sacramento, B.L.D.; Azevedo, A.D.D.; Alves, A.T.; Moura, S.C.; Ribas, R.F. Photosynthetic parameters as physiological indicators of tolerance to cadmium stress in sunflower genotypes. *Rev. Caatinga* **2018**, *31*, 907–916. [CrossRef]
39. Dghim, F.; Abdellaoui, R.; Boukhris, M.; Neffati, M.; Chaieb, M. Physiological and biochemical changes in *Periploca angustifolia* plants under withholding irrigation and rewatering conditions. *S. Afr. J. Bot.* **2018**, *114*, 241–249. [CrossRef]
40. Guo, W.H.; Li, B.; Huang, Y.M.; Zhao, H.X.; Zhang, X.S. Effects of different water stresses on eco-physiological characteristics of *Hippophae rhamnoides* seedlings. *Acta Bot. Sin.* **2003**, *45*, 1238–1244.
41. Djanaguiraman, M.; Nair, R.; Giraldo, J.P.; Prasad, P.V.V. Cerium oxide nanoparticles decrease drought-induced oxidative damage in sorghum leading to higher photosynthesis and grain yield. *ACS Omega* **2018**, *3*, 14406–14416. [CrossRef]
42. Sourour, A. A review: Morphological, physiological, biochemical and molecular plant responses to water deficit stress. *Int. J. Eng. Sci.* **2017**, *6*, 1–4. [CrossRef]
43. Buckley, T.N. The control of stomata by water balance. *New Phytol.* **2005**, *168*, 275–292. [CrossRef]
44. Yang, X.H.; Chen, X.Y.; Ge, Q.Y.; Li, B.; Tong, Y.P.; Zhang, A.M.; Li, Z.S.; Kuang, T.Y.; Lu, C.M. Tolerance of photosynthesis to photoinhibition, high temperature and drought stress in flag leaves of wheat: A comparison between a hybridization line and its parents grown under field conditions. *Plant Sci.* **2006**, *171*, 389–397. [CrossRef] [PubMed]
45. Sterratt, D. Nernst Equation. In *Encyclopedia of Computational Neuroscience*; Springer Science & Business Media: New York, NY, USA, 2014; pp. 1–3.
46. Reitner, J. *Encyclopedia of Earth Sciences Series*; Springer Science & Business: Berlin/Heidelberg, Germany, 2011; pp. 227–238.
47. Debono, M.W. Dynamic protoneural networks in plants: A new approach of spontaneous extracellular potential variations. *Plant Signal. Behav.* **2013**, *8*, e24207. [CrossRef]
48. Chatterjee, C.; Gopal, R.; Dube, B.K. Impact of iron stress on biomass, yield, metabolism and quality of potato (*Solanum tuberosum* L.). *Sci. Hortic.* **2006**, *108*, 1–6. [CrossRef]
49. Qin, J.; He, N.J.; Wang, Y.; Xiang, Z.H. Ecological issues of Mulberry and sustainable development. *J. Resour. Ecol.* **2012**, *3*, 330–339.
50. Javed, Q.; Wu, Y.Y.; Xing, D.K.; Azeem, A.; Ullah, I.; Zaman, M. Re-watering: An effective measure to recover growth and photosynthetic characteristics in salt-stressed *Brassica napus* L. *Chil. J. Agric. Res.* **2017**, *77*, 78–86. [CrossRef]
51. Kertész, Á.; Hlaváčová, Z.; Vozáry, E.; Staroňová, L. Relationship between moisture content and electrical impedance of carrot slices during drying. *Int. Agrophys.* **2015**, *29*, 61–66. [CrossRef]
52. Gil, P.M.; Gurovich, L.; Schaffer, B.; Alcayaga, J.; Rey, S.; Iturriaga, R. Root to leaf electrical signaling in avocado in response to light and soil water content. *J. Plant Physiol.* **2008**, *165*, 1070–1078. [CrossRef]

horticulturae

MDPI

Article

Interactive Impacts of Temperature and Elevated CO_2 on Basil (*Ocimum basilicum* L.) Root and Shoot Morphology and Growth

T. Casey Barickman [1,*], Omolayo J. Olorunwa [1], Akanksha Sehgal [2], C. Hunt Walne [2], K. Raja Reddy [2] and Wei Gao [3]

1 North Mississippi Research and Extension Center, Mississippi State University, Verona, MS 38879, USA; ojo26@msstate.edu
2 Department of Plant and Soil Sciences, Mississippi State University, Mississippi State, MS 39762, USA; as5002@msstate.edu (A.S.); chw148@msstate.edu (C.H.W.); krreddy@pss.msstate.edu (K.R.R.)
3 USDA UVB Monitoring and Research Program, Natural Resource Ecology Laboratory, Department of Ecosystem Science and Sustainability, Colorado State University, Fort Collins, CO 80523, USA; wei.gao@colostate.edu
* Correspondence: t.c.barickman@msstate.edu; Tel.: +1-662-566-2201

Abstract: Recent evidence suggests that the effects of temperature significantly affect the growth and development of basil plants with detrimental impacts on yield. The current research investigated the interactive effects of varying temperature and CO_2 levels on the shoot and root morphology and growth of early and late-season basil plants. Basil plants were subjected to control (30/22 °C), low (20/12 °C), and high (38/30 °C) temperature under ambient (420 μL L^{-1}) and elevated (720 μL L^{-1}) CO_2 concentrations. Decreasing the temperature to 20/12 °C caused more adverse effects on the morphological traits of the early-season basil. Relative to the control treatments, low- and high-temperature stresses decreased 71 and 14% in marketable fresh mass, respectively. Basil exhibited an increase in plant height, node and branch numbers, specific leaf area, anthocyanin and nitrogen balance index, root tips, and root crossings when subjected to high-temperature stress. Furthermore, elevated CO_2 affected many morphological features compared to ambient CO_2 concentrations. The findings of this study suggest that varying the growth temperature of basil plants would more significantly impact the shoot and root morphologies and growth rates of basil than increasing the CO_2 concentrations, which ameliorated the adverse impacts of temperature stress.

Keywords: genovese; leaf area; root length; nitrogen balance index; anthocyanin; epicuticular leaf waxes

Citation: Barickman, T.C.; Olorunwa, O.J.; Sehgal, A.; Walne, C.H.; Reddy, K.R.; Gao, W. Interactive Impacts of Temperature and Elevated CO_2 on Basil (*Ocimum basilicum* L.) Root and Shoot Morphology and Growth. *Horticulturae* **2021**, 7, 112. https://doi.org/10.3390/horticulturae7050112

Academic Editor: Yanyou Wu

Received: 6 April 2021
Accepted: 7 May 2021
Published: 14 May 2021

Publisher's Note: MDPI stays neutral with regard to jurisdictional claims in published maps and institutional affiliations.

1. Introduction

Climate change is increasingly recognized as a serious, global agricultural concern affecting plant growth and development with detrimental impacts on the yields of many important crops. Over the past century, there have been a dramatic increase in atmospheric carbon dioxide (CO_2) concentrations with a corresponding rise in global temperatures [1]. Global atmospheric CO_2 is rising (above 417 μL L^{-1} in 2021) [2] and is projected by climate models to reach 540 to 970 μL L^{-1} by 2100 because of human activities, declining carbon sinks, and natural global cycles [3,4]. As delineated in the fourth U.S. climate assessment, global temperature is projected to rise in the range of 1.5 °C and 4.5 °C in the next century due to the levels of atmospheric CO_2 and other greenhouse gases increasing at an alarming rate [1]. Atmospheric CO_2 and temperature are critical in the photosynthesis, physiological, and developmental processes that occur in many crops, especially C3 crops, including basil (*Ocimum basilicum* L.) [5,6]. Since climate change induces multiple abiotic stressors that affect crop yield worldwide, the impact of elevated CO_2 and temperature stress on basil growth and development has been distinguished as a vital area for additional studies. A great deal of the research to date, nonetheless, has focused on individual abiotic stresses and not their interactions with less consideration on the morphology and growth parameters of basil roots and shoots. Hence, it is imperative to understand the interactive effects of elevated CO_2 and temperature stress on basil growth and morphology to ensure sustainable crop production.

Basil is an essential herbaceous aromatic plant with a noteworthy contribution to enhancing cuisine nutrition, healthy living, and landscape aesthetics. Globally, a large proportion of high-quality basil is cultivated for its essential oil, dry leaves, and flowers [7,8]. Generally, basil is widely adapted to various climates and regions, and therefore it is cultivated throughout the globe. However, recent evidence suggests that basil growth and development can be seriously impaired by low-temperature stress and is susceptible to growth temperatures below 10 °C [9,10]. Low-temperature stress can be deleterious to basil growth and morphology, especially during its seedling and vegetative stage, with resultant effects on reduced productivity [10]. Chilling causes brown discoloration of interveinal leaf areas (LA), increases leaf-blade thickening, decreases plant growth, and deteriorates quality and marketability [9,10]. Moreover, low-temperature stress decreases plant height (PH) and fresh mass (FM) of basil by 36% and 63%, respectively, after 15 days of treatment [11]. However, according to previous studies, basil is a thermophilic plant that can sustain growth at a temperature in the range of 29 °C and 35 °C [9,12]. Corroborating this information, Walters and Currey [13] recorded increased biomass, PH, FM, dry mass accumulation (DM), node numbers (NN), and internode length of basil when the growth temperatures were increased to 29 °C.

The response of basil plants to elevated atmospheric CO_2 has not been extensively explored. However, previous research has indicated that elevated CO_2 will significantly impact basil growth and development primarily because of the significant role CO_2 plays in the respiration and photosynthesis of C3 plants [5,14]. Al Jaouni et al. [15] reported that biomass production increased by 40% along with the photosynthetic and respiratory rate of basil, which significantly improved by 80% when atmospheric CO_2 was increased from 360 to 620 $\mu L\ L^{-1}$. The improved photosynthetic rate was attributed to the role of elevated atmospheric CO_2 in repressing the oxygenation reaction of Rubisco, leading to improved carbon gains [14].

Individual and multiple interacting abiotic stresses have been noted to significantly affect plant roots' growth and morphology due to the pivotal role the root system plays in plant growth [16,17]. Root systems are instrumental in providing anchorage, water, and nutrient acquisition for plant growth. Recent evidence suggests that the plant root system is more critical than the above-ground traits to adapting to abiotic stress [17–19]. Varying temperature levels have been noted to have either beneficial or detrimental impacts on plant roots [20,21]. Lahti et al. [22] reported that spruce seedlings subjected to high-temperature stress increased their total root biomass and length growth, indicating the beneficial growth role of rising soil temperatures. Conversely, low-temperature stress constrained plant root morphology, specifically length, depth, and width, when growth temperature decreased to 18/13 °C (day/night), signifying the plant root's sensitivity to chilling stress [17,23]. Previous studies on the interactive effects of multiple abiotic stresses on plants have shown that increasing CO_2 from the ambient concentrations ameliorates other abiotic stressors' adverse effects by increasing the carbon gains in the plant roots [21]. Accordingly, further investigation is crucial to thoroughly understand the interactive impacts of temperature stress and elevated CO_2 on basil root growth and morphology to determine promotive or inhibitive role.

Under these considerations, it is imperative to note that understanding plant root and shoot response to various abiotic stresses is vital in increasing crop productivity while adapting to harsh environmental conditions. Moreover, limited information has been provided on basil roots in response to individual and multiple abiotic stress. Therefore, this study aims to investigate the individual and interactive impacts of elevated CO_2 and temperature stress on the growth and morphology of basil shoots and roots.

2. Materials and Methods

Basil 'Genovese' (Johnny's Selected Seeds, Winslow, ME, USA) seeds were planted in polyvinyl-chloride pots (15.2 cm diameter by 30.5 cm height). Each pot was filled up with 500 g gravel and then filled with a mixture of sand and topsoil (3:1 VV) in the soil-plant-atmosphere-research (SPAR) units at the Rodney Foil Plant Science research facility of Mississippi State University, Mississippi State, MS, USA, June–July 2019. The SPAR units can control environmental conditions, including temperature and CO_2 concentration levels, at predetermined set points. More information on the SPAR chamber's subtleties is portrayed by Reddy et al. [24] and Wijewardana et al. [25].

Six seeds previously selected by size and quality were planted in each pot, and approximately 14 days after sowing (DAS), the plants were thinned to one plant per pot, and temperature and CO_2 treatments were initiated. Throughout the experiment, basil plants were irrigated with full-strength Hoagland's nutrient solution [26] three times daily (7 a.m., 12 p.m., and 5 p.m.) via an automated computer-controlled drip system. Irrigation amounts were based on the evapotranspiration of the basil plants within each chamber. Irrigation was then applied at 120% of the amount of water lost the previous day and split between each irrigation cycle.

The experiment was organized in a randomized complete block design within a three by two factorial arrangement with temperature and CO_2 treatments. A total of six SPAR chambers represents three blocks with ten replications. Each SPAR chamber consisted of three rows of pots with ten pots per row in each SPAR chamber. All environmental growing conditions except for temperature and CO_2 were kept the same throughout the experiment.

2.1. Temperature and CO_2 Treatments

Basil plants were randomly assigned to each chamber consisting of 20/12, 30/22, and 38/30 °C temperature treatments in combination with ambient (420 $\mu L\ L^{-1}$) or elevated (720 $\mu L\ L^{-1}$) CO_2 concentrations (Table 1). The day- and night-time temperatures were, respectively, initiated at dawn and one hour after nightfall. Table 1 shows the average environmental conditions in which the experiment was conducted. During the experiment, three temperature treatments, 20/12, 30/22, and 38/30 °C, were regarded as low, optimum, and high temperatures, respectively, for basil growth and development.

Table 1. Temperature-stress treatments based on the percentage of daily evapotranspiration (ET) imposed at 14 days after sowing, mean and standard error day/night temperature, mean and standard error day chamber CO_2 concentration, mean and standard error day/night vapor pressure deficit (VPD), and mean and standard error day/night evapotranspiration (ET) during the experimental period of 38 days for each treatment.

Treatments		Measured Temperature (°C)	CO_2 ($\mu L\ L^{-1}$)	VPD (kPa)	Mean ET ($L\ H_2O\ d^{-1}$)
		Day/night	Day	Day/night	Day/night
Control	30/22 °C, 420 $\mu L\ L^{-1}$	26.27 ± 0.02	430.47 ± 0.98	1.82 ± 0.01	14.64 ± 1.41
Control + High CO_2	30/22 °C, 720 $\mu L\ L^{-1}$	26.34 ± 0.01	731.21 ± 1.52	1.98 ± 0.01	12.60 ± 1.27
High Temperature	38/30 °C, 420 $\mu L\ L^{-1}$	32.16 ± 0.49	434.19 ± 1.21	2.80 ± 0.07	8.74 ± 0.64
Low Temperature	20/12 °C, 420 $\mu L\ L^{-1}$	19.53 ± 0.56	431.08 ± 0.66	0.89 ± 0.08	8.59 ± 0.47
High Temperature + High CO_2	38/30 °C, 720 $\mu L\ L^{-1}$	32.09 ± 0.49	728.79 ± 0.83	2.87 ± 0.07	18.41 ± 1.86
Low Temperature + High CO_2	20/12 °C, 720 $\mu L\ L^{-1}$	19.56 ± 0.57	724.78 ± 0.35	0.95 ± 0.09	6.39 ± 0.37

2.2. Morphophysiological Measurements

At 17 and 38 days after treatment (DAT), basil plants from each treatment combination were harvested to assess their phenotype and to obtain growth data on early- and late-season effects of temperature and CO_2. Basil phenotypic data of plant height (PH), node number (NN), branch number (BN), and marketable FM were measured for each treatment combination. LA was measured using the LI-3100 leaf-area meter (Li-Cor Bioscience,

Lincoln, NE). Plant component FM was obtained from all basil plants using a weighing scale. The plant FM samples were then dried in a forced-air oven at 75 °C for two days to obtain basil dry mass (DM). Specific leaf area (SLA) is the measure of the leaf area formed per unit of leaf biomass, and plants usually use variation in SLA as a means of adapting to suboptimal conditions [27]. The measured DM and LA were utilized to estimate SLA ($cm^2\ g^{-1}$).

2.3. Root Image Acquisition and Analysis

The basil plants were severed at the soil surface and divided into stem and roots. The roots were carefully washed to remove excess soil from the root system and ensure clean measurements. The total root length (TRL) was measured using a meter ruler. Next, the cleansed roots were soaked in a 5 mm Plexiglass tray filled with water, where individual roots were straightened out and set apart for root imaging. A specialized dual-scan optical scanner (Regent Instruments, Inc., Québec, Canada) connected to a PC was used to capture gray-scaled root images according to the method described by Wijewardana et al. [25]. Acquired images were analyzed for the total root length (TRL), root surface area (RSA), average root diameter (RAD), root volume (RV), number of roots (RN), number of roots having laterals (RNL), number of tips (RNT), number of forks (RNF), and number of crossings (RNC) using WinRHIZO Pro software (Regent Instruments, Québec, Canada). More information on the root parameters can be found at www.regentinstruments.com (1 March 2021).

2.4. Physiological Measurements

At 17 DAT, a Dualex® Scientific Polyphenols (FORCE-A, Orsay, France) device was clipped on the second most fully developed basil leaf across treatments to obtain total chlorophyll (TCI) in the mesophyll, flavonoids (Flav), anthocyanin (Anth) in the epidermis, and a nitrogen balance index (NBI). The NBI shows the plants' nitrogen status by utilizing the proportion of chlorophyll and flavonoid units in the leaves. The TCI was evaluated as the ratio of the leaf transmission of near-infrared and red wavelengths. The Flav and Anth index is based on the measurement of chlorophyll fluorescence, while the NBI is a ratio between chlorophyll and flavonol indexes.

$$NBI = Chl/Flav$$

2.5. Epicuticular Wax Content Determination

The epicuticular leaf waxes were extracted and quantitively analyzed in accordance with the method of Ebercon et al. [28] with minor modifications as described by Singh and Reddy [29].

2.6. Data Analysis

The experimental design was a randomized complete block in a factorial arrangement with three temperature treatments, two CO_2 treatments, three-block, and ten replications. Data were analyzed using the PROC GLIMMIX analysis of variance (ANOVA) followed by mean separation. Statistical analysis of the data was performed using SAS (version 9.4; SAS Institute, Cary, NC, USA). The standard errors were based on the pooled error term from the ANOVA table. Duncan's multiple range test ($p \leq 0.05$) was used to differentiate between treatment classifications when F values were significant for the main effects. Model-based values were reported rather than the unequal standard error from a data-based calculation because pooled errors reflected the statistical testing. Diagnostic tests, such as Shapiro–Wilk in SAS, were conducted to ensure that treatment variances were statistically equal before pooling.

3. Results

3.1. Shoot Growth and Morphology

The analysis of variance revealed that temperature and CO_2 independently affect the morphological traits of basil (Table 2). The temperature treatments significantly affected ($p < 0.001$) the PH of both the early-season (17 DAT) and late-season (38 DAT) basil. However, CO_2 only affected ($p < 0.01$) the PH of the early-season basil (17 DAT) and no significant difference ($p > 0.05$) of the late-season basil (38 DAT). Also, there was no interaction between temperature and CO_2 effects on the PH of basil. At 17 DAT, low-temperature stress decreased PH of basil plants by 55% and 46% when subjected to elevated CO_2 compared to the control temperature at ambient CO_2. At 38 DAT, basil PH decreased by 17% and 22% at the high- and low-temperature stresses, respectively, compared to the control temperature at ambient CO_2 (Table 3).

Table 2. The mean plant height (PH), node number (NN), branch number (BN), leaf area (LA), leaf dry mass (LDM), shoot dry mass (SH DM), stem dry mass (ST DM), root dry mass (RDM), and root-to-shoot ratio (RS-Ratio) of basil plants grown without temperature stress (Control), with low-temperature stress, and with high-temperature stress at 420 and 720 $\mu L\ L^{-1}$ of CO_2 concentration after 17 days of treatment.

Treatment	PH [a]	NN	BN	LA	LDW	SH DW	ST DW	RDW	RS Ratio [b]
				420 $\mu L\ L^{-1}$					
Control	36.5 ^{a}	7.1 ^{b}	15.3 ^{c}	1223.6 ^{a,b}	4.479 ^{b}	6.667 ^{b}	2.188 ^{c}	0.941 ^{a}	0.140 ^{b}
High Temperature	36.8 ^{a}	8.5 ^{a}	24.0 ^{b}	1044.9 ^{b}	4.366 ^{b}	6.893 ^{b}	2.528 ^{b,c}	0.891 ^{a}	0.128 ^{b}
Low Temperature	16.5 ^{c}	4.6 ^{d}	6.4 ^{e}	403.8 ^{c}	1.909 ^{d}	2.263 ^{d}	0.354 ^{d}	0.411 ^{b}	0.198 ^{a}
				720 $\mu L\ L^{-1}$					
Control	36.6 ^{a}	7.0 ^{b}	15.3 ^{c}	1321.1 ^{a}	5.779 ^{a}	8.568 ^{a}	2.789 ^{a,b}	1.021 ^{a}	0.119 ^{b}
High Temperature	38.2 ^{a}	8.9 ^{a}	26.7 ^{a}	1139.41 ^{a,b}	5.227 ^{a,b}	8.340 ^{a}	3.113 ^{a}	0.966 ^{a}	0.121 ^{b}
Low Temperature	19.6 ^{b}	5.0 ^{c}	9.0 ^{d}	413.0 ^{c}	2.948 ^{c}	3.544 ^{c}	0.596 ^{d}	0.532 ^{b}	0.155 ^{b}
Treatment [c,d]	***	***	***	***	***	***	***	***	**
CO_2	**	*	**	NS	***	***	***	NS	*
Trt * CO_2	NS	NS	NS	NS	NS	NS	NS	NS	NS

[a] Plant-height units in centimeters (cm); node number and branch number on a per-plant basis; leaf area units in centimeters squared; remaining parameter units are on a gram-per-plant basis. [b] RS ratio, root-to-shoot ratio (root dry mass/shoot dry mass). [c] SE, standard error of the mean; PH = 1.2848; NN = 0.1487; BN = 1.0653; LA = 586.3; LDW = 2.2082; SH DW = 4.8444; ST DW = 2.7489; RDW = 0.8512; RS ratio = 0.00981; RDW = 0.06842; RS ratio = 0.01446. [d] NS represents non-significant $p > 0.05$. *, **, *** represent significance levels at $p \leq 0.05$, $p \leq 0.01$, and $p \leq 0.001$, respectively.

At 17 DAT, low- and high-temperature stresses decreased the LA of basil by 67% and 15%, respectively, compared to the control temperature at ambient CO_2 (Table 2). While at 38 DAT, low- and high-temperature stresses decreased the LA of basil by 40% and 34%, respectively, compared to the control temperature at ambient CO_2. However, the LA of basil was not different ($p > 0.05$) when subjected to elevated CO_2. The basil plants increased in SLA by 21% under high-temperature stress, whereas they decreased SLA by 27% at low-temperature stress (Figure 1). Elevated CO_2 decreased the SLA of basil by 9% when compared to the control treatments. However, SLA's response to both CO_2 and temperature was not significant ($p > 0.05$).

At 17 and 38 DAT, the DM of basil leaf and stem were significantly different ($p > 0.001$) when subjected to both low- and high-temperature stresses. Basil plants under low-temperature stress showed a more decreased measure of biomass per plant than basil under control temperature at ambient CO_2. At 17 DAT, high temperature showed a 3% reduction in leaf DM of basil, and this decline was ameliorated by elevated CO_2 to increase basil leaf DM by 17% (Figure 2C). However, at 38 DAT, high-temperature stress decreased basil leaf DM by 16% compared to basil under control temperature at ambient CO_2 (Figure 3C). It is imperative to state that elevated CO_2 significantly ($p > 0.001$) increased both leaf and stem DM of basil at 17 DAT. However, elevated CO_2 did not show a significant effect on leaf DM ($p > 0.05$) but stem DM ($p < 0.05$) at 38 DAT.

Table 3. The mean plant height (PH), node number (NN), branch number (BN), leaf area (LA), leaf dry mass (LDM), shoot dry mass (SH DM), stem dry mass (ST DM), root dry mass (RDM), and root-to-shoot ratio (RS-Ratio) of basil plants grown without temperature stress (Control), with low-temperature stress, and with high-temperature stress at 420 and 720 μL L^{-1} of CO_2 concentration after 38 days of treatment.

Treatment	PH [a]	NN	BN	LA	LDW	SH DW	ST DW	RDW	RS Ratio [b]
					420 μL L^{-1}				
Control	61.7 [a]	10.0 [b]	29.9 [a]	6946.3 [a]	25.03 [a,b]	58.081 [a,b]	33.049 [a,b]	6.841 [a]	0.116 [c]
High Temperature	51.4 [b]	11.6 [a]	24.5 [b]	4616.3 [b,c]	18.63 [c]	43.274 [c]	24.645 [c]	7.387 [a]	0.164 [a]
Low Temperature	47.9 [c,d]	7.2 [d]	15.7 [c]	4149.3 [b,c]	16.62 [c]	25.206 [d]	8.589 [d]	3.049 [b]	0.123 [b,c]
					720 μL L^{-1}				
Control	60.9 [a]	10.1 [b]	29.7 [a]	8078.9 [a]	28.39 [a]	67.126 [a]	38.733 [a]	8.511 [a]	0.128 [b,c]
High Temperature	50.9 [b,c]	11.4 [a]	24.7 [b]	5215.7 [b]	20.65 [b,c]	52.123 [b,c]	31.469 [b,c]	7.518 [a]	0.142 [a,b]
Low Temperature	47.3 [d]	7.7 [c]	17.1 [c]	3582.0 [c]	18.95 [c]	29.113 [d]	10.167 [d]	4.599 [b]	0.157 [a]
Treatment [c,d]	***	***	***	***	***	***	***	***	**
CO_2	NS	NS	NS	NS	NS	NS	*	NS	NS
Trt * CO_2	NS	NS	NS	NS	NS	NS	NS	NS	*

[a] Plant height units in centimeters (cm); node number and branch number on a per-plant basis; leaf area units in centimeters squared; remaining parameter units are on a gram-per-plant basis. [b] RS ratio, root-to-shoot ratio (root dry mass/shoot dry mass). [c] SE, standard error of the mean; PH = 1.2848; NN = 0.1487; BN = 1.0653; LA = 586.3; LDW = 2.2082; SH DW = 4.8444; ST DW = 2.7489; RDW = 0.8512; RS ratio = 0.00981. [d] NS represents non-significant $p > 0.05$. *, **, *** represent significance levels at $p \leq 0.05$, $p \leq 0.01$, and $p \leq 0.001$, respectively.

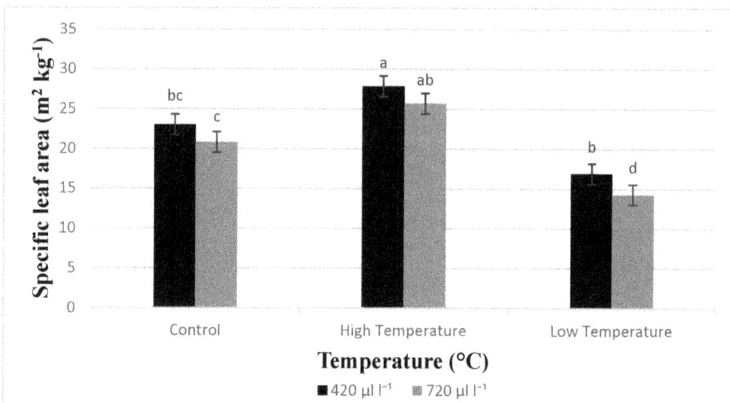

Figure 1. Average specific leaf area (SLA) for basil plants grown without temperature stress (control; 30/22 °C), with low-temperature (20/12 °C) stress, and with high-temperature (38/30 °C) stress at 420 and 720 μL L^{-1} of CO_2 concentration after 38 days of treatment. The standard error mean for SLA was 1.2941. Different lower-case letters indicate a significant difference at $p = < 0.05$ by the least significant difference.

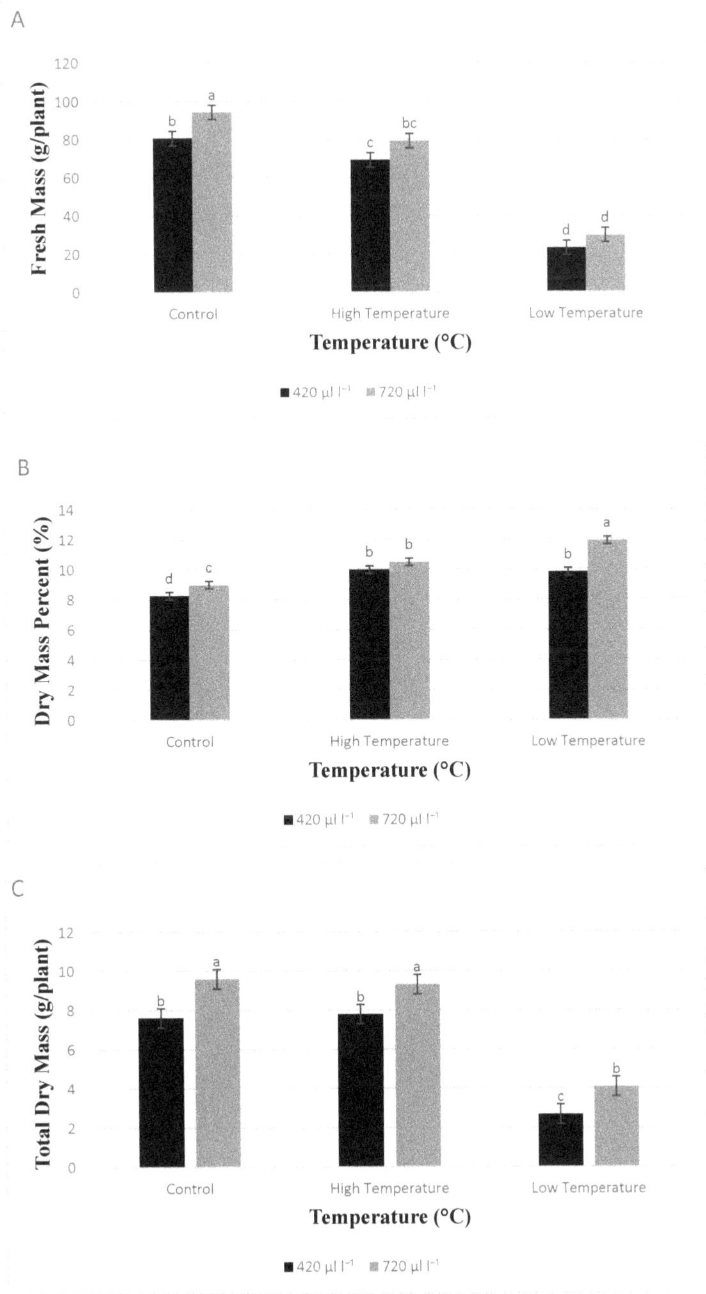

Figure 2. (**A**) Fresh mass (FM), (**B**) total dry mass (total DM), and (**C**) dry mass percent (DM %) for basil plants grown without temperature stress (control; 30/22 °C), with low-temperature (20/12 °C) stress, and with high-temperature (38/30 °C) stress at 420 and 720 μL L^{-1} of CO_2 concentration after 17 days of treatment. The standard error mean was FM = 3.7485, total DM = 0.497, and DM percent = 0.2487. Different lower-case letters indicate a significant difference at $p = < 0.05$ by the least significant difference.

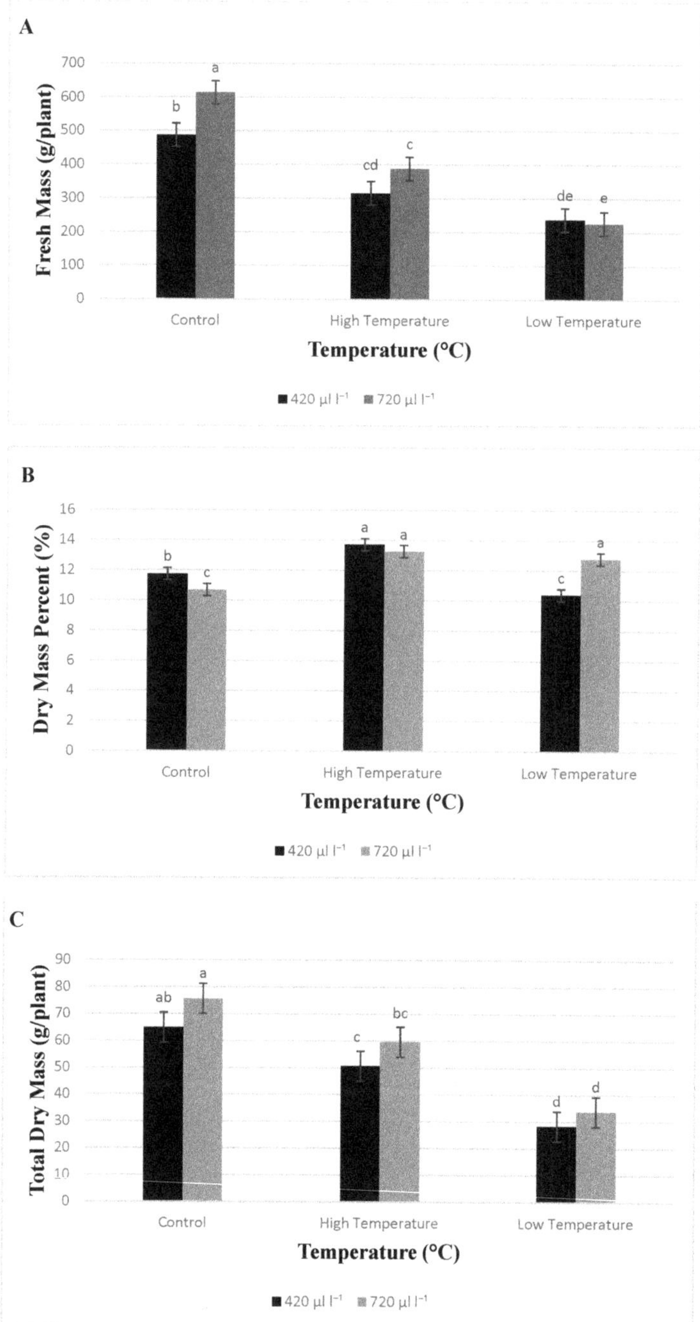

Figure 3. (**A**) Fresh mass (FM), (**B**) total dry mass (total DM), and (**C**) dry mass percent (DM %) for basil plants grown without temperature stress (control; 30/22 °C), with low-temperature (20/12 °C) stress, and with high-temperature (38/30 °C) stress at 420 and 720 μL L^{-1} of CO_2 concentration after 38 days of treatment. The standard error mean was FM = 34.65, total DM = 5.5682, and DM percent = 0.4008. Different lower-case letters indicate a significant difference at $p = < 0.05$ by the least significant difference.

Contrary to the control temperature at ambient CO_2, basil plant marketable FM decreased by 71% and 14% when exposed to cold and heat stresses, respectively (Figure 2A). Elevated CO_2 ameliorated the adverse effects and decreased marketable FM by up to 63% more at low temperatures than basil grown at the control temperature. At 38 DAT, similar results were discovered for decreasing marketable FM (Figure 3A).

The basil plants showed a significant reduction in leaf wax content when subjected to both low- and high-temperature stresses and elevated CO_2 (Figure 4). However, there was no interaction effect between CO_2 and temperature treatments on basil leaf wax content.

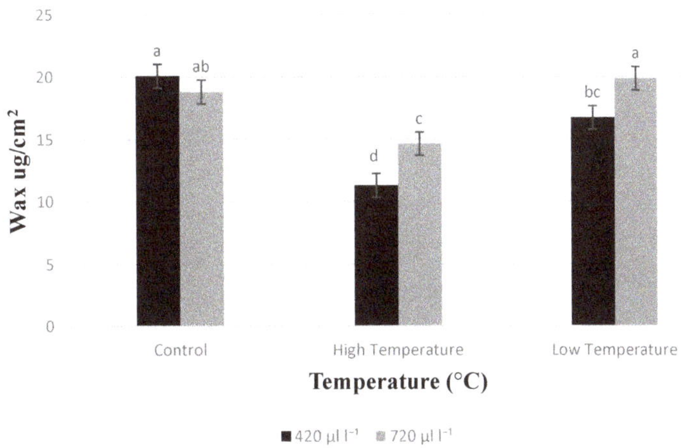

Figure 4. Average epicuticular wax content for basil plants grown without temperature stress (control; 30/22 °C), with low-temperature (20/12 °C) stress, and with high-temperature (38/30 °C) stress at 420 and 720 μL L^{-1} of CO_2 concentration after 34 days of treatment. The standard error mean for wax was 0.9463. Different lower-case letters indicate a significant difference at $p = < 0.05$ by the least significant difference.

3.2. Root Development Parameters

Amongst the root traits, TRL and RSA were more sensitive to the interactions between temperature and CO_2 than other root traits (Table 4). Basil RL, RAD, RV, RNT, RNF, and RNC were also significantly affected by the main effects of temperature and CO_2 (Table 4). At 17 DAT, TRL decreased by 41% and 17% under low- and high-temperature stresses, respectively, compared to the control temperature at ambient CO_2. It is interesting to note that the elevated CO_2 mitigated the detrimental effects and decreased basil TRL by 27% and 8% at low- and high-temperature treatments, respectively, compared to the control temperature at ambient CO_2. Similar results were observed with decreasing RSA of basil. Under high-temperature stress, basil plants exhibited 31%, 27%, and 12% reduction in RSA, RAD, and RV, respectively, compared to the control treatment. RNT, RNF, and RNC, which influence root's architecture, were observed to be considerably higher and lower under high and low-temperature stress, respectively, compared with the control treatments.

3.3. Physiological Parameters

The results showed that temperature, CO_2, and its interactions on basil's epidermal anthocyanin index (Anth) were discovered to be significant. The concentration of Anth in basil increased by 7% and 10% under high-temperature stress at both ambient and elevated CO_2, respectively (Figure 5A). However, the Anth concentration of basil decreased by 21% and 2% under low temperature at both ambient and elevated CO_2, respectively. The results also showed a higher index of flavonoids under low-temperature stress, while under high-temperature stress, it decreased significantly (Figure 5B). Elevated CO_2 significantly increased the index of basil flavonoids under both high- and low-temperature stress. The basil leaves showed significantly higher TCI when subjected to low-temperature stress than the control treatments (Figure 5C). Comparing the basil TCI to the control treatments, high-temperature treatments significantly increased the basil TCI. However, elevated CO_2 significantly decreased the NBI of the basil plant at both low- and high-temperature stresses (Figure 5D). Similar results were observed of TCI when subjected to elevated CO_2 at low temperatures.

Table 4. The mean longest root length (RL), total root length (TRL), root surface area (RSA), root average diameter (RAD), root volume (RV), root tips (RNT), root forks (RNF), and root crossings (RNC) of basil plants grown without temperature stress (control), with low-temperature stress, and with high-temperature stress at 420 and 720 µL L^{-1} of CO_2 concentration after 17 days of treatment.

Treatment	RL [a]	TRL	RSA	RAD	RV	RNT	RNF	RNC
				420 µL L^{-1}				
Control	45.1 [a]	4572.9 [a]	854.3 [a]	0.598 [a]	14.00 [a]	10052 [a]	38545 [a,b]	2412.6 [b]
High Temperature	40.6 [b,c]	3792.2 [b,c]	623.4 [b,c]	0.524 [c]	9.68 [b]	12271 [a]	33856 [b]	2475.1 [b]
Low Temperature	36.9 [c]	2701.2 [d]	497.3 [d]	0.584 [a,b]	7.35 [b]	5448 [b]	17831 [c]	1115.2 [c]
				720 µL L^{-1}				
Control	46.7 [a]	4159.1 [a,b]	738.6 [a,b]	0.561 [a,b,c]	15.45 [a]	12477 [a]	46580 [a]	3287.8 [a]
High Temperature	43.0 [a,b]	4194.4 [a,b]	715.4 [b,c]	0.541 [b,c]	9.78 [b]	10898 [a]	31358 [b]	2428.6 [b]
Low Temperature	39.9 [b,c]	3354.1 [c]	602.1 [c,d]	0.576 [a,b]	8.65 [b]	9624 [a]	22960 [c]	1480.0 [c]
Treatment [b,c]	***	***	***	*	***	**	***	***
CO_2	NS	NS	NS	NS	NS	NS	NS	*
Trt*CO_2	NS	*	*	NS	NS	NS	NS	NS

[a] RL, TRL, RAD on a centimeter-per-plant basis; RSA, RV on a cubic-centimeter basis; RNT, RNF, and RNC on a number-per-plant basis. [b] SE, standard error of the mean; RL = 1.4215; TRL = 212.96; RSA = 43.7643; RAD = 0.01825; RV = 1.0304; RNT = 1260.17; RNF = 2952.16; RNC = 218.76. [c] NS represents non-significant $p > 0.05$. *, **, *** represent significance levels at $p \leq 0.05$, $p \leq 0.01$, and $p \leq 0.001$, respectively; within columns, values followed by the same letter are not significantly different.

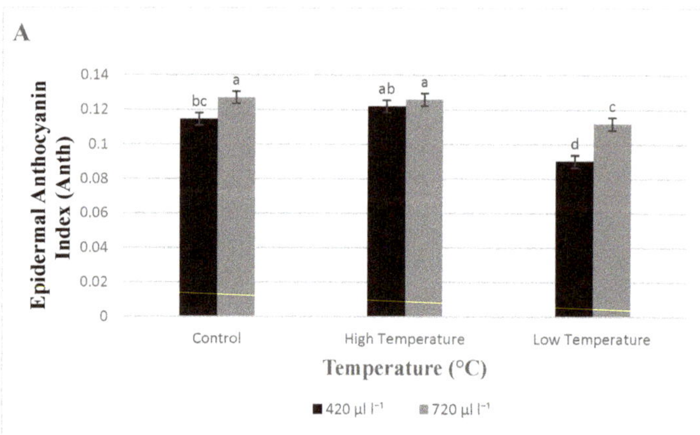

Figure 5. *Cont.*

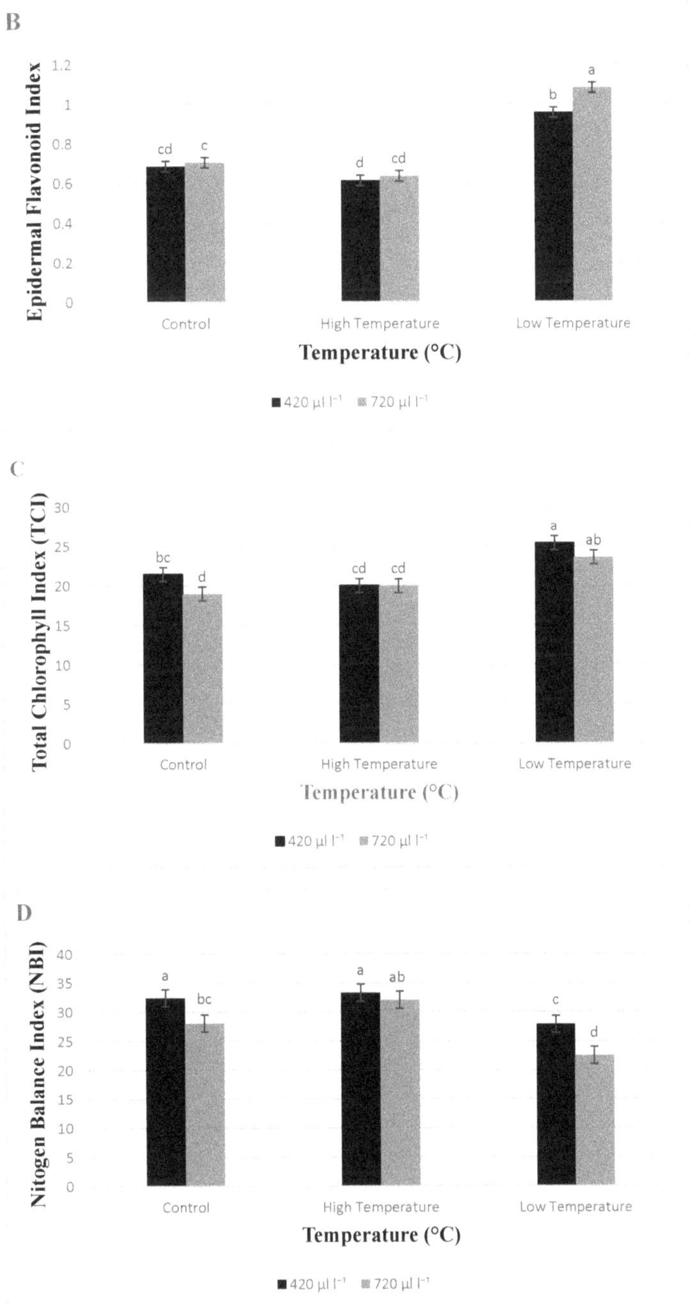

Figure 5. (**A**) Epidermal anthocyanin index, (**B**) epidermal flavonoid index, (**C**) total chlorophyll index, and (**D**) nitrogen balance index of basil leaf tissue subjected to no temperature stress (control; 30/22 °C), with low-temperature (20/12 °C) stress, and with high-temperature (38/30 °C) stress at 420 and 720 µL L^{-1} of CO_2 concentration. The standard error mean was anthocyanin = 0.003594, flavonoids = 0.02642, TCI = 0.88, and NBI = 1.4829. Different lower-case letters indicate significant difference at $p = < 0.05$ by least significant difference.

4. Discussion

Temperature and elevated CO_2 remain important factors that significantly affect the growth and development of C3 plants, including basil [13,30]. Hence, exploring the interactive effects of multiple abiotic stressors on the growth and development traits associated with basil roots and shoots is imperative. Understanding crop performance to temperature stress and elevated CO_2 is crucial during the early seedling stage because it affects developing a uniform and healthy plant canopy.

In this current research, increasing the temperature to 38/30 °C at ambient CO_2 concentration caused less adverse effects on the early season's morphological features (17 DAT) of basil. Previous research indicated an increase in the PH, NN, and BN primarily because basil plants prefer warmer temperatures [31]. For instance, in this current study, high-temperature stress caused a significant decrease in late-season basil PH, NN, and BN by 17%, 16%, 18%, respectively, compared to the control treatments. However, in contrast to the control treatments, there was a significant increase in the early-season basil PH, NN, and BN by 1%, 20%, and 57%, respectively. These findings suggest that growing basil under high-temperature stress is beneficial to basil at its early stage and detrimental to the late-season basil.

On the other hand, growing both the early- and late-season basil under low-temperature stress significantly decreased basil PH, NN, and BN by 55%, 35%, and 58%, respectively. These results are in line with previous studies [9,13]. They reported a reduction in basil PH when exposed to low temperature and increased PH of basil grown under high-temperature stress. It is important to note that CO_2 only affected the PH of basil at 17 DAT, and there was no significant difference in basil at 38 DAT. Thus, these results signified the role of elevated CO_2 in ameliorating the adverse impacts of temperature stress due to the higher production of carbon available for increased rates of photosynthesis. Several studies have surmised the beneficial effects of elevated CO_2 while varying the temperature levels [5,20,21,27].

Altering plants' growth temperature could result in fewer basil leaves, perhaps due to the reduction in the rate of emergence of plant NN and increased rate of leaf senescence [32]. Basil plants revealed the highest decrease in the LA when exposed to low-temperature stress, indicating their sensitivity to cold stress compared to heat stress. There was a 27% decline in the SLA of basil in response to cold stress because the LA is directly linked to SLA. Correspondingly, Bannayan et al. [27] and Caliskan et al. [32] posited that increasing basil's growth temperature would increase SLA and vice versa. In this study, when SLA declined in response to elevated CO_2, there was no significant difference in the LA considering elevated CO_2. The present exploration also revealed a positive effect of elevated CO_2 on basil's total biomass, which is usually expected of C3 plants. The adverse impacts of both low- and high-temperature stresses on the FM and DM of basil were ameliorated by elevated CO_2. Thus, these results suggest that basil plants could cope with sub-optimal temperature conditions because of enhanced photosynthetic rates when CO_2 is elevated. Al Jaouni et al. [15] also reported that basil biomass production increased by 40% under elevated CO_2.

The plant root system is composed of various sorts of roots that change in morphology and function. The root architecture illustrates the root system's spatial organization in the soil and is crucial for plants in obtaining water and nutrients required for growth and development [33,34]. Recent evidence suggests that the root traits required for obtaining resources from the soil are also linked with plants' adaptive characteristics to reduce the adverse effects of variation in growth temperatures [16,18]. The adaptive traits of the plant's roots to environmental changes indicated that TRL and RSA were more sensitive to both temperature- and CO_2 -change. Luo et al. [17] revealed root width and depth to be susceptible to temperature changes. However, RNT and RNC, which influence the root's architecture, were considerably larger under high-temperature stress. Increasing the CO_2 levels also contributed to increasing the RNC. These results are consistent with previous studies on C3 crops, such as members of genus *Brassica* and order *Fabales* [17,35]. Previous studies additionally revealed that high RNT and RNC contribute to a thinner RAD [36].

In support of our research, basil plants showed decreased RAD and RV when subjected to heat stress. This implies that a reduced RAD under temperature stress would obtain additional soil nutrients and increase nutrient absorption, suggesting basil's tolerance to warmer conditions [17,37].

Analogous to shoot and root traits, basil plants' physiological traits were also affected by temperature stress. Growing basil plants under low-temperature stress significantly increased the concentration of flavonoids index in basil leaves. While under high-temperature treatment, the index of flavonoids of basil decreased considerably. These findings contradict many studies [31,38,39] because increased flavonoids are usually associated with thermophilic plant defense mechanisms against heat stress. Moreover, increased growth temperature from 30/22 °C to 38/30 °C for basil under elevated CO_2 produced more Anth content. These results suggest that subjecting basil to temperature stress and elevated CO_2 does not cause a drastic loss in Anth, which matches those observed in earlier studies [15]. Moreover, epicuticular wax decreased significantly both under hot and cold stress conditions, which is unexpected because increased wax content is always utilized as a physiological trait for selecting thermophilic plants [40].

In contrast with the present results, previous studies have demonstrated that decreasing plant growth temperature significantly increased the chlorophyll index [41,42]. Many studies have utilized the plant chlorophyll index as a metric for characterizing the plant's tolerance to multiple abiotic stresses, especially temperature stress in grain and vegetables [41,43,44]. Low temperature significantly increased the non-destructive TCI of basil leaves by 18%. However, a reduction of TCI was observed for basil leaves when the temperature was increased to 38/30 °C. It is important to note that the lowest basil TCI was observed under elevated CO_2 at 30/22 °C, while the maximum basil TCI was recorded when basil was subjected to low temperatures at ambient CO_2. In addition to using TCI as an essential tool for selecting plants for adapting to multiple abiotic stressors, NBI has been noted to determine in vivo the plant nitrogen status [42,45]. It can also be used to measure the ratio of carbon and nitrogen capacity of plants. In this study, the NBI of basil leaves was observed at maximum when subjected to 30/22 °C at ambient CO_2, indicating basil tolerance to heat stress. However, the lowest basil NBI values were recorded under low-temperature stress, which further supports the previous information on the sensitivity of basil to chilling stress.

5. Conclusions

Elevated CO_2 and temperature stress independently affect the growth and morphology of basil roots and shoots. Decreasing the basil's growth temperature to 20/12 °C was the major determining factor in both the early- and late-season basil's morphological features. Low-temperature stress also resulted in the significant reduction of physiological parameters, thus repressing basil plants' growth. Furthermore, the accelerated concentration of chlorophyll and flavonoid pigment degradation of basil plants was observed when elevated CO_2 interacted with low-temperature stress. These results further proved the susceptibility of basil plants to chilling stress. Contrarily, elevated CO_2 remarkably ameliorated the damage caused by low-temperature stress on the morphology and growth of basil roots, shoots, and physiological parameters. However, basil, being a thermophilic plant, was observed to increase its plant height, node numbers, branch numbers, net photosynthesis, specific leaf area, anthocyanin, and nitrogen-balance index when subjected to high-temperature stress. The outcomes of this research suggest that altering the growth temperature of basil plants would more significantly impact the growth and development rates of basil than increasing the CO_2 concentrations, which ameliorated the adverse impacts of temperature stress.

Author Contributions: T.C.B.: conceptualization, methodology, validation, formal analysis, investigation, resources, data curation, writing—original draft, writing—review & editing, visualization, supervision, project administration, funding acquisition. O.J.O.: formal analysis, writing—original draft, writing—review & editing. A.S.: methodology, validation, investigation. C.H.W.: methodology, validation, formal analysis, investigation. K.R.R.: conceptualization, methodology, validation, formal analysis, investigation, resources, data curation, writing—review & editing, visualization, supervision, project administration, funding acquisition. W.G.: conceptualization, methodology, validation, resources, funding acquisition. All authors have read and agreed to the published version of the manuscript.

Funding: This material is based on the work supported by the USDA-NIFA Hatch Project under accession number 149210, and the National Institute of Food and Agriculture, 2019-34263-30552, and MIS 043050 funded this research.

Institutional Review Board Statement: Not applicable.

Informed Consent Statement: Not applicable.

Data Availability Statement: The data presented in this study are available on request from the corresponding author.

Acknowledgments: We thank David Brand for technical assistance and graduate students at the Environmental Plant Physiology Laboratory for their help during data collection.

Conflicts of Interest: The authors declare that they have no known competing financial interest or personal relationships that could have appeared to influence the work reported in this paper.

Abbreviations

ANTH, epidermal anthocyanin index; BN, branch number; DAS, days after sowing; DAT, days after treatment; DM, dry mass; Flav, epidermal flavonoids index; FM, fresh mass; LA, leaf area; NBI, nitrogen balance index; NN, node numbers; PH, plant height; RAD, average root diameter; RDW, root dry mass; RN, number of roots; RNC, number of crossings; RNF, number of forks; RNL, number of roots having laterals; RNT, number of tips; RSA, root surface area; RV, root volume; SLA, specific leaf area; SPAR, soil-plant-atmosphere-research; TCI, total chlorophyll index; and TRL, total root length.

References

1. USGCRP. *Impacts, Risks, and Adaptation in the United States: Fourth National Climate Assessment, Volume II*; Reidmiller, D.R., Avery, C.W., Easterling, D.R., Kunkel, K.E., Lewis, K.L.M., Maycock, T.L., Stewart, B.C., Eds.; US Global Change Research Program: Washington, DC, USA, 2018; p. 1515.
2. Baker, H. Atmospheric CO2 Will Pass an Alarming Milestone in 2021. Available online: https://www.livescience.com/co2-concentration-rising-past-alarming-threshold.html (accessed on 8 February 2021).
3. IPCC. *Climate Change 2014: Impacts, Adaptation, and Vulnerability. Part A: Global and Sectoral Aspects. Contribution of Working Group II to the Fifth Assessment Report of the Intergovernmental Panel on Climate Change*; Field, C.B., Barros, V.R., Dokken, D.J., Mach, K.J., Mastrandrea, M.D., Bilir, T.E., Chatterjee, M., Ebi, K.L., Estrada, Y.O., Genova, R.C., et al., Eds.; Cambridge University Press: Cambridge, UK; New York, NY, USA, 2014; p. 1132.
4. Stocker, T.F.; Qin, D.; Plattner, G.-K.; Tignor, M.; Allen, S.K.; Boschung, J.; Nauels, A.; Xia, Y.; Bex, V.; Midgley, P.M. *Climate Change 2013: The Physical Science Basis. Intergovernmental Panel on Climate Change, Working Group I Contribution to the IPCC Fifth Assessment Report (AR5)*; Cambridge University Press: New York, NY, USA, 2013; p. 25.
5. Reddy, A.R.; Reddy, K.R.; Hodges, H.F. Interactive effects of elevated carbon dioxide and growth temperature on photosynthesis in cotton leaves. *Plant Growth Regul.* **1998**, *26*, 33–40. [CrossRef]
6. Dong, J.; Gruda, N.; Li, X.; Tang, Y.; Zhang, P.; Duan, Z. Sustainable vegetable production under changing climate: The impact of elevated CO_2 on yield of vegetables and the interactions with environments-A review. *J. Clean. Prod.* **2020**, *253*, 119920. [CrossRef]
7. Simon, J.E.; Quinn, J.; Murray, R.G. Basil: A Source of Essential Oils. In *Adv. New Crops*; Janick, J., Simon, J.E., Eds.; Timber Press: Portland, OR, USA, 1990; pp. 484–489.
8. Kopsell, D.A.; Kopsell, D.E.; Curran-Celentano, J. Carotenoid and chlorophyll pigments in sweet basil grown in the field and greenhouse. *Hort. Sci.* **2005**, *40*, 1230–1233. [CrossRef]
9. Chang, X.; Alderson, P.; Wright, C. Effect of temperature integration on the growth and volatile oil content of basil (*Ocimum Basilicum L.*). *J. Hortic. Sci. Biotechnol.* **2005**, *80*, 593–598. [CrossRef]
10. Ribeiro, P.; Simon, J.E. Breeding sweet basil for chilling tolerance. In *Issues in New Crops and New Uses*; Janick, J., Whipkey, A., Eds.; ASHS Press: Alexandria, VA, USA, 2007; pp. 302–305.
11. Mortensen, L.M. The effect of air temperature on growth of eight herb species. *Amer. J. Plant Sci.* **2014**, *5*, 1542–1546. [CrossRef]
12. Hiltunen, R.; Holm, Y. Essential of oil of *Ocimum*. In *Basil: The Genus Ocimum*; Hiltunen, R., Holm, Y., Eds.; Harwood academic publishers: Amsterdam, The Netherlands, 1999; pp. 113–135.

13. Walters, K.J.; Currey, C.J. Growth and development of basil species in response to temperature. *Hort. Sci.* **2019**, *54*, 1915–1920. [CrossRef]
14. Watanabe, C.K.; Sato, S.; Yanagisawa, S.; Uesono, Y.; Terashima, I.; Noguchi, K. Effects of elevated CO_2 on levels of primary metabolites and transcripts of genes encoding respiratory enzymes and their diurnal patterns in *Arabidopsis thaliana*: Possible relationships with respiratory rates. *Plant Cell Physiol.* **2014**, *55*, 341–357. [CrossRef] [PubMed]
15. Jaouni, S.A.; Saleh, A.M.; Wadaan, M.A.M.; Hozzein, W.N.; Selim, S.; AbdElgawad, H. Elevated CO_2 induces a global metabolic change in basil (*Ocimum Basilicum L.)* and peppermint (*Mentha Piperita L.*) and improves their biological activity. *J. Plant Physiol.* **2018**, *224–225*, 121–131. [CrossRef]
16. de Dorlodot, S.; Forster, B.; Pagès, L.; Price, A.; Tuberosa, R.; Draye, X. Root system architecture: Opportunities and constraints for genetic improvement of crops. *Trends Plant Sci.* **2007**, *12*, 474–481. [CrossRef] [PubMed]
17. Luo, H.; Xu, H.; Chu, C.; He, F.; Fang, S. High temperature can change root system architecture and intensify root interactions of plant seedlings. *Front. Plant Sci.* **2020**, *11*, 160. [CrossRef] [PubMed]
18. Bardgett, R.D.; Mommer, L.; De Vries, F.T. Going underground: Root traits as drivers of ecosystem processes. *Trends Ecol. Evol.* **2014**, *29*, 692–699. [CrossRef] [PubMed]
19. Fitter, A.H.; Caldwell, M.M.; Pearcy, R.W. Architecture and biomass allocation as components of the plastic response of root systems to soil heterogeneity. In *Exploitation of Environmental Heterogeneity by Plants: Ecophysiological Processes Above-And Belowground*; Academic Press: Cambridge, MA, USA, 1994; pp. 305–323.
20. Brand, D.; Wijewardana, C.; Gao, W.; Reddy, K.R. Interactive effects of carbon dioxide, low temperature, and ultraviolet-b radiation on cotton seedling root and shoot morphology and growth. *Front. Earth Sci.* **2016**, *10*, 607–620. [CrossRef]
21. Wijewardana, C.; Henry, W.B.; Gao, W.; Reddy, K.R. Interactive effects on CO_2, drought, and ultraviolet-b radiation on maize growth and development. *J. Photochem. Photobiol. B.* **2016**, *160*, 198–209. [CrossRef] [PubMed]
22. Lahti, M.; Aphalo, P.J.; Finér, L.; Ryyppö, A.; Lehto, T.; Mannerkoski, H. Effects of soil temperature on shoot and root growth and nutrient uptake of 5-year-old norway spruce seedlings. *Tree Physiol.* **2005**, *25*, 115–122. [CrossRef]
23. Balliu, A.; Sallaku, G. The environment temperature affects post-germination growth and root system architecture of pea (*Pisum Sativum L*) plants. *Sci. Hortic.* **2021**, *278*, 109858. [CrossRef]
24. Reddy, K.; Read, J.J.; McKinion, J.M. Soil-Plant-Atmosphere-Research (SPAR) facility: A Tool for plant research and modeling. *Biotronics* **2001**, *30*, 27–50.
25. Wijewardana, C.; Hock, M.; Henry, B.; Reddy, K.R. Screening corn hybrids for cold tolerance using morphological traits for early-season seeding. *Crop Sci.* **2015**, *55*, 851–867. [CrossRef]
26. Hoagland, D.R.; Arnon, D.I. *The Water-Culture Method for Growing Plants without Soil*, 2nd ed.; Circular 347; California Agricultural Experiment Station: Berkeley, CA, USA, 1950; p. 347.
27. Bannayan, M.; Tojo Soler, C.M.; Garcia y. Garcia, A.; Guerra, L.C.; Hoogenboom, G. Interactive effects of elevated [CO_2] and temperature on growth and development of a short- and long-season peanut cultivar. *Clim. Chang.* **2009**, *93*, 389–406. [CrossRef]
28. Ebercon, A.; Blum, A.; Jordan, W.R. A rapid colorimetric method for epicuticular wax contest of sorghum leaves. *Crop Sci.* **1977**, *17*, 179–180. [CrossRef]
29. Singh, S.K.; Reddy, K.R. Regulation of photosynthesis, fluorescence, stomatal conductance and water-use efficiency of cowpea (*Vigna Unguiculata [L.] Walp.*) under drought. *J. Photochem. Photobiol. B Biol.* **2011**, *105*, 40–50. [CrossRef]
30. Yuan, L.; Yuan, Y.; Liu, S.; Wang, J.; Zhu, S.; Chen, G.; Hou, J.; Wang, C. Influence of high temperature on photosynthesis, antioxidative capacity of chloroplast, and carbon assimilation among heat-tolerant and heat-susceptible genotypes of nonheading chinese cabbage. *HortScience* **2017**, *52*, 1464–1470. [CrossRef]
31. Al-Huqail, A.; El-Dakak, R.M.; Sanad, M.N.; Badr, R.H.; Ibrahim, M.M.; Soliman, D.; Khan, F. Effects of climate temperature and water stress on plant growth and accumulation of antioxidant compounds in sweet basil (*Ocimum basilicum L.*) leafy vegetable. *Scientifica* **2020**, 3808909. [CrossRef] [PubMed]
32. Caliskan, O.; Odabas, M.S.; Cirak, C. The modeling of the relation among the temperature and light intensity of growth in *Ocimum basilicum L. J. Med. Plants Res* **2009**, *3*, 965–977. [CrossRef]
33. Yamauchi, A.; Pardales, J.R., Jr.; Kono, Y. Root system structure and its relation to stress tolerance. In *Dyanamics of Roots and Nitrogen in Cropping Systems of the Semi-Arid Tropics*; Ito, O., Johansen, C., Adu-Gyamfi, J.J., Katayama, K., Kumar, J.V.D.K., Rego, T.J., Eds.; Japan International Research Center for Ag Sciences: Tsukuba, Japan, 1996; pp. 211–233.
34. Aidoo, M.K.; Bdolach, E.; Fait, A.; Lazarovitch, N.; Rachmilevitch, S. Tolerance to high soil temperature in foxtail millet (*Setaria Italica L.*) Is related to shoot and root growth and metabolism. *Plant Physiol. and Biochem.* **2016**, *106*, 73–81. [CrossRef] [PubMed]
35. Nagel, K.A.; Kastenholz, B.; Jahnke, S.; van Dusschoten, D.; Aach, T.; Mühlich, M.; Truhn, D.; Scharr, H.; Terjung, S.; Walter, A.; et al. Temperature responses of roots: Impact on growth, root system architecture and implications for phenotyping. *Funct. Plant Biol.* **2009**, *36*, 947–959. [CrossRef] [PubMed]
36. Kaspar, T.C.; Bland, W.L. Soil temperature and root growth. *Soil Sci.* **1992**, *154*, 290–299. [CrossRef]
37. De Kroon, H.; Mommer, L.; Nishiwaki, A. Root competition: Towards a mechanistic understanding. In *Root Ecology*; de Kroon, H., Visser, E.J.W., Eds.; Ecological Studies (Analysis and Synthesis); Springer: Berlin/Heidelberg, Germany, 2003; Volume 168, pp. 215–234, ISBN 978-3-662-09784-7.

38. Shamloo, M.; Babawale, E.A.; Furtado, A.; Henry, R.J.; Eck, P.K.; Jones, P.J.H. Effects of genotype and temperature on accumulation of plant secondary metabolites in canadian and australian wheat grown under controlled environments. *Sci. Rep.* **2017**, *7*, 9133. [CrossRef] [PubMed]
39. Sublett, W.L.; Barickman, T.C.; Sams, C.E. Effects of elevated temperature and potassium on biomass and quality of dark red 'lollo rosso' lettuce. *Horticulturae* **2018**, *4*, 11. [CrossRef]
40. Jumrani, K.; Bhatia, V.S. Interactive effect of temperature and water stress on physiological and biochemical processes in soybean. *Physiol. Mol. Biol. Plant.* **2019**, *25*, 667–681. [CrossRef] [PubMed]
41. Reddy, K.R.; Seghal, A.; Jumaa, S.; Bheemanahalli, R.; Kakar, N.; Redoña, E.D.; Wijewardana, C.; Alsajri, F.A.; Chastain, D.; Gao, W.; et al. Morpho-Physiological characterization of diverse rice genotypes for seedling stage high- and low-temperature tolerance. *Agronomy* **2021**, *11*, 112. [CrossRef]
42. Ronga, D.; Rizza, F.; Badeck, F.-W.; Milc, J.; Laviano, L.; Montevecchi, G.; Pecchioni, N.; Francia, E. Physiological responses to chilling in cultivars of processing tomato released and cultivated over the past decades in southern europe. *Sci. Hortic.* **2018**, *231*, 118–125. [CrossRef]
43. Agati, G.; Tuccio, L.; Kusznierewicz, B.; Chmiel, T.; Bartoszek, A.; Kowalski, A.; Grzegorzewska, M.; Kosson, R.; Kaniszewski, S. Nondestructive optical sensing of flavonols and chlorophyll in white head cabbage (*Brassica Oleracea L. Var. Capitata Subvar. Alba*) grown under different nitrogen regimens. *J. Agric. Food Chem.* **2016**, *64*, 85–94. [CrossRef] [PubMed]
44. de Freitas, G.M.; Thomas, J.; Liyanage, R.; Lay, J.O.; Basu, S.; Ramegowda, V.; do Amaral, M.N.; Benitez, L.C.; Bolacel Braga, E.J.; Pereira, A. Cold Tolerance response mechanisms revealed through comparative analysis of gene and protein expression in multiple rice genotypes. *PLoS ONE* **2019**, *14*, e0218019. [CrossRef] [PubMed]
45. Cerovic, Z.G.; Masdoumier, G.; Ghozlen, N.B.; Latouche, G. A New optical leaf-clip meter for simultaneous non-destructive assessment of leaf chlorophyll and epidermal flavonoids. *Physiol. Plant.* **2012**, *146*, 251–260. [CrossRef] [PubMed]

horticulturae

Article

Effect of Saline–Alkali Stress on Sugar Metabolism of Jujube Fruit

Yan Wang [1,2], Yifeng Feng [2,3], Min Yan [2], Ju Yu [1,2], Xiaofeng Zhou [1,2], Jingkai Bao [2], Qiaoqiao Zhang [2,3] and Cuiyun Wu [2,3,*]

1 College of Life Science and Technology, Tarim University, Alar 843300, China; wy_371446391@163.com (Y.W.); smileyuxiaofei@163.com (J.Y.); zhxf112233@163.com (X.Z.)
2 National-Local Joint Engineering Laboratory of High Efficiency and Superior-Quality Cultivation and Fruit Deep Processing Technology on Characteristic Fruit Trees, Alar 843300, China; f15909073303@163.com (Y.F.); yanmin961106@163.com (M.Y.); baomouren1997@163.com (J.B.); zhangqq0621@163.com (Q.Z.)
3 College of Horticulture and Forestry, Tarim University, Alar 843300, China
* Correspondence: wcyby@163.com

Abstract: Sugar, an osmoregulatory substance used by plants to adapt to abiotic stresses such as drought and salinity, is one of the most important indexes of fruit quality. In this study, 0–150 mM saline–alkali solutions ($NaCl:NaHCO_3$ = 3:1) were used to irrigate the roots of 10-year-old "Junzao" fruit trees during the growth period to explore the regulation mechanism of different concentrations of saline–alkali stress on sugar and reactive oxygen metabolism in jujube fruit at maturity. The results showed that under low stress (0~90 mM), the contents of sucrose, glucose, and fructose in the jujube fruit and the activities of sucrose phosphate synthase (SPS), sucrose synthase decomposition direction (SS-I), and sucrose synthase synthesis direction (SS-II) increased with increases in stress concentration, results that were consistent with the relative expression trends of the SPS and SS genes; however, the results were reversed under high concentrations (120 and 150 mM). The soluble acid invertase (S-AI) activity decreased with increases in stress concentration under low stress, and the results were reversed with high stress, which was consistent with the relative expression trends of the *ZjcINV3*, *ZjnINV1*, and *ZjnINV3*. Research regarding the response of antioxidant enzymes in fruits under saline–alkali stress showed that only the differences in peroxidase (POD) activity under saline–alkali stress were consistent with sugar accumulation; the proline (PRO), catalase (CAT) decreased and the malondialdehyde (MDA) superoxide dismutase (SOD) increased with increases in saline–alkali stress. These results indicate that the sugar metabolism and antioxidase jointly promote and regulate sugar accumulation in jujube fruits in a low saline–alkali environment.

Keywords: jujube; saline–alkali stress; sugar metabolism; antioxidant enzyme; RT-qPCR

Citation: Wang, Y.; Feng, Y.; Yan, M.; Yu, J.; Zhou, X.; Bao, J.; Zhang, Q.; Wu, C. Effect of Saline–Alkali Stress on Sugar Metabolism of Jujube Fruit. *Horticulturae* **2022**, *8*, 474. https://doi.org/10.3390/horticulturae8060474

Academic Editor: Costanza Ceccanti

Received: 26 April 2022
Accepted: 24 May 2022
Published: 26 May 2022

1. Introduction

Xinjiang is a national high-quality fruit production area and the most advantageous production area of the country for jujube cultivation. In 2020, the cultivation area comprised about one-third of cultivation areas nationwide, and the annual yield comprised more than 50% of national production [1]; however, jujube is mostly grown in non-cultivated areas in which soil salinization is a serious problem that affects the growth of jujube trees and jujube quality.

Sugar is an important factor of fruit quality that is often affected by external environments such as light, water, salinity, and temperature [2]. Some natural phenomena in production and previous studies have shown that [3–6] mild salt stress can improve sugar accumulation in fruits and severe salt stress can decrease the sugar content. In a previous study, after sprayings with different concentrations of an NaCl solution, the fructose, glucose, and sucrose in grapefruit increased under low salinity (20~60 mM) and decreased under high salinity (100~150 mM) [6,7]. Differences in the sugar contents of fruits caused by different concentrations of salt stress have also been reported for tomatoes [4,8,9], Lycium chinensis [5], and figs [10]. These differences are often related to the activities of sucrose-metabolizing enzymes and gene expression levels [11]. Tomatoes [9] were treated with 25~75 mM NaCl, and the salt-stress-induced AI and SS-I enzyme activities promoted the accumulation of hexose. The up-regulation of SS, AI, and neutral invertase (NI) genes under 75 mM NaCl stress was shown to lead to increases in the contents of glucose, fructose, and sucrose in figs [10]. NaCl stress was shown to inhibit NI and AI and promoted SS, which reduced the hexose and soluble sugar contents in wolfberry fruit; low salt (0.3% and 0.6%) stress was shown to promote SPS activity and increase sucrose accumulation in fruit, and high salt (0.9%) stress led to the opposite result. Salt stress has a concentration-dependent effect on sugar metabolism [12,13] and leads to changes in oxidative stress in plants [14–16]. Research has shown that decreases in the total soluble sugar of kiwi fruit leaves [17] under NaCl stress are related to significantly increased proline levels. Begonia [18,19] was shown to adapt to saline–alkali stress by increasing the activities of SOD and POD and accumulating sugar to participate in osmotic regulation. The regulation and adaptation of sugar to stress in plants is an integrated response that involves gene expression regulation, post-transcriptional regulation, and the metabolite feedback of sugar metabolism at multiple biological levels [20,21].

At present, there have been few reports on the effect of saline–alkali stress on the sugar metabolism in jujube fruit. One report showed that the soluble sugar content in jujube fruit first increased and then decreased with increases in saline–alkali stress (0~0.6%), specifically demonstrating that 0.4% saline–alkali stress could significantly increase the soluble sugar in jujube fruit by 13% [22]; however, there is still no extensive research on the response of jujube fruit sugar accumulation in enzymatic and physiological of regulation under saline–alkali stress that uses a combination of physiology, molecular and oxidase [23]. Taking into account all these facts, the objectives of this study were to understand how jujube adjusts osmolytes and antioxidase activity to increase carbohydrate metabolic under the conditions of saline–alkali stress and to identify which genes are involved in the regulation of sugar metabolism in jujube.

2. Materials and Methods

2.1. Experimental Materials

The experiment was conducted in a jujube orchard of Alar City 10 Regiment, Aksu Prefecture, Xinjiang Region (40°59′ N 81°28′ E), located on the southern edge of the Taklamakan desert; it has a continental arid desert climate with an annual average temperature of 10.4 °C, an annual average precipitation of 40.1~82.5 mm, and an annual average evaporation of 1876.6~2558.9 mm. The materials used in the experiment were 10-year-old "Junzao" jujube trees, with wild jujube as rootstock. The row spacing was 4 m, and the plant spacing was 2 m (1250 trees/hm^2). A rain shelter was arranged above the test site to prevent the impact of rainwater on treatment (Figure 1A). Connections between treatments were intercepted by digging a 100 cm deep ditch (Figure 1B) filled with a double-layer plastic film to prevent leakage and isolate plots (Figure 1C) so that the treatments of each plot did not affect each other. Four drip lines (two emitters with 0.3 m intervals; 3.2 L/h for each emitter) located 0.4 and 0.8 m away from the tree row provided the salt–alkali solution on both sides of the trees (Figure 1D). The soil was sandy loam, and the jujube trees were distributed in a 0–60 cm soil layer.

2.2. Experimental Design

The experiment had a single-factor randomized block design. According to measurements of the saline–alkali soil in the Alar jujube orchard, the ratio of NaCl to $NaHCO_3$ was 3:1. The concentrations of saline–alkali used in the study were 0 (CK), 30, 60, 90, 120, and 150 μM. Each sample group comprised three plots of four plants, with a total of 72 jujube trees; each group was irrigated with the saline solution every 14 days after the young fruiting stage for 3 irrigations in total.

Figure 1. (**A**) Rain shelter to prevent the impact of rainwater; (**B**) connections between treatments were intercepted by digging a 100 cm deep ditch; (**C**) the ditch was insulated with double-layer plastic; (**D**) the drip lines used to provide the salt–alkali solution.

2.3. Sample Collection

2.3.1. Plant Sample Collection

Samples were collected at fruit maturation (120 days after flowering), and 30 fruits were selected from a different part of each tree. The fruit samples were immediately put into an incubator and brought back to the laboratory before the fruit pulp was extracted and chopped. The pulp was quickly divided into three parts; one part was dried in a vacuum-freeze lyophilizer (GOLD-SIM, FD5-3) for 72 h, pulverized, extracted as powder, and stored at −20 °C in a cryogenic refrigerator for sugar content determination, while the other two parts were quickly frozen with liquid nitrogen and stored at −80 °C in a cryogenic refrigerator for enzymatic activity and enzyme gene determination.

2.3.2. Soil Sample Collection

Soil samples for each treatment were collected from three points under the trees: 0–20, 20–40, and 40–60 cm soil layers. Next, they were subjected to natural air drying, pushed through a 2 mm soil sieve, and stored in sealed bags.

2.4. Experimental Method

2.4.1. Measurement of the Soil Salt Contents

For the preparation of leachate, we first weighed 10 g of drying soil sample and placed it into a bottle, stirred in 50 mL of deionized water, and shook the mixture for 5 min. We then filtered the mixture with filter paper to obtain a clear 5:1 water–soil immersion filter solution.

For the measurement of soil salt contents, we poured 20 mL of the immersion solution into a dry evaporation dish and then placed it into a steam oven until the temperature rose from 105 to 110 °C and the solution reached a constant weight.

2.4.2. Extraction and Determination of Sugars

Sugars were extracted and measured according to the protocol of Pu [24]. We weighed 1 g of freeze-dried jujube powder and performed ultrasonic extraction with distilled water. After centrifugation, the supernatant was passed through a 0.45 μm microporous membrane. Sugars were detected with high-performance liquid chromatography (HPLC) using an Agilent 1206 HPLC system (Agilent Technologies, Waldbronn, Germany) with an evaporative light-scattering detector cell maintained at 60 °C. The column was a Waters XBridgeTM BEH Amide column (4.6 × 250 mm, 5 μm), the column temperature was 30 °C, the atomization tube temperature was 60 °C, and the drift tube temperature was 60 °C. The gas flow was 1.6 L/min, the gain value was 1.0, the mobile phase was acetonitrile/water (volume ratio of 76:24) to which 0.2% triethylamine was added, the injection volume was

10 μL, the running time was 18 min, and each sample was measured 3 times. We performed linear regression analysis on peak area (Y) with mass concentration (X); developed standard curves of sucrose (100 mg/mL), fructose (100 mg/mL), and glucose (100 mg/mL); and established a regression equation (Table 1). Then, we calculated the contents of the sugar components according to the peak area of the sample and the standard curve.

Table 1. Linear equation and correlation coefficients of constituent sugars in mixed standard solution determined with HPLC–ELSD.

Sugar Constituent	Regression Equation	Correlation Coefficient
Fructose	$y = 749.23\ X^{1.6239}$	$R^2 = 0.9993$
Glucose	$y = 1046.9\ X^{1.5509}$	$R^2 = 0.9991$
Sucrose	$y = 1055.0\ X^{1.6852}$	$R^2 = 0.9989$

2.4.3. Enzyme Extraction and Activity Assays

Jujube and liquid nitrogen were placed in a mortar and ground into powder; then, 0.1 g of the powder was weighed and placed in a 2 mL centrifuge tube. The enzyme activity was determined with a sucrose phosphate synthase kit (SPS), a soluble acid convertase kit (S-AI), a neutral translate kit (NI), a sucrose synthetase kit (SS-I), a sucrose synthetase kit (SS-II), a proline kit (PRO), a malondialdehyde kit (MDA), a superoxide dismutase kit (SOD), a catalase kit (CAT), and a peroxidase kit (POD) produced by Suzhou Keming Biotechnology Co., Ltd., Suzhou, China. We used a microassay determination method.

2.4.4. RNA Extraction

We used the TransGen Biotech TransZol Plant kit to extract RNA. The mortar, medicine spoon, centrifuge tube, and gun head used for RNA extraction were treated with DEPC water and operated in a low-temperature, enzyme-free environment.

2.4.5. cDNA

cDNA synthesis was performed using the TransGen Biotech EasyScript® One-Step gDNA Removal and cDNA Synthesis SuperMix kits.

2.4.6. RT-qPCR

Fluorescence quantification was performed using the TransGen Biotech TransStart® Green qPCR SuperMix kit. The PCR reaction was conducted with the synthesized cDNA as the template. The reaction system comprised 10.5 L of the cDNA template, 0.2 L of the left and right primers, two sets of 5 μL of PerfectStart™ Green qPCR SuperMix, 0.2 L of Passive Reference Dye, and 3.9 L of ddH_2O. The fluorescence quantification PCR (Applied BiosystemsT M QuantStudio TM 5, Thermo Fisher Scientific, Waltham, MA, USA) response procedure was set to 42 cycles of 94 °C degeneration for 30 s, 94 °C denaturation for 5 s, 55 °C annealing for 15 s, 72 °C extension for 10 s; finally, the system was heated at 1.6 °C/s to 95 °C s and then maintained for 15 s, allowed to cool down for 60 s at 1.6 °C/s and then held at 0.075 °C/s. The amplification of the selected and reference genes was conducted using three biological replicates each. The gene expression of the control group was used as the reference gene for analysis. The relative abundance of transcripts was calculated by using the $2^{-\Delta\Delta Ct}$ method. Enzyme gene from 11 gene by Zhang Chunmei [25], and primer sequence information is shown in Table 2.

2.5. Statistical Analysis

DPS7.05 statistical analysis was performed. Values are presented as the mean ($n = 3$) ± standard error (SE). Duncan's multiple range tests were conducted to determine whether there were significant differences between individual treatments at $p < 0.05$.

Table 2. The primer sequence of qRT-PCR.

Gene Symbol	Primer Sequence (5′–3′)	Primer Sequence (3′–5′)
ZjSPS1	AGTCCCACTCGCTACTTCGT	TCCAAATCCTCCAGCACATA
ZjSPS2	TCCCAAGCCCTCAGGTATTT	GTAGTTTCTGTTTGCGTGTAG
ZjSPS4	GCTATGACAGCAACGGAGAT	AACAGCACAAAGCCTACACG
ZjSS1	AAGTCATAAGATCCGCACAG	AACACGAACATATTCCCAAA
ZjSS2	ACTGTCTATTTCCCTTTCACG	TCATTGTTATCCTCCCTGCT
ZjSS3	ATTGGGTGTAACGCAGTGTA	TGTGGTTCATGGCTATAAGAT
ZjnINV1	TTTCTCGATGTTGACCCTGTT	TTATGCAAGCCTTCCCTTCT
ZjnINV3	AACAGAGGAATACTCCCACA	CATGAAATCGAATACCCAAT
ZjcINV3	TTTCTCGATGTTGACCCTGTT	TTATGCAAGCCTTCCCTTCT
ZjvINV1	ATTCCAAAGGGTCCCAAAGC	TGGTTAAGCCAGGGTCAGTG
ZjvINV2	ACCCGATAACCCGAAGGAAG	GTCTGTACGGACGCCCAACC
UBQ2	CACCCGTTACTTGCTTTC	CTCTTCCCATTGTCCTCC

3. Results

3.1. Differences in Soil Salt Content under Different Saline–Alkali Treatments

As shown in Figure 2, the salt content in different soil layers gradually increased with increases in the saline–alkali stress. With increases in the saline–alkali stress, the salinity in the 0–20, 20–40, and 40–60 cm soil layers under saline–alkali treatment increased by 35.31~54.70%, 31.36~50.11%, and 19.66~39.62%, respectively, compared to the control. The fitting curve of salt content in the 0–60 soil layer under saline–alkali treatment was $y = 0.0275X + 5.5005$, $R^2 = 0.9860$, and it increased by 2.02, 2.37, 3.31, 4.13, and 4.98 g/kg compared to the control total salt of 4.65 g/kg. The fitted growth curve was $y = 0.0256X + 1.0611$, $R^2 = 0.9842$, and the total salt content of the treatment gradually increased.

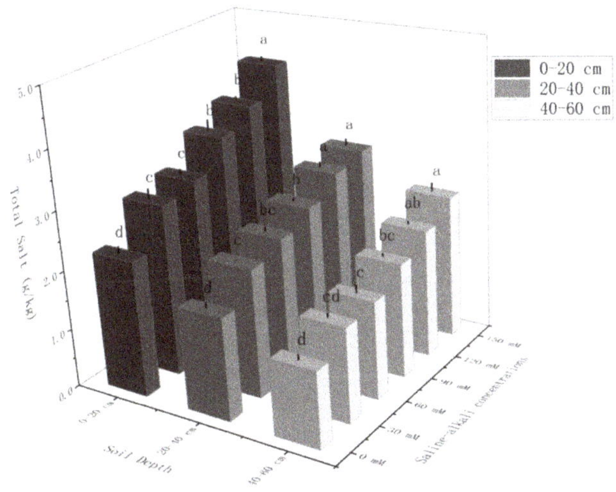

Figure 2. Total salinity in 0–20, 20–40, and 40–60 cm soil layers under different saline–alkali treatments. Note: Small letters indicate significant differences at a 5% level ($p \leq 0.05$).

3.2. Effects of Different Saline–Alkali Stress Treatments on the Contents of Sugar Components in Jujube Fruit

The differences in the sugar components and total sugar content of the jujube fruit under saline–alkali treatment are shown in Table 3. Under low stress (0~90 mM), the fructose and glucose contents in the fruit increased with increases in the saline–alkali stress, and under high stress (120~150 mM), it decreased with increases in stress concentration; when the stress was 90 mM, the fructose and glucose contents were the highest (99.03 and

86.34 mg/g, respectively), and significantly higher than those of the CK treatment by 23.4% and 24.8%, respectively, and significantly higher than those of 120 mM by 11.0% and 15.1%, respectively. The sucrose content increased with increases in stress concentration under 0~60 mM and was the highest under 60 mM (206.11 mg/g), which was significantly higher than the values of the control and 30 mM by 22.6 and 10.1%, respectively. Under 90~150 mM of stress, the sucrose content decreased with increases in stress concentration, but there were no significant differences among treatments. The content of soluble sugar was considered to be the sum of the three sugar components, and its content changed with the stress of saline–alkali in a trend consistent with those of fructose and glucose. At 90 mM, the soluble sugar content was the highest (387.52 mg/g), which was significantly higher than that of the CK treatment by 22.1%. At 30 mM of stress, the total soluble sugar content was higher than that at 60, 120, and 150 mM by 7.8%, but the difference was not significant. The sugar content changes in jujube fruit first increased and then decreased with increases in saline–alkali stress.

Table 3. Differences in sugar component contents in jujube fruit under different saline stress conditions.

Saline–Alkali Stress	Fructose (mg/g)	Glucose (mg/g)	Sucrose (mg/g)	Sum of Soluble Sugars (mg/g)
0 mM	80.25 ± 3.10 ^{c}	69.17 ± 2.63 ^{c}	168.05 ± 5.38 ^{c}	317.47 ± 11.03 ^{c}
30 mM	92.95 ± 2.61 ^{a,b}	79.43 ± 1.34 ^{a,b}	187.25 ± 4.42 ^{b}	359.62 ± 7.52 ^{b}
60 mM	95.16 ± 0.46 ^{a,b}	80.94 ± 0.64 ^{a,b}	206.11 ± 1.74 ^{a}	382.20 ± 2.77 ^{a,b}
90 mM	99.03 ± 2.49 ^{a}	86.34 ± 2.29 ^{a}	202.15 ± 2.24 ^{a}	387.52 ± 6.94 ^{a}
120 mM	89.23 ± 2.16 ^{b}	75.01 ± 1.66 ^{b,c}	200.15 ± 2.08 ^{a}	364.38 ± 3.13 ^{a,b}
150 mM	86.82 ± 3.97 ^{b,c}	77.98 ± 4.40 ^{b}	199.91 ± 2.30 ^{a}	364.71 ± 9.64 ^{a,b}

Note: values indicate the means ± SE, n = 3. Small letters indicate significant differences at a 5% level ($p \leq 0.05$). Different lowercase letters in the same column indicate significant differences in measurement factors.

3.3. Comparison of Key Enzyme Activities of Sucrose Metabolism under Different Saline–Alkali Stress Treatments

The differences in the activities of five sucrose-metabolizing enzymes in jujube fruits under saline–alkali treatment are shown in Figure 3. Under low stress (0~90 mM), the activities of SPS, SS-I, and SS-II significantly increased with increases in stress. The activities of SPS and SS-I showed significant decreasing trends and SS-II showed a downward trend under 120 to 150 mM of saline–alkali stress, but the differences were not significant. The activity of SS-I was higher than that of SS-II. The NI activity increased with the increased stress, and there were no significant differences between treatments except for the CK treatment. The activity S-AI first decreased and then increased with increasing stress. When the stress was 90 mM, the activity of S-AI was the lowest (10.21 μg/min/g).

3.4. Effects of Different Saline–Alkali Stress Treatments on Gene Expression of Key Enzymes of Sucrose Metabolism in Jujube Fruit

11 genes and a reference gene (UBQ2) selected by Zhang [25] were used for study sugar accumulation in the jujube fruits under the saline–alkali treatment. As shown in Figure 4, the relative expression levels of 10 genes were all lower than those of the CK treatment, and only the expression level of the *ZjSPS4* gene was higher than that of the CK treatment, which first increased and then decreased with increases in stress. The expression levels of the sucrose synthase genes *ZjSS1*, *ZjSS2*, and *ZjSS3* and the invertase genes first increased and then decreased with increases in concentration (except for in the CK treatment), and when the stress was 90 mM, the expression levels were significantly higher than in other treatments. The expression of the *ZjvINV2*, *ZjcINV3*, *ZjnINV1*, and *ZjnINV3* genes first decreased and then increased, and when the stress was 60 mM, the expression levels were significantly higher than in other saline treatments.

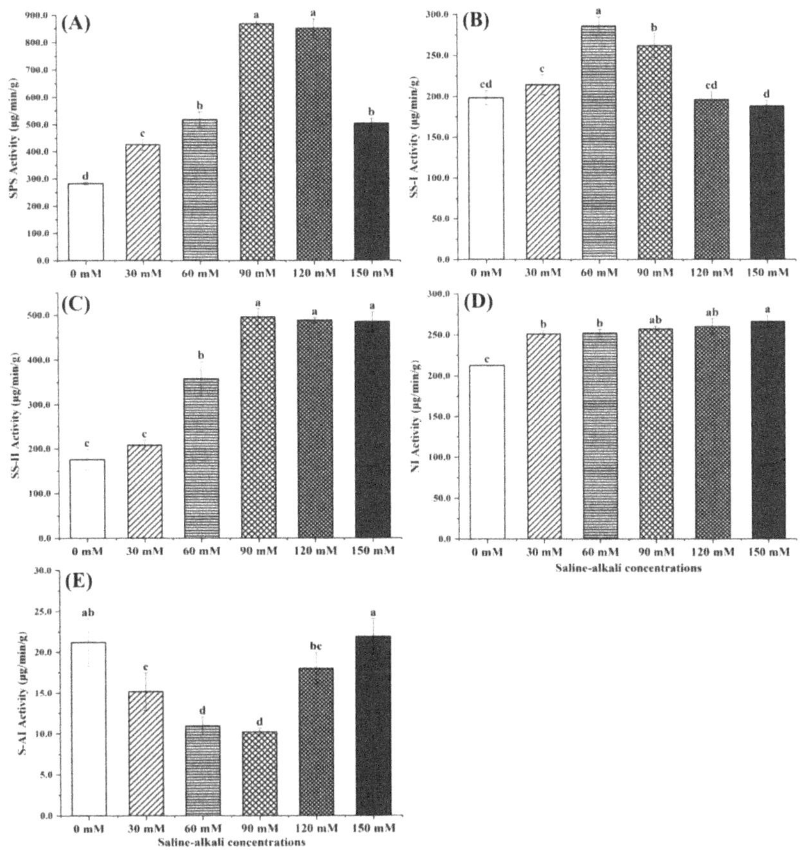

Figure 3. Differences of the sucrose metabolic enzyme activities in jujube fruits under different saline–alkali stress conditions. (**A**) SPS activity; (**B**) SS-I activity; (**C**) SS-II activity; (**D**) NI activity; (**E**) S-AI activity. Note: values indicate the means ± SE, n = 3. Small letters indicate significant differences at a 5% level ($p \leq 0.05$).

3.5. Effects of Different Saline–Alkali Treatments on Proline, Malondialdehyde, and Antioxidant Enzymes in Jujube Fruit

The effect of saline–alkali treatment on the antioxidant enzyme activity of jujube fruits is shown in Figure 5. The PRO content decreased with increases in saline–alkali stress and was significantly lower than that of the CK treatment (3.84 mg/g). The content at 30 mM (2.56 mg/g) was 38.7% higher than that of the CK treatment, but there were no significant differences among treatments. The SOD and MDA trends were similar, and there were no significant differences between the 30 mM (480.01 U/g) and CK treatments; the SOD content under 60~150 mM of stress was significantly higher than that under the 30 mM and CK treatments. The CAT content had the same trend as the PRO content, and its activity (223.57 umol/min/g) under the 30 mM treatment was 35.3% higher than that under the 150 mM treatment. The POD content first increased and then decreased with increases in salt stress. The POD content under the 60 mM treatment (46.59 U/g) was significantly higher than that under the CK and 150 mM treatments by 30.2% and 133.2%, respectively.

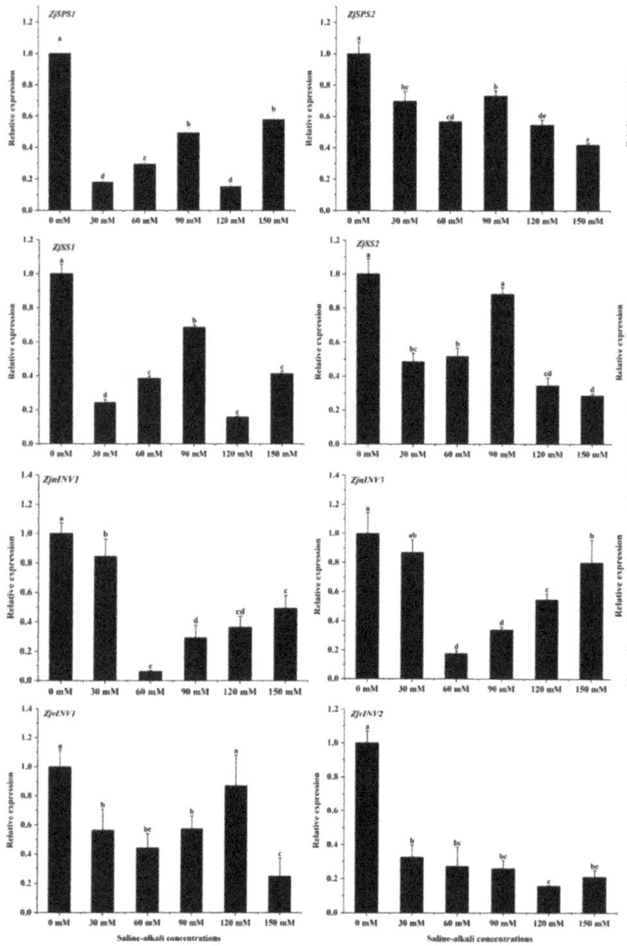

Figure 4. Differences in key enzyme genes of sucrose metabolism in jujube fruits under saline–alkali stress. Note: values indicate the means ± SE, $n = 3$. Small letters indicate significant differences at a 5% level ($p \leq 0.05$).

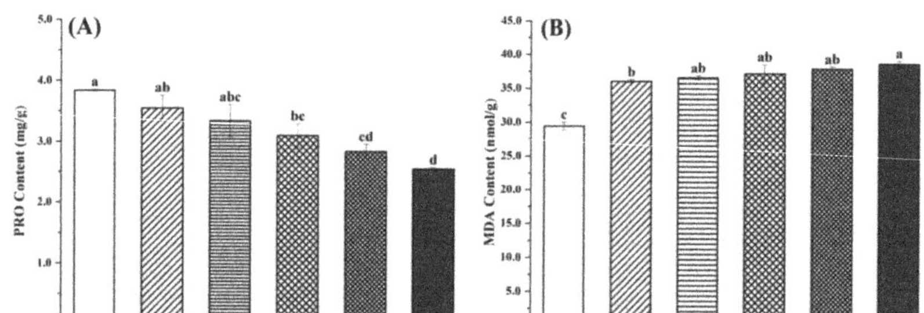

Figure 5. *Cont.*

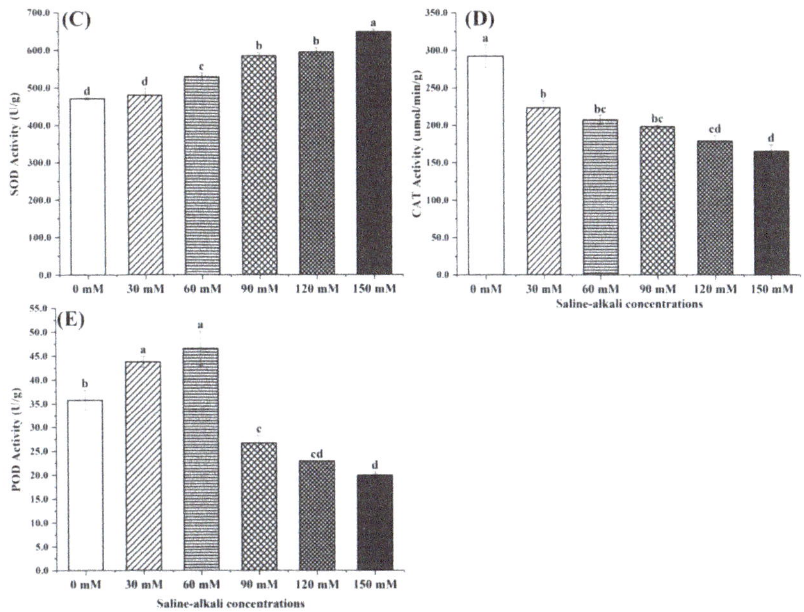

Figure 5. Effects of different saline–alkali treatments on the contents of PRO (**A**), MDA (**B**), SOD (**C**), CAT (**D**), and POD (**E**) in jujube fruit. Note: values indicate the means ± SE, n = 3. Small letters indicate significant differences at a 5% level ($p \leq 0.05$).

4. Discussion

As an osmotic regulator, sugar plays an important role in maintaining osmotic balance in response to saline–alkali stress. Sugar is an important indicator of fruit quality and an important osmotic regulator used by plants to resist stress, and their accumulation in plants experiencing salinity stress is an adaptive mechanism [26,27]. Sucrose, glucose, and fructose are the main soluble sugar components in fruit; they act as signaling molecules in plant metabolic processes and defense responses, and respond differently to stress. According to previous studies [28,29], in order to resist stress, plants can accumulate sugar to increase intracellular solute concentration, maintain osmotic balance and consequently normal cell turgor, prevent the excessive dehydration of protoplasts, and reduce the harm caused by stress, this is consistent with the response of this study under low salt stress (0~90 mM); however, excessive stress will lead to an imbalance of permeability and the inhibition of sugar accumulation [6,7,21], this is consistent with the inhibition of sugar accumulation under high salinity stress (120~150 mM) in this study.

Effects of SPS and SS on sugar synthesis and decomposition by expression of related genes under saline–alkali stress. The regulation of metabolic enzyme gene expression and metabolic enzyme activity determines the amount of sugar accumulation in fruit [30–32]. Under saline–alkali stress, the changes in sucrose content were mainly affected by SPS and SS-I activities, and the changes in fructose and glucose were mainly affected by SS-II activities, which shows that sugar accumulation during fruit ripening is closely related to SPS and SS under saline-alkali stress. A similar change was also observed in tomatoes [4,8], wolfberries [5], and figs [10]; meanwhile, the relative expression levels of genes also changed. The relative expression level of *ZjSS1*, *ZjSS2*, and *ZjSS3* was consistent with the change of SS-I enzyme activity under saline–alkali stress, but *ZjSPS1*, *ZjSPS2* and *ZjSPS4* did not show significant trends consistent with the SPS enzyme; it may be that the regulation of SPS activity by SPS genes may be controlled by both positive and negative regulation of multiple genes in saline-alkali stress [30,33–35].

Invertases and related genes' response to saline-alkali stress regulates sugar content in jujube fruit. As invertases, S-AI and NI also regulate sugar accumulation in response to abiotic stress [36]. In this study, the activity of NI were no significant differences under different treatments, and all significantly higher than control, indicating that NI was significantly affected by saline–alkali stress, but there was no significant concentration-effect [12,37,38]. The changes of S-AI activity with the stress concentration, contrary to the accumulation of fructose and glucose. These results show that saline–alkali stress regulates the conversion level of sucrose to fructose and glucose by S-AI, as well as controlling intracellular sugar accumulation and osmotic concentration to maintain osmotic balance [5,39]. In addition, the responses of *ZjcINV3*, *ZjnINV1*, and *ZjnINV2* to stress were consistent with the trends of S-AI activity, which suggests that they could be crucial in regulating S-AI activity in response to saline–alkali stress; this is consistent with the result that the change of *Hf VIN1* gene expression and VIN enzyme activity with a decrease in temperature stress [40], which indicates that plants need more invertases to decompose sucrose into hexose under low temperature, salinity and other stresses, to provide more energy for cells to maintain increased respiratory consumption and enhance resistance to stress [36,37,39]. Hydrolyze sucrose by IN in vacuole for tissue utilization, and down-regulation of gene expression can respond to the needs of sucrose decomposition under stress [39,40].

Response of antioxidant enzymes to salinity stress and regulation of sugar content in jujube fruits. Sugar metabolism is often accompanied by biological oxidation in plants with stress, and the protective mechanism of active oxygen in cells plays an important role in responding to stress [14]. Salinity stress can cause an imbalance of reactive oxygen species; (ROS), PRO, MDA, SOD, POD, and CAT are important indexes of the plant stress response, and they act as potent salinity mitigators [15,41]. In the present study, PRO and CAT presented similar activity decreases with increases in stress concentration. It has been reported that proline is an osmoprotectant and scavenger of free radicals in plant cells, accumulating in cells to protect osmotic balance [42,43]. In this connection, PRO decreased indicating that osmotic balance gradually weakened in saline–alkali stress. It is possible that with the increase of saline–alkali stress concentration, the membrane permeability is destroyed, the proline anabolism in the jujube fruit is unbalanced, and the resistance to salinity gradually weakens [44,45]. The inhibition of CAT activity during saline stress is related to the destruction of cell metabolism and structure [46]. In this study, CAT activity was found to be restrained by the different saline–alkali stress concentrations in jujube fruit, which may be caused irreversible damage to the structure and function of CAT proteins [47].

The MDA level directly reflects the degree of cell membrane damage and the level of lipid peroxidation [47,48]; this elevation of MDA occurrence might be due to the impairment of the cell membrane, reduction of lipid peroxidation level insufficient activity of enzymatic antioxidant [49]. In the current experiment, the MDA content increased gradually in a concentration-dependent manner in the "Junzao" jujube fruit under saline–alkali stress, which indicates that membranes were increasingly severely damaged with increasing saline–alkali stress. A similar increase in MDA was also observed in Arabidopsis [50], tomato [51,52], wheat [53], maize [54], tobacco [55] with salt stress.

SOD content could reflect the antioxidant capacity of plants, and increases in activity are usually related to plant tolerance [41,47]. In order to resist saline–alkali stress, the jujube fruit can prevent cell damage by enhancing SOD activity, scavenging free radicals, and maintaining the metabolic balance of reactive oxygen [56,57]. The MDA and SOD presented similar activity increases with increasing saline–alkali stress, which indicates that the level of SOD under salt-alkali stress is closely related to the damage degree reflected by MDA to stress [47,56,58].

As a defensive enzyme, POD is mainly involved in ROS scavenging, and its activity is related to plant tolerance [41]. In this study, the change in POD activity was consistent with the response of sucrose in jujube fruits. Under low saline–alkali stress, the ability of POD to scavenge excessive ROS in cells may be closely related to sugar accumulation; however, the protective effect of the antioxidant enzyme system was limited [38,57]. Saline–alkali stress can lead to decreased enzyme activity when it reaches the minimum tolerance limit of plants (60 mM), at which point their defense mechanism is destroyed and POD enzyme activity decreases [38,41].

The concentration effect of stress on sugar accumulation is reflected in the enzyme activity and antioxidase. The responses of MDA, SOD, and NI to different saline–alkali stresses were consistent in this study, indicating that the hydrolysis of sucrose by NI to fructose and glucose was closely related to the degree of lipid peroxidation damage and the stress resistance of cell membrane [37,38]; this could be because that the responses of these antioxidase and soluble sugars to saline–alkali stress play roles in maintaining cellular redox homeostasis, promoting sugar accumulation, and maintaining osmotic material balance [44,56].

Stress has a concentration effect on sugar accumulation in jujube fruit. Previous studies have demonstrated that salt stress affects carbohydrate contents in the fruit, depending on the magnitude of stress [10,17]. Moderate salt stress can improve the soluble solids, or soluble sugar content in fruit [4–9], but the sugar content of grapes (100~150mM) [6,7], figs (100 mM) [10], tomatoes (0.3%) [8], and Lycium chinensis (0.9%) [5] were inhibited at the high saline stress. Similar results were obtained in this study, and under 90 mM saline–alkali stress treatment, the content of fructose, glucose, soluble total sugar, and the activity of SPS, SS-II, and POD in jujube fruit reaches the highest; meanwhile, there is still no extensive research on that activity of enzymes and expression of genes in response to saline–alkali stress and affect sugar accumulation in fruit [9,10,13], especially in jujube fruit. The change of S-AI and *ZjcINV3, ZjnINV1, ZjnINV2* showed an opposite trend compared with the increase in sugar in this study. It has been evidenced that the transcript levels and regulation of invertases play an important role in regulating sugar accumulation in response to stress [10,11,13,37–40], which suggests a different transcriptional level of invertase-encoding genes resulting in sugar accumulation in the fruit.

5. Conclusions

In summary, this research explored the effects of enzyme activities, gene expression, and oxidase markers on sugar accumulation in jujube fruit under saline–alkali stress. In "Junzao" fruit, low saline–alkali stress promoted the increase in soluble sugar content, and high saline–alkali stress inhibited the accumulation of sugar. Increases in SPS and SS-II activities and the inhibition of AI activities are the main reasons for the accumulation of sucrose, glucose, and fructose, while down-regulation of *ZjcINV3, ZjnINV1*, and *ZjnINV3* may be the key genes on sugar accumulation in fruit. The activity of POD reflects the ability of "Junzao" fruit to resist saline–alkali stress by promoting sugar accumulation and maintaining osmotic balance. Studying the relationship between environmental factors and sugar accumulation of jujube could provide a theoretical basis for the ecological regulation of fruit sugars' metabolism, as well as provide relevant references for explaining the influence of salinized soil on jujube fruit quality in Xinjiang.

Author Contributions: Y.W.: data curation, methodology, formal analysis, writing—original draft, and writing—review and editing. Y.F.: methodology and writing—review and editing. M.Y.: investigation, and formal analysis. J.Y. and X.Z.: methodology, project administration, and supervision. J.B.: writing—review, and editing. Q.Z.: funding acquisition and supervision. C.W.: funding acquisition, conceptualization, project administration, supervision, writing—original draft, and writing—review and editing. All authors have read and agreed to the published version of the manuscript.

Funding: This research was funded by Major scientific and technological projects of XPCC (2017DB006); Innovation and Entrepreneurship Platform and Base Construction Project of XPCC (2019CB001); and Scientific Research Innovation Project of postgraduate in Xinjiang Autonomous Region (XJ2020G270).

Institutional Review Board Statement: Not applicable.

Informed Consent Statement: Not applicable.

Data Availability Statement: All the data generated or analyzed during this study are included in this published article.

Acknowledgments: The authors would like to thank the University of Tarim for supporting the project.

Conflicts of Interest: The authors declare no conflict of interest.

Abbreviations

SPS	sucrose phosphate synthase
SS-I	Sucrose synthase decomposition direction
SS-II	Sucrose synthase synthesis direction
NI	Neutral invertase
S-AI	Soluble acid invertase
PRO	Proline
MDA	Malondialdehyde
SOD	Superoxide dismutase
CAT	Catalase
POD	Peroxidase
ROS	Reactive oxygen species

References

1. Wei, S.; Hongtao, D.; Yanfeng, Z.; Cun, W.; Hairong, L.; Dengke, L. Problems and countermeasures of Chinese jujube industry in Aksu area. *J. Fruit Resour.* **2021**, *2*, 84–87. (In Chinese)
2. Chen, T.; Zhang, Z.; Li, B.; Qin, G.; Tian, S. Molecular basis for optimizing sugar metabolism and transport during fruit development. *aBIOTECH* **2021**, *2*, 330–340. [CrossRef]
3. Hao, Y. Developing and utilizing saline-alkali land to produce high quality fruit. *Chin. Fruit Ind. Inf.* **2013**, *30*, 30–31. (In Chinese)
4. Ladewig, P.; Trejo-TéLlez, L.I.; Servin-Juarez, R.; Contreras-Oliva, A.; Gomez-Merino, F.C. Growth, yield and fruit quality of Mexican tomato landraces in response to salt stress. *Not. Bot. Horti Agrobot. Cluj-Napoca.* **2021**, *49*, 12005. [CrossRef]
5. Juan, Y.; Xing, X.; Yuqing, W.; Rongxia, Z. Sugar sand sucrose-metabolizing enzymesin fruits of *Lycium barbarum* under salt stress. *J. Ningxia Agric. Coll.* **2004**, *25*, 28–31. (In Chinese)
6. Li, X.L.; Wang, C.R.; Li, X.Y.; Yao, Y.X.; Hao, Y.J. Modifications of Kyoho grape berry quality under long-term NaCl treatment. *Food Chem.* **2013**, *139*, 931–937. [CrossRef]
7. Sun, H.; Sun, T.-Y.; Xu, L.-L.; Du, Y.-P. Effects of the long-term treatment of low-concentrated salt on grape berry quality and transcriptome. *Plant Physiol. J.* **2017**, *53*, 2197–2205. (In Chinese) [CrossRef]
8. Wang, W.; Cai, L.; Long, Z.; Zhang, X.; Zhao, F. Effects of non-uniform salt stress on growth, yield, and quality of tomato. *Soil Sci. Plant Nutr.* **2021**, *67*, 545–556. [CrossRef]
9. Shaowei, L.; Fei, Q.; Tianlai, L. Effects of NaCl stress on photosynthetic characteristics and sucrose metabolism of tomato leaves. *North. Hortic.* **2012**, *3*, 14–18. (In Chinese) [CrossRef]
10. Mascellani, A.; Natali, L.; Cavallini, A.; Mascagni, F.; Caruso, G.; Gucci, R.; Havlik, J.; Bernardi, R. Moderate salinity stress affects expression of main sugar metabolism and transport genes and soluble carbohydrate content in ripe fig fruits (*Ficus carica* L. cv. Dottato). *Plants* **2021**, *10*, 1861. [CrossRef]
11. Guanglian, L.; Min, Z.; Chunhui, H.; Dongfeng, J.; Xiaobiao, X. Progress in research on sugar metabolism and related enzyme genes in fruit. *Acta Agric. Univ. Jiangxiensis* **2020**, *42*, 187–195. [CrossRef]
12. Qijun, M. Molecular Mechanism by which Apple Sucrose Transporter MdSUT2.2 Involves in Regulating Sugar Content in Response to Drought and Salt Stresses. Doctor's Thesis, Shandong Agricultural University, Taian, China, 2018. (In Chinese).
13. Gao, Z.; Sagi, M.; Lips, S.H. Carbohydrate metabolism in leaves and assimilate partitioning in fruits of tomato (*Lycopersicon esculentum* L.) as affected by salinity. *Plant Sci.* **1998**, *135*, 149–159. [CrossRef]
14. Alhasnawi, A.N.; Kadhimi, A.A.; Isahak, A.; Mohamad, A.; Doni, F.; Mohtar, W.; Yusoff, W.; Zain, C.R.B.M. Salinity stress in plant and an important antioxidant enzyme. *Life Sci. J.* **2014**, *11*, 913–920. [CrossRef]
15. Ali, A.; Yun, D. Salt stress tolerance; what do we learn from halophytes? *J. Plant Biol.* **2017**, *60*, 431–439. [CrossRef]
16. Arif, Y.; Singh, P.; Siddiqui, H.; Bajguz, A.; Hayat, S. Salinity induced physiological and biochemical changes in plants: An omic approach towards salt stress tolerance. *Plant Physiol. Biochem.* **2020**, *156*, 64–77. [CrossRef] [PubMed]
17. Abid, M.; Zhang, Y.J.; Li, Z.; Bai, D.F.; Zhong, Y.P.; Fang, J.B. Effect of salt stress on growth, physiological and biochemical characters of four kiwifruit genotypes. *Sci. Hortic.* **2020**, *271*, 109473. [CrossRef]
18. Jia, X.; Wang, H.; Svetla, S.; Zhu, Y.F.; Hu, Y.; Cheng, L.; Zhao, T.; Wang, Y.X. Comparative physiological responses and adaptive strategies of apple Malus halliana to salt, alkali and saline-alkali stress. *Sci. Hortic.* **2019**, *245*, 154–162. [CrossRef]
19. Jia, X.; Zhu, Y.; Zhang, R.; Zhu, Z.; Zhao, T.; Cheng, L.; Gao, L.; Liu, B.; Zhang, X.; Wang, Y. Ionomic and metabolomic analyses reveal the resistance response mechanism to saline-alkali stress in Malus halliana seedlings. *Plant Physiol. Biochem.* **2020**, *147*, 77–90. [CrossRef]

20. Roitsch, T.; González, M. Function and regulation of plant invertases: Sweet sensations. *Trends Plant Sci.* **2004**, *9*, 606–613. [CrossRef]
21. Hu, L.; Zhou, K.; Liu, Y.; Yang, S.; Zhang, J.; Gong, X.; Ma, F. Overexpression of MdMIPS1 enhances salt tolerance by improving osmosis, ion balance, and antioxidant activity in transgenic apple. *Plant Sci.* **2020**, *301*, 110654. [CrossRef]
22. Jie, W.; He-Li, W.; Cui-Yun, W.; Zhang, Q.; Jiang, Y.; Xiang-Yu, L. Effects of mixed salt-alkali stress on the internal quality of Zizyphus jujuba 'Huizao'. *Agric. Res. Arid Areas* **2015**, *33*, 144–147. (In Chinese)
23. Soltabayeva, A.; Ongaltay, A.; Omondi, J.O.O.; Srivastava, S. Morphological, Physiological and Molecular Markers for Salt-Stressed Plants. *Plants* **2021**, *10*, 243. [CrossRef] [PubMed]
24. Pu, Y.; Ding, T.; Wang, W.; Xiang, Y.; Ye, X.; Li, M.; Liu, D. Effect of harvest, drying and storage on the bitterness, moisture, sugars, free amino acids and phenolic compounds of jujube fruit (*Zizyphus jujuba* cv. Junzao). *J. Sci. Food Agric.* **2018**, *98*, 628–634. [CrossRef] [PubMed]
25. Chunmei, Z. Molecular Mechanism Related to Themetabolism of Sugar, Acid and Domestication for Ziziphus Jujuba Mill. Doctor's Thesis, Northwest Agriculture & Forestry University, Yangling, Shanxi, China, 2016.
26. Tao, H.; Gexiang, Z.; Fuchao, Z.; Yu, C. Research progress in plant salt stress response. *Mol. Plant Breed.* **2018**, *16*, 3006–3015. (In Chinese) [CrossRef]
27. Munns, R.; Tester, M. Mechanisms of salinity tolerance. *Annu. Rev. Plant Biol.* **2008**, *59*, 651–681. [CrossRef] [PubMed]
28. Zhu, J. Abiotic stress signaling and responses in plants. *Cell* **2016**, *167*, 313–324. [CrossRef]
29. Muchate, N.S.; Nikalje, G.C.; Rajurkar, N.S.; Suprasanna, P.; Nikam, T.D. Plant salt stress: Adaptive responses, tolerance mechanism and bioengineering for salt tolerance. *Bot. Rev.* **2016**, *82*, 371–406. [CrossRef]
30. Lastdrager, J.; Hanson, J.; Smeekens, S. Sugar signals and the control of plant growth and development. *J. Exp. Bot.* **2014**, *65*, 799–807. [CrossRef] [PubMed]
31. Fang, J.; Zhu, X.; Jia, H.; Wang, C. Research advances on physiological function of plant sucrose synthase. *J. Nanjing Agric. Univ.* **2017**, *40*, 759–768. (In Chinese)
32. Li, J.; Wu, L.; Foster, R.; Ruan, Y.L. Molecular regulation of sucrose catabolism and sugar transport for development, defence and phloem function. *J. Integr. Plant Biol.* **2017**, *59*, 322–335. [CrossRef]
33. Lu, S.; Li, T.; Jiang, J. Effects of salinity on sucrose metabolism during tomato fruit development. *Afr. J. Biotechnol.* **2010**, *9*, 842–849. [CrossRef]
34. Lu, S.; Li, T.; Jiang, J. Tomato key sucrose metabolizing enzyme activities and gene expression under NaCl and PEG iso-osmotic stresses. *Agric. Sci. China.* **2009**, *8*, 1046–1052. [CrossRef]
35. Na, C.; DongDong, H.; Daoyuan, W.; Lijuan, P.; Xiaoyuan, C.; Ming, C.; Tong, W.; Mian, W.; Zhen, Y.; Shanlin, Y. Expression analysis of the sucrose synthase gene AhSuSy in different tissue and under abiotic stresses in peanut. *J. Peanut Sci.* **2013**, *42*, 25–32. (In Chinese)
36. Jietang, Z. Advances in research on invertase in plant development and response to abiotic and biotic stresses. *J. Trop. Subtrop. Bot.* **2016**, *24*, 352–358. (In Chinese)
37. Dahro, B.; Wang, F.; Peng, T.; Liu, J.-H. PtrA/NINV, an alkaline/neutral invertase gene of Poncirus trifoliata, confers enhanced tolerance to multiple abiotic stresses by modulating ROS levels and maintaining photosynthetic efficiency. *BMC Plant Biol.* **2016**, *16*, 1–8. [CrossRef]
38. Huang, Y.-W.; Nie, Y.-X.; Wan, Y.-Y.; Chen, S.-Y.; Sun, Y.; Wang, X.-J.; Bai, J.-G. Exogenous glucose regulates activities of antioxidant enzyme, soluble acid invertase and neutral invertase and alleviates dehydration stress of cucumber seedlings. *Sci. Hortic.* **2013**, *162*, 20–30. [CrossRef]
39. Koch, K. Sucrose metabolism: Regulatory mechanisms and pivotal roles in sugar sensing and plant development. *Curr. Opin. Plant Biol.* **2004**, *7*, 235–246. [CrossRef]
40. Lu, B.; Zhiguo, Z.; Shijie, Z.; Dongmei, H.; Qiaoping, Q. Solation of three types of invertase genes from Hemerocallis fulva and their responses to low temperature and osmotic stress. *Acta Hortic. Sin.* **2021**, *48*, 300–312. (In Chinese) [CrossRef]
41. Yang, S.Y.; Chen, X.Y.; Hui, W.K.; Ren, Y.; Ma, L. Progress in responses of antioxidant enzyme systems in plant to environmental stresses. *J. Fujian Agric. For. Univ.* **2016**, *45*, 481–489. (In Chinese) [CrossRef]
42. Arteaga, S.; Yabor, L.; Díez, M.J.; Prohens, J.; Boscaiu, M.; Vicente, O. The use of proline in screening for tolerance to drought and salinity in common bean (*Phaseolus vulgaris* L.) genotypes. *Agronomy* **2020**, *10*, 817. [CrossRef]
43. Yu, Z. Effects of Exogenous Proline on Growth and Proline Metabolism of Trifoliate Orange Rootstock under Boron Stress. Master's Thesis, Huazhong Agricultural University, Wuhan, China, 2021.
44. Ghosh, U.K.; Islam, M.N.; Siddiqui, M.N.; Cao, X.; Khan, M.A.R. Proline, a multifaceted signalling molecule in plant responses to abiotic stress: Understanding the physiological mechanisms. *Plant Biol.* **2021**, *24*, 227–239. [CrossRef] [PubMed]
45. Naliwajski, M.; Skłodowska, M. The Relationship between the Antioxidant System and Proline Metabolism in the Leaves of Cucumber Plants Acclimated to Salt Stress. *Cells* **2021**, *10*, 609. [CrossRef]
46. Sanoubar, R.; Cellini, A.; Gianfranco, G.; Spinelli, F. Osmoprotectants and antioxidative enzymes as screening tools for salinity tolerance in radish (*Raphanus sativus*). *Hortic. Plant J.* **2020**, *6*, 14–24. [CrossRef]
47. Mansour, M.M.F. The plasma membrane transport systems and adaptation to salinity. *J. Plant Physiol.* **2014**, *171*, 1787–1800. [CrossRef] [PubMed]

48. Singhal, R.K.; Saha, D.; Skalicky, M.; Mishra, U.N.; Chauhan, J.; Behera, L.P.; Lenka, D.; Chand, S.; Kumar, V.; Dey, P.; et al. Crucial Cell Signaling Compounds Crosstalk and Integrative Multi-Omics Techniques for Salinity Stress Tolerance in Plants. *Front. Plant Sci.* **2021**, *12*, 670369. [CrossRef]
49. Tahjib-Ul-Arif, M.; Sohag, A.A.M.; Afrin, S.; Bashar, K.K.; Afrin, T.; Mahamud, A.G.M.S.U.; Polash, M.A.S.; Hossain, M.T.; Sohel, M.A.T.; Brestic, M.; et al. Differential Response of Sugar Beet to Long-Term Mild to Severe Salinity in a Soil–Pot Culture. *Agriculture* **2019**, *9*, 223. [CrossRef]
50. Ellouzi, H.; Ben Hamed, K.; Cela, J.; Munné-Bosch, S.; Abdelly, C. Early effects of salt stress on the physiological and oxidative status of Cakile maritima (halophyte) and Arabidopsis thaliana (glycophyte). *Physiol. Plant.* **2011**, *142*, 128–143. [CrossRef]
51. Gapińska, M.; Skłodowska, M.; Gabara, B. Effect of short- and long-term salinity on the activities of antioxidative enzymes and lipid peroxidation in tomato roots. *Acta Physiol. Plant.* **2008**, *30*, 11–18. [CrossRef]
52. Gong, B.; Wen, D.; Vanden Langenberg, K.; Wei, M.; Yang, F.; Shi, Q.; Wang, X. Comparative effects of NaCl and NaHCO3 stress on photosynthetic parameters, nutrient metabolism, and the antioxidant system in tomato leaves. *Sci. Hortic.* **2013**, *157*, 1–12. [CrossRef]
53. Esfandiari, E.; Gohari, G. Response of ROS-Scavenging systems to salinity stress in two different wheat (*Triticum aestivum* L.) cultivars. *Not. Bot. Horti Agrobot. Cluj-Napoca.* **2017**, *45*, 287–291. [CrossRef]
54. Chen, Y.E.; Mao, J.J.; Sun, L.Q.; Huang, B.; Ding, C.B.; Gu, Y.; Liao, J.Q.; Hu, C.; Zhang, Z.W.; Yuan, S.; et al. Exogenous melatonin enhances salt stress tolerance in maize seedlings by improving antioxidant and photosynthetic capacity. *Physiol. Plant.* **2018**, *164*, 349–363. [CrossRef] [PubMed]
55. Wang, H.; Wang, M.; Xia, Z. Overexpression of a maize SUMO conjugating enzyme gene (ZmSCE1e) increases Sumoylation levels and enhances salt and drought tolerance in transgenic tobacco. *Plant Sci.* **2019**, *281*, 113–121. [CrossRef] [PubMed]
56. Aleem, M.; Aleem, S.; Sharif, I.; Wu, Z.; Aleem, M.; Tahir, A.; Atif, R.M.; Cheema, H.M.N.; Shakeel, A.; Lei, S.; et al. Characterization of SOD and GPX Gene Families in the Soybeans in Response to Drought and Salinity Stresses. *Antioxidants* **2022**, *11*, 460. [CrossRef]
57. Wang, C.; Wen, D.; Sun, A.; Han, X.; Zhang, J.; Wang, Z.; Yin, Y. Differential activity and expression of antioxidant enzymes and alteration in osmolyte accumulation under high temperature stress in wheat seedlings. *J. Cereal Sci.* **2014**, *60*, 653–659. [CrossRef]
58. Rasel, M.; Tahjib-Ul-Arif, M.; Hossain, M.A.; Hassan, L.; Farzana, S.; Brestic, M. Screening of Salt-Tolerant Rice Landraces by Seedling Stage Phenotyping and Dissecting Biochemical Determinants of Tolerance Mechanism multidimensional roles in salt-stressed plants. *J. Plant Growth Regul.* **2020**, *40*, 1853–1868. [CrossRef]

horticulturae

Article

Translocation and Utilization Mechanisms of Leaf Intracellular Water in Karst Plants *Orychophragmus violaceus* (L.) O. E. Schulz and *Brassica napus* L.

Deke Xing [1], Weixu Wang [1], Yanyou Wu [2,*], Xiaojie Qin [1], Meiqing Li [1], Xiaole Chen [1] and Rui Yu [1]

[1] School of Agricultural Engineering, Jiangsu University, Zhenjiang 212013, China
[2] State Key Laboratory of Environmental Geochemistry, Institute of Geochemistry, Chinese Academy of Sciences, Guiyang 550081, China
* Correspondence: wuyanyou@mail.gyig.ac.cn; Tel.: +86-851-84391746

Abstract: *Orychophragmus violaceus* (L.) O. E. Schulz adapts to karst environments through a variety of adaptability mechanisms. However, the leaf intracellular water translocation and utilization mechanism is still unknown. This study hypothesizes that plants adapt to dehydration by synergistically adjusting the leaf anatomy, cell elasticity and intracellular water translocation. Leaf structure, elastic modulus (Em), physiological capacitance (CP), impedance (Z), water potential (Ψ_L), leaf tensity (LT) and chlorophyll fluorescence parameters of the detached leaves in plants of *O. violaceus* and *Brassica napus* L. were measured at each water loss time (0, 1, 2, 3, 4 and 5 h). The uniform leaves were randomly selected from five different plants for each species. The cell vacuole volume and translocation resistance of intracellular water could be represented by the electrophysiological parameters, such as CP and Z. The results indicated that timely shrinkage of *O. violaceus* leaves and mesophyll cells together with the increased water translocation resistance retained the intracellular water and maintained the turgor pressure. Water within sponge parenchyma could also be translocated into palisade parenchyma. The PSII reaction center was kept stable, and the photosynthetic activity of *O. violaceus* was clearly inhibited at 3 h. Palisade parenchyma of *B. napus* leaves increased quickly to improve the intercellular water translocation due to the strong cell stiffness. Gradually increasing intracellular water translocation resistance and recovery of the cell elasticity slowed down the leaf water loss, which, however, could not timely stop the damage on the PSII reaction center and the photochemical efficiency. The photochemical efficiency was seriously inhibited at 4 h and 5 h. The response mechanism of intracellular water to dehydration can be investigated with the help of leaf electrophysiological traits. However, the direct determination of plant drought resistance using electrophysiological information can still not be realized at present and needs further research.

Keywords: electrophysiology; anatomical structure; water potential; cell elasticity; water status

Citation: Xing, D.; Wang, W.; Wu, Y.; Qin, X.; Li, M.; Chen, X.; Yu, R. Translocation and Utilization Mechanisms of Leaf Intracellular Water in Karst Plants *Orychophragmus violaceus* (L.) O. E. Schulz and *Brassica napus* L. *Horticulturae* **2022**, *8*, 1082. https://doi.org/10.3390/horticulturae8111082

Academic Editors: Rossano Massai and Othmane Merah

Received: 5 September 2022
Accepted: 14 November 2022
Published: 16 November 2022

1. Introduction

The soils in karst areas are characterized by karst drought, high pH, low nutrients and high bicarbonate. Droughts with a high degree of spatial heterogeneity in these areas tend to occur increasingly frequently, which is the key factor limiting plant growth and deteriorating the fragile karst environment [1]. *Orychophragmus violaceus* (L.) O. E. Schulz belongs to Cruciferae, and is commonly called Chinese violet cress [2]. Some scholars strongly recommend this typical karst adaptable plant as a marginal raw material for terrestrial biomass [3]. This species is also cultivated as a medicine or ornamental plant and has wide market prospects. Cultivation of this plant helps to improve the economic income of local farmers in karst areas. *O. violaceus* is always taken as a model plant for studying the adaptive mechanisms of plants under karst adversities. It has been reported that *O. violaceus* adapts to karst adversities through a variety of adaptability mechanisms, i.e., photosynthetic adjusting, carbonic anhydrase regulation [4] and inorganic nutrient utilization [5].

Brassica napus L. is also a cruciferous plant with good karst drought resistance. *B. napus* is an important oil crop in southern China, as this type of crop can extract edible oil and fuel, thus having great economic value [6]. Besides, studies have reported that *O. violaceus* and *B. napus* are all suitable as pioneer plants for ecological restoration in karst areas, but they are different in adaptive mechanisms and drought resistance. Consequently, the *B. napus* plants were selected as a comparative species for conducting the investigations on the leaf intracellular water of *O. violaceus*. The studies on plant adaptive mechanisms aim to find methods for effectively evaluating plant stress resistance and matching the heterogeneous karst adversities during the ecological restoration. However, the determining methods established based on the above-mentioned adaptive mechanisms are time consuming and cannot determine the dynamic adaptability of a plant. In fact, plants adapt to adversities, especially drought stress, by the timely regulation of photosynthesis and growth, which is directly related to the intracellular water. Therefore, this study aims at investigating the translocation and utilization mechanisms of leaf intracellular water in karst adapted plants, and providing a basis for establishing a new method for effectively evaluating plant drought resistance in the short-term.

Water stress alters metabolisms in plants, thereby reducing photosynthesis and limiting plant growth [7]. Most (about 97%) of the water absorbed by plant roots dissipates through transpiration, but only a small amount (1~3%) is retained in leaf cells to support plant photosynthesis, growth and other physiological and biochemical processes. Researchers have measured the drought resistance of plants by abscisic acid and the indicators related to metabolism or osmoregulation. Abscisic acid (ABA) can induce stomatal closure and reduce water loss in plants under water deficit [8]. To avoid cell damage, plants produce substances such as phenolic compounds, proline, sugars, anthocyanins and glycosides, which have protective effects on osmoregulation [9]. The intracellular water required for photosynthesis is also regulated by carbonic anhydrase [10]. In fact, leaf intracellular water regulated by the above-mentioned strategies exhibits complex changes, which makes the determination of leaf intracellular water more difficult. Karst plants have made adaptive changes in the anatomical structure and cell behavior for surviving, which can be reflected by the leaf mechanical and electrophysiological traits. These leaf physical traits can be easily determined and are responsive.

Plant leaves are the most sensitive organs to adversities. Leaves can balance water gain and loss by rapidly adjusting anatomical structure and mechanical properties [11]. A water deficit can cause the leaf to shrink, and lead to changes in leaf density (LD) and water movement within leaves, which can reduce water loss [12]. High leaf water storage capacity is related to the increase in total leaf thickness, palisade tissue thickness, and spongy tissue thickness [13]. Leaf elastic modulus (Em) varies with LD and can reflect the variation of leaf anatomy. However, the leaf anatomical and mechanical properties only reflect the static water status of plant. Leaf water potential (Ψ_L) follows a circadian rhythm parallel to atmospheric evapotranspiration demand [14], and leaves can improve their absorption ability of water after a certain drop in Ψ_L. Water loss also induces mesophyll cells to generate and maintain a certain turgor pressure, thus changing the variation in Ψ_L [15]. Leaf electrophysiological information is increasingly used for detecting plant water status. A mesophyll cell can be modeled as a concentric sphere capacitor due to the special composition and structure [16]. A cell membrane with strictly selective permeability will influence the concentration of intracellular electrolytes. The water metabolism in leaves alters the electrolyte concentration and changes the corresponding electrophysiological parameters [12]. Intracellular water status can be obtained by measuring the electrophysiological indicators such as physiological capacitance (CP), impedance (Z) and leaf tensity (LT) of plant by a self-made parallel-plate capacitor [16]. Measurements of Ψ_L and electrophysiological information can investigate the dynamic leaf water status. It is apparent that the leaf intracellular water dynamic as well as the static status is of equal importance for regulating the water availability. However, the synergistic mechanism of leaf anatomy, mechanical strength, Ψ_L and intracellular water translocation on intracellular water status has not been reported yet.

Leaf dehydration is easy to be controlled, so experiments can be repeated multiple times, data are more reliable and changes recorded during dehydration are not affected by other parts of the plant [17]. In this study, the detached leaves of *O. violaceus* and *B. napus* were used as experimental materials, they were soaked in double-distilled water for 30 min and then quickly dehydrated. By comparing the corresponding changes of chlorophyll fluorescence parameters such as maximum photosystem II (PSII) quantum efficiency (F_v/F_m), electron transport rate (ETR), photochemical quenching (qP) and non-photochemical quenching (NPQ) at different times of water loss, the photosynthetic characteristics of *O. violaceus* and *B. napus* were studied. The synergistic influence of leaf anatomical structure, mechanical strength, Ψ_L and intracellular water translocation on the water status were determined, and the photosynthetic adaptive mechanisms were investigated. This study hypothesized that plants can adapt to dehydration by synergistically adjusting the leaf anatomy, cell elasticity and intracellular water translocation. The results of this study can provide a new method for determining the leaf intracellular water and provide a basis for improving the evaluating efficiency of drought resistance of pioneer plants for ecological restoration in karst areas.

2. Materials and Methods

2.1. Plant Materials

The experiment was conducted in the lab at Jiangsu University, Jiangsu Province (32.20° N, 119.45° E), China. The leaves of *Orychophragmus violaceus* (L.) O. E. Schulz and *Brassica napus* L. were selected as the experimental materials in this study. The study area receives a mean annual air temperature of about 15.6 °C. The fourth and fifth fully expanded uniform leaves were completely randomly taken from five different plants for each species at 09:00–10:00 in the morning and immediately soaked into double-distilled water for 30 min, in order to make sure all the leaves were in a uniform initial state (water-saturated). As such, an accurate comparison could be made between *O. violaceus* and *B. napus*. Thereafter, the water on the leaf's surface was wiped off, and the leaves were placed on a dry and ventilated table (26 °C for 5 h). Next, the measurements were taken at 0 (baseline), 1, 2, 3, 4 and 5 h after dehydration, and the determination of each parameter was repeated five times with five different randomly selected leaves [10]. The photosynthetic photon flux density (PPFD) in the lab was 160 $\mu mol/m^2 \cdot s$, and the relative air humidity was 40 ± 5%.

2.2. Determination of Leaf Water Potential, Leaf Area and Water Content

Leaf water potential (Ψ_L) was determined by using a dew point microvoltmeter in a universal sample room (*C-52-SF, Psypro*, Wescor, Logan, UT, USA). The leaves were scanned with a broad-leafed image analysis system (*WinFOLIA, Regent Instruments Inc.*, Quebec, Canada) to obtain the leaf area (LA, cm^2). The fresh weight of the leaves (FW, g) was recorded. The leaves were then dried in an oven at 80 °C to constant weight (DW, g). The leaf water content (WC, %) was calculated by the following [18]:

$$WC = \frac{FW - DW}{FW} \times 100\% \quad (1)$$

2.3. Leaf Elastic Modulus Measurement

The increased stresses (F_s, N) with increasing deformation rates (ΔX, %) of leaf at each water loss time were recorded with the texture analyzer TA.XtplusC (Stable Micro Systems, Godalming, Surrey, UK) using the P/2n probe with a diameter of 2 mm. The working parameters and test mode were set up according to Xing et al. [10], and then the leaf elastic modulus (Em, N per unit of deformation) was calculated according to the following equation:

$$F_s = Em \times \Delta X \quad (2)$$

2.4. Determination of Leaf Anatomy and Leaf Density

Leaf anatomy was observed by using the paraffin sectioning method [19]. Leaf pieces (0.5×0.5 cm) were cut off between the main veins and immersed under the formalin-acetic acid-alcohol (FAA) fixing solution [10]. These paraffin sections were stained with safranin and fast green dye and permanently mounted on slides. Samples were observed by using inverted light microscopes (DMi8, Leica, Wetzlar, Germany), and images were taken. The leaf's total thickness (D_t, μm), upper and lower epidermis thickness, palisade parenchyma thickness, sponge parenchyma thickness and palisade-sponge ratio (%) were measured by the ImageJ software (National Institutes of Health—NIH, Bethesda, MD, USA). The tightness degree of leaf tissue structure (CTR, %) is the ratio of palisade parenchyma thickness to leaf thickness, and loose degree of leaf tissue structure (SR, %) is the ratio of sponge parenchyma thickness to leaf thickness.

Leaf density (LD, g/cm^3) was calculated as follows:

$$LD = \frac{DW}{D_t \times LA} \times 10^4 \tag{3}$$

where DW (g) is the leaf dry weight, D_t (μm) is the leaf total thickness and LA (cm^2) is the leaf area.

2.5. Determination of Physiological Capacitance, Impedance and Leaf Tensity

The physiological capacitance (CP, pF) and impedance (Z, MΩ) at each water loss time was determined by using the LCR HiTester (model 3532-50, Hioki, Nagano, Japan) with a frequency and voltage of 3 kHz and 1 V, respectively [20]. Each leaf was clipped onto the custom-made parallel-plate capacitor. The value of LT (cm^2/cm) was calculated according to the following equation [21]:

$$LT = \frac{A_{CP}}{d_L} = \frac{CP}{\varepsilon_0}\left[\frac{1000iRT}{81,000iRT + (81 - a)M\Psi_L}\right] \tag{4}$$

where A_{CP} (cm^2) is the effective area of the leaf in contact with the capacitor plates, d_L (cm) is the leaf effective thickness, ε_0 is the vacuum dielectric constant (with value of 8.854×10^{-12} F/m); I is the dissociation coefficient (with value of 1), R is the gas constant (with value of 8.30×10^{-3} L·MPa/mol·K), T is the thermodynamic temperature (T = 273 + t °C, K), 81 is the relative dielectric constant of water at normal temperature, a is the relative dielectric constant of the cytosol solute, M is the relative molecular mass of the cytosol solute (g/mol), and Ψ_L (MPa) is the leaf water potential. In this study, the sucrose $C_{12}H_{22}O_{11}$ was identified as the solute in the cytosol, therefore, a was 3.30, M was 342 g/mol, and t was 20 °C [10].

2.6. Chlorophyll Fluorescence Parameters Measurement

Chlorophyll fluorescence (ChlF) parameters were determined by using an IMAGING-PAM modulated chlorophyll fluorescence imaging system (PAM-2000, Walz, Germany). Leaves were dark adapted for 30 min to ensure complete relaxation of all reaction centers before the measurements. The minimum fluorescence (F_o) was determined using a measuring beam, whereas the maximum ChlF (F_m) was recorded after a 0.8 s saturating light pulse (6000 μmol/m^2·s). Maximum PSII quantum yield (F_v/F_m) was calculated as $(F_m - F_o)/Fm$. Then the action light was applied, the minimum (F_o') and maximum (F_m') fluorescence under light and the steady state fluorescence (F_s) were recorded after the fluorescence value was stable. The electron transport rate (ETR) is calculated as follows: $ETR = PPFD \times \Phi_{PSII} \times 0.85 \times 0.5$, where the 0.5 represents the two-quantum absorption per electron transport, and 0.85 represents the absorbed part of the incident photon is 85%, PPFD is the photosynthetic photon flux density. The qP is calculated as follows: $qP = \frac{F'_m - F_s}{F'_m - F'_o}$, and the NPQ is calculated as follows: $NPQ = \frac{F_m - F'_m}{F'_m}$.

2.7. Statistical Analysis

Data are presented as means of at least five replicates (five uniform leaves from five randomly selected plants for each species). The results were analyzed by one-way ANOVA with the Duncan's multiple comparison at $p \leq 0.05$ with the SPSS22.0 software (*SPSS, IBM, Armonk,* New York, NY, USA). The statistically significant differences between different treatments for each species were determined, respectively. The data are shown as the means ± SE (n = 5).

3. Results

3.1. Leaf Water Potential, Leaf Area and Water Content

For O. violaceus, Ψ_L at 5 h was significantly lower than that at 0 h and 1 h, but there was no significant difference between the values at 2~5 h. LA decreased significantly with the increase of water loss time. The WC at 5 h was significantly lower than that at 0~3 h, but there was no significant difference between the values at 5 h and 4 h. For B. napus, Ψ_L and WC did not decrease significantly during the water loss, but the LA at 5 h was significantly lower than that at 0 h and 1 h (Table 1).

Table 1. Leaf water potential (Ψ_L, MPa), leaf area (LA, cm^2) and leaf water content (WC, %) at different dehydration times.

Species	Water Loss Time (h)	Leaf Water Potential	Leaf Area	Leaf Water Content
O. violaceus	0	−0.81 ± 0.05 a	20.10 ± 0.24 a	86.01 ± 1.20 a
	1	−1.12 ± 0.11 ab	19.31 ± 0.14 b	82.74 ± 0.95 b
	2	−1.20 ± 0.15 bc	18.77 ± 0.19 c	81.32 ± 0.82 bc
	3	−1.29 ± 0.14 bc	18.13 ± 0.12 d	79.05 ± 1.03 cd
	4	−1.49 ± 0.13 bc	17.45 ± 0.06 e	77.12 ± 0.96 de
	5	−1.58 ± 0.13 c	16.81 ± 0.02 f	75.75 ± 0.55 e
B. napus	0	−0.83 ± 0.08 a	22.97 ± 0.38 a	81.52 ± 1.35 a
	1	−1.00 ± 0.08 ab	22.68 ± 0.36 ab	79.94 ± 1.02 ab
	2	−1.02 ± 0.09 ab	22.39 ± 0.34 abc	79.10 ± 0.81 ab
	3	−1.10 ± 0.10 ab	21.98 ± 0.29 abc	78.49 ± 0.63 b
	4	−1.02 ± 0.08 ab	21.66 ± 0.30 bc	77.87 ± 0.60 b
	5	−1.20 ± 0.11 b	21.38 ± 0.29 c	77.11 ± 0.32 b

Note: Means (n = 5) in the same column for each plant species followed by different letters differ significantly at $p \leq 0.05$, according to one-way ANOVA.

3.2. Changes of Elastic Modulus and Leaf Anatomical Structure

The Em values of *O. violaceus* at 0, 1 and 3 h were significantly lower than those at other levels. There was no significant difference between the values at 2 h and 5 h (Figure 1A). The values of Em of *B. napus* at 0, 2 and 3 h were remarkably lower than that at 1 h but higher than that at 4 h, there was no significant difference between the values at 4 h and 5 h (Figure 1B).

With the extension of dehydration time, the values of total leaf thickness, sponge parenchyma and lower epidermis of *O. violaceus* at 3~5 h were significantly lower than those at 0 and 1 h, respectively. Additionally, the values of each parameter at 3~5 h exhibited no clear difference. The value of the upper epidermis at 4 h was clearly lower than those at 0~2 h, and the value of palisade parenchyma at 0 h was clearly higher than those at 1, 3 and 4 h (Table 2).

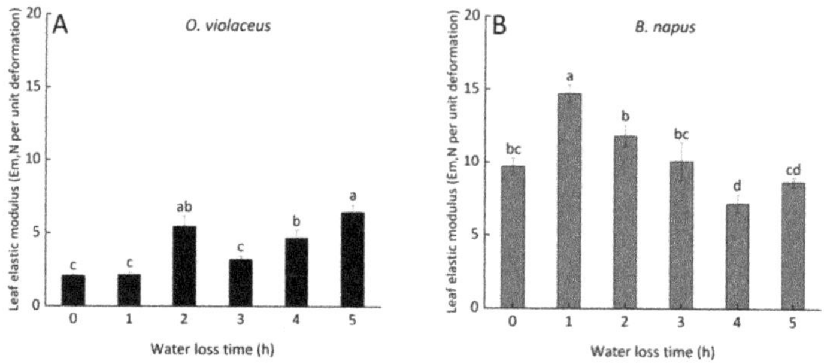

Figure 1. Changes of elastic modulus (Em, N per unit deformation) of *O. violaceus* (**A**) and *B. napus* (**B**). (Note: Different letters appear above the error bars of the same plant species when subsequent values differ significantly at $p \leq 0.05$, according to one-way ANOVA).

Table 2. Leaf anatomical parameters of *O. violaceus*.

Water Loss Time (h)	Thickness (μm)				
	Total Leaf Thickness	Upper Epidermis	Palisade Parenchyma	Sponge Parenchyma	Lower Epidermis
0	150.63 ± 7.57 a	13.97 ± 0.81 a	27.66 ± 1.69 a	95.02 ± 5.70 a	13.98 ± 0.69 a
1	142.42 ± 0.54 ab	13.15 ± 0.38 a	20.88 ± 2.80 b	94.54 ± 2.05 a	13.85 ± 0.04 a
2	130.18 ± 2.56 b	13.20 ± 1.17 a	26.13 ± 1.36 ab	79.71 ± 2.80 b	11.15 ± 0.46 b
3	100.59 ± 5.61 c	12.06 ± 0.62 ab	20.77 ± 1.30 b	57.87 ± 3.03 c	9.90 ± 0.73 bc
4	98.42 ± 3.85 c	10.09 ± 0.56 b	20.28 ± 1.09 b	58.62 ± 3.11 c	9.44 ± 0.24 bc
5	106.20 ± 2.49 c	11.93 ± 0.63 ab	23.93 ± 2.12 ab	61.59 ± 2.22 c	8.74 ± 0.67 c

Note: Means (n = 5) in the same column followed by different letters differ significantly at $p \leq 0.05$, according to one-way ANOVA.

Table 3. Leaf anatomical parameters of *B. napus*.

Water Loss Time (h)	Thickness (μm)				
	Total Leaf Thickness	Upper Epidermis	Palisade Parenchyma	Sponge Parenchyma	Lower Epidermis
0	154.41 ± 2.94 ab	17.09 ± 0.85 a	33.19 ± 0.47 c	95.59 ± 2.96 a	8.54 ± 0.33 b
1	155.90 ± 1.59 ab	16.95 ± 0.43 a	37.71 ± 1.03 b	90.55 ± 1.42 a	10.69 ± 0.85 a
2	157.80 ± 3.89 a	16.53 ± 0.88 a	40.54 ± 0.74 a	89.05 ± 3.85 a	11.68 ± 0.52 a
3	138.78 ± 2.27 c	15.67 ± 0.62 a	30.80 ± 0.79 cd	80.61 ± 1.01 b	11.70 ± 0.74 a
4	153.46 ± 4.17 ab	15.32 ± 0.31 ab	32.12 ± 0.73 c	94.90 ± 3.31 a	11.13 ± 0.38 a
5	146.88 ± 1.29 bc	13.59 ± 0.46 b	29.04 ± 0.64 d	94.32 ± 1.96 a	9.93 ± 0.76 ab

Note: Means (n = 5) in the same column followed by different letters differ significantly at $p \leq 0.05$, according to one-way ANOVA.

For *O. violaceus*, the palisade-sponge ratio at 1 h was significantly lower than those at 2~5 h, but showed no clear difference with that at 0 h. The CTR at 1 h was significantly lower than those at other water loss times, but there was no clear difference between those values at 2~5 h. The SR at 1 h was remarkably higher than those at 2~5 h, but had no significant difference with that at 0 h. For *B. napus*, the palisade-sponge ratio at 2 h was significantly higher than those at 0 h and 3~5 h, but showed no significant difference with that at 1 h, there was no significant difference between the values at 0, 4 and 5 h. The CTR values at 1 h and 2 h were significantly higher than those at other levels, but there was no significant difference between the values at 4 h and 5 h. The SR values at 1~3 h were significantly lower than those at 0, 4 and 5 h (Table 4).

Table 4. Comparison of leaf tissue characteristics between *O. violaceus* and *B. napus*.

Species	Water Loss Time (h)	Palisade-Sponge Ratio (%)	CTR (%)	SR (%)
O. violaceus	0	29.14 ± 1.05 bc	18.35 ± 0.40 b	63.04 ± 1.09 ab
	1	22.24 ± 3.49 c	14.65 ± 1.92 c	66.39 ± 1.62 a
	2	32.85 ± 2.07 ab	20.05 ± 0.74 ab	61.23 ± 1.76 bc
	3	35.86 ± 0.76 ab	20.63 ± 0.32 ab	57.55 ± 0.35 c
	4	34.70 ± 1.85 ab	20.60 ± 0.72 ab	59.50 ± 1.25 bc
	5	39.09 ± 4.26 a	22.52 ± 1.75 a	58.01 ± 1.85 c
B. napus	0	34.76 ± 0.74 cd	21.50 ± 0.19 b	61.88 ± 0.79 a
	1	41.69 ± 1.64 ab	24.19 ± 0.63 a	58.09 ± 0.83 b
	2	45.74 ± 2.53 a	25.73 ± 0.90 a	56.38 ± 1.14 b
	3	38.20 ± 0.51 bc	22.19 ± 0.21 b	58.09 ± 0.29 b
	4	33.88 ± 0.41 cd	20.94 ± 0.11 bc	61.81 ± 0.48 a
	5	30.83 ± 1.12 d	19.77 ± 0.49 c	64.20 ± 0.83 a

Note: Means ($n = 5$) in the same column for each plant species followed by different letters differ significantly at $p \leq 0.05$, according to one-way ANOVA.

3.3. Changes of Physiological Capacitance and Impedance

The electrophysiological parameters CP and Z could represent the cell vacuole volume and resistance of intracellular water translocation [21,22]. The values of CP of *O. violaceus* at 1, 5 h were lower than that at 0 h but higher than those at 3, 4 h (Table 5). The CP of *O. violaceus* at 4 h was lower than those at other levels. The Z value of *O. violaceus* at 4 h was clearly higher than those at 0~2 h and 5 h, those at 1, 2 and 3 h showed no clear difference and were higher than the values at 0 and 5 h. The CP value of *B. napus* at 1 h was significantly higher than those at other levels, and the value at 5 h was clearly lower than those at other levels, and there was no clear difference between the values at 3 h and 4 h. The CP of *B. napus* at 2 h was higher than that at 3 h or 4 h but lower than that at 0 h. The Z value of *B. napus* at 5 h was higher than those at other levels, but the values at 0 h and 1 h were clearly lower than those at 2, 4 and 5 h, and the value at 4 h was remarkably higher than those at 0~3 h (Table 5).

Table 5. Leaf physiological capacitance (CP, pF) and impedance (Z, MΩ) at different dehydration times.

Species	Water Loss Time (h)	Physiological Capacitance	Physiological Impedance
O. violaceus	0	165.829 ± 3.203 a	0.686 ± 0.026 c
	1	106.725 ± 0.112 bc	1.131 ± 0.088 b
	2	95.636 ± 0.694 cd	1.256 ± 0.049 b
	3	87.890 ± 7.766 d	1.328 ±0.092 ab
	4	65.463 ± 4.508 e	1.486 ± 0.086 a
	5	114.747 ± 3.202 b	0.875 ± 0.052 c
B. napus	0	278.716 ± 10.921 b	0.526 ± 0.027 de
	1	322.141 ± 10.274 a	0.412 ± 0.019 e
	2	161.440 ± 11.663 c	0.711 ± 0.032 c
	3	133.522 ± 3.965 d	0.641 ± 0.024 cd
	4	130.789 ± 1.907 d	1.017 ± 0.060 b
	5	99.254 ± 2.585 e	1.165 ± 0.082 a

Note: Means ($n = 5$) in the same column for each plant species followed by different letters differ significantly at $p \leq 0.05$, according to one-way ANOVA.

3.4. Changes of Leaf Tensity and Leaf Density

The LT values of *O. violaceus* at 1 and 2 h showed no significant difference, the value at 5 h was higher than those at 1~4 h but lower than that at 0 h. The LT of *O. violaceus* at 4 h was clearly lower than those at other levels (Figure 2A). The LT value of *B. napus* at 0 h was higher than those at 2~5 h but lower than that at 1 h, the value at 5 h was significantly lower than those at other levels (Figure 2B). The LD values of *O. violaceus* increased significantly at 3, 4, and 5 h compared to those at 0, 1 and 2 h, there was no significant difference between the values at 3, 4 and 5 h (Figure 2C). The LD values of *B. napus* had no significant change at 0~2 h. The LD value of *B. napus* at 3 h was clearly higher than those at 0~2 h and 4 h (Figure 2D).

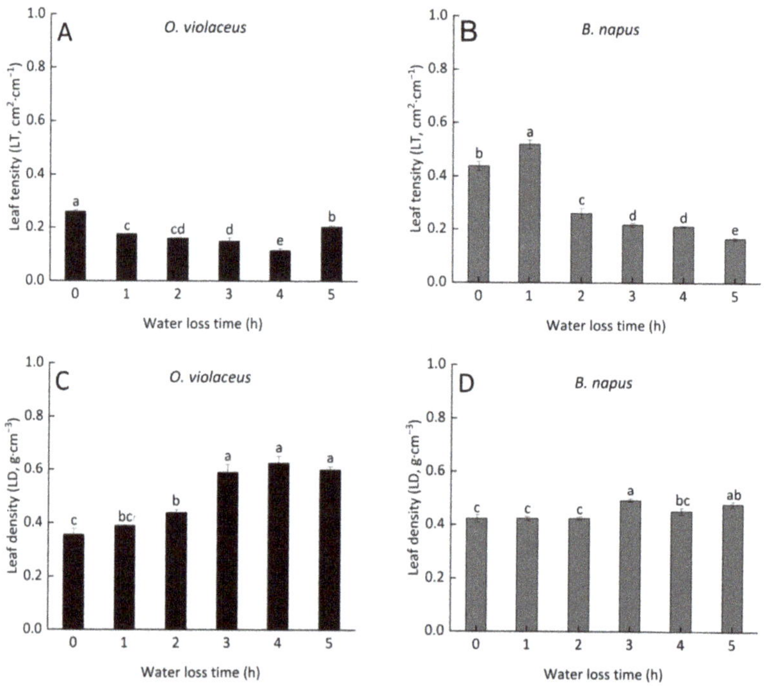

Figure 2. Leaf tensity (LT, $cm^2 \cdot cm^{-1}$) and leaf density (LD, $g \cdot cm^{-3}$) of *O. violaceus* and *B. napus* at different dehydration times. (Note: (**A**) LT of *O. violaceus*; (**B**) LT of *B. napus*; (**C**) LD of *O. violaceus*; (**D**) LD of *B. napus*. Different letters appear above the error bars of the same parameter in the same plant species when subsequent values differ significantly at $p \leq 0.05$, according to one-way ANOVA).

3.5. Chlorophyll Fluorescence Parameters

The F_v/F_m values of *O. violaceus* did not change significantly at 0~2 h, then decreased as dehydration time increased (Figure 3A). Lower F_v/F_m value of *B. napus* was associated with increasing dehydration time (Figure 3B). The ETR values of *O. violaceus* showed no clear difference at 1 and 2 h, and the values of *O. violaceus* at 3 and 4 h also showed no remarkable difference but were lower than those at 1 and 2 h. The ETR of *O. violaceus* at 0 h was remarkably higher than those at other levels, while that at 5 h was lower than the values at other levels (Figure 3C). The ETR values of *B. napus* kept stable at 2~4 h, which were clearly lower than that at 0 h but higher than that at 5 h (Figure 3D). The qP of *O. violaceus* at 3 h was higher than that at 1 h but lower than those at 0 and 2 h, and the values at 0 h and 2 h were significantly higher than those at other levels. The qP of *O. violaceus* at 5 h was clearly lower than those at other levels (Figure 3E). The qP of *B. napus* at 1~3 h showed no clear difference, they were significantly lower than the value at 0 h but higher than those at 4 h and 5 h (Figure 3F). The NPQ of *O. violaceus* increased remarkably at 1 h compared to that at 0 h, then exhibited no clear difference between the values at 1~4 h, the value at 5 h was significantly higher than those at 0~3 h, but showed no significant difference with that at 4 h (Figure 3G). The NPQ values of *B. napus* at 4 h and 5 h were significantly higher than those at 0~2 h, but showed no significant difference with that at 3 h, and that at 0 h was clearly lower than the values at 2~5 h (Figure 3H).

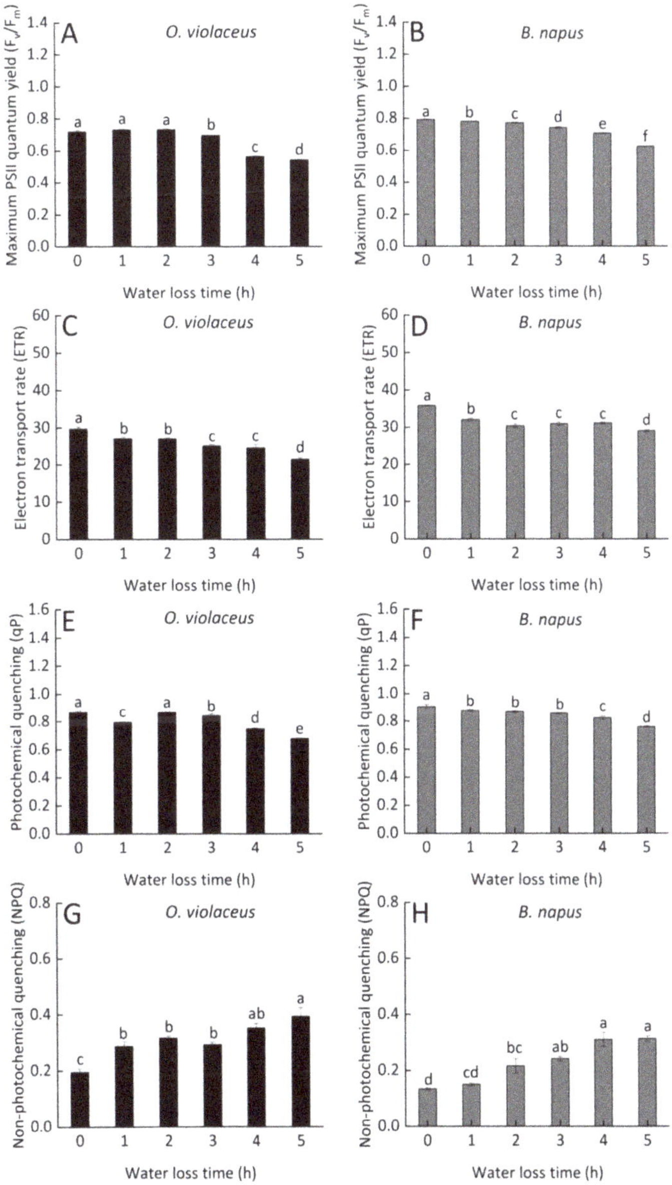

Figure 3. The F_v/F_m, ETR, qP and NPQ of *O. violaceus* and *B. napus* at different dehydration times (Note: (**A**,**C**,**E**,**G**) are the F_v/F_m, ETR, qP, NPQ of *O. violaceus*, respectively; (**B**,**D**,**F**,**H**) are the F_v/F_m, ETR, qP, NPQ of *B. napus*, respectively. Different letters appear above the error bars of the same parameter in the same plant species when subsequent values differ significantly at $p \leq 0.05$, according to one-way ANOVA).

4. Discussion

4.1. Leaf Intracellular Water Translocation vs. Anatomical Structure and Electrophysiology

The present study aimed to investigate the responses of leaf anatomical and physical traits of *O. violaceus* and *B. napus* to dehydration. Changes of palisade or spongy parenchyma in *O. violaceus* at different water loss times altered the intracellular water distribution. Water in spongy parenchyma of *O. violaceus* maintained stable at 1 h but was obviously lost at 3 h, while the water in palisade parenchyma maintained stable at 1~5 h (Figure 4). Most importantly, the electrophysiological indices, i.e., CP, Z, have been successfully used to determine the dynamic traits and metabolism of the intracellular water [16]. CP is closely related to the change of vacuole volume [21]. By analyzing variations of CP and Z of *O. violaceus*, we found that water translocation occurred within mesophyll cell or between palisade and spongy parenchyma at 1~5 h. Water was mainly translocated from spongy parenchyma into palisade parenchyma of *O. violaceus* at 3 h. Spongy parenchyma of *B. napus* shrank at 1~3 h and recovered at 5 h (Figure 4), the water was mainly translocated from spongy parenchyma into palisade parenchyma at 1~3 h, and the resistance of water translocation increased with increasing water loss times. *B. napus* leaf also exhibited lower water loss rate than *O. violaceus* during the dehydration period.

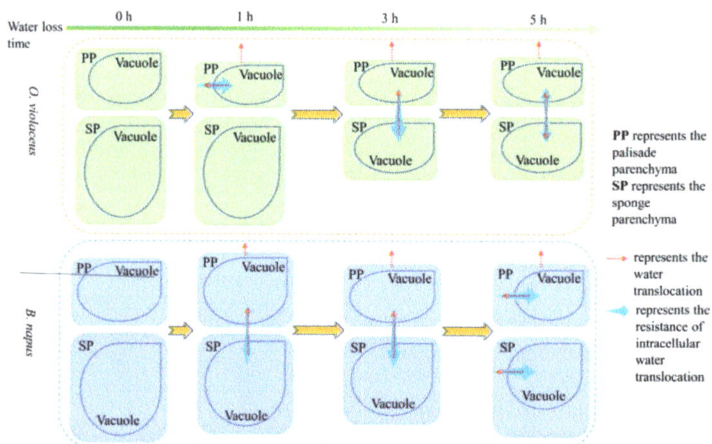

Figure 4. Leaf intracellular water translocation in *O. violaceus* and *B. napus*.

Anatomical structure, WC could just reflect the static leaf water status in a moment, and the determination of anatomical structure was destructive. However, plant electrophysiology determination was non-destructive and could be used to investigate the dynamic traits of intracellular water, which helped to study the water metabolism [23]. Therefore, it has the potential to quickly determine the dynamic adaptability of *O. violaceus* and *B. napus*.

4.2. Dynamic Leaf Water Status under Dehydration

Responses of leaf traits to dehydration differ between *O. violaceus* and *B. napus* but are all aimed at adjusting the leaf intracellular water and coping with water deficit environments [24]. At 1 h, spongy parenchyma contributed to the improvement of gas exchange [25], which kept the transpiration and water loss in *O. violaceus*. Due to the elasticity of mesophyll cells of *O. violaceus*, the palisade parenchyma was prone to shrink as further water losing [26], which, however, maintained the intracellular turgor pressure and kept the Ψ_L. Leaf Z represents the resistance to current, which is generated by the transport of dielectric materials including inorganic and organic ions. It is negatively correlated with the intracellular water transport rate [22]. Increased Z at 1 h enhanced the translocation resistance of intracellular water and maintained the intracellular substances.

Water loss at 1 h decreased the electron transport and photochemical efficiency, but did not affect the integrity of the PSII reaction center of *O. violaceus*. Mesophyll cells of *B. napus* with high stiffness helped to keep the leaf morphology and intracellular water. Increased palisade parenchyma might be attributed to the translated water from intracellular HCO_3^-, which was regulated by carbonic anhydrase in *B. napus* leaves [27]. Increased palisade parenchyma can improve the intercellular water transport efficiency and increase the mesophyll cell superficial area, therefore improving the leaf water holding capacity [28]. As a result, the intracellular water translocation resistance of *B. napus* was slightly reduced. However, PSII of *B. napus* was more sensitive to dehydration than *O. violaceus*.

At 2~3 h, mesophyll cells of *O. violaceus* with high elasticity were prone to shrink, caused by the increasing water loss. Sponge parenchyma occupies more spaces than palisade parenchyma in leaves [29]. Shrinking sponge parenchyma in *O. violaceus* leaves clearly decreased the mesophyll cell volume and reduced the transpired dissipation [25]. Meanwhile, water within the sponge parenchyma was translocated into palisade parenchyma, but the stable translocation resistance of intracellular water, which was kept by the increased CTR and slightly decreased Ψ_L, mitigated the water loss and maintained the leaf water status of *O. violaceus*. However, damage of dehydration on the PSII reaction center of *O. violaceus* became obvious at 3 h. No obvious water loss was observed due to the increased translocation resistance of intracellular water in *B. napus*. However, recovery of the cell elasticity of *B. napus* caused decrease in mesophyll cell volume but remarkably increased the LD and also maintained the Ψ_L. Meanwhile, the decreased CTR was conducive to the water movement among intercellular spaces. As a result, although the PSII reaction center of *B. napus* suffered from the damage, it could still maintain stable light transport and photochemical efficiency.

At 4~5 h, the mesophyll cells were prone to perform stiffness, and there was no obvious change in leaf anatomy of *O. violaceus*. The intra- and inter-cellular water movement declined due to the less WC and high LD, which alleviated the leaf water loss, kept the Ψ_L and reduced the translocation resistance of intracellular water. The latter one might also be attributed to the water regulation caused by carbonic anhydrase in *O. violaceus* leaves, since the carbonic anhydrase of *O. violaceus* would be activated under water deficit conditions [4]. However, water deficit at this period significantly inhibited the photochemical efficiency and damaged the PSII reaction center of *O. violaceus*. Previous studies have shown that the destruction of plant leaf epidermis leads to the reduction of photosynthesis and increase of water loss [30]. The upper epidermis of *B. napus* is thicker than that of *O. violaceus*, so it can prevent further water loss from leaves and kept the Ψ_L. The remaining intracellular water and cell elasticity recovered the volume of sponge parenchyma. The clearly increased intracellular water translocation resistance of *B. napus* indicated that the water movement within cells occurred. The photochemical efficiency was influenced by dehydration and was seriously inhibited at 4–5 h, which was attributed to the slow water loss in *B. napus* leaves.

5. Conclusions

This study explained the different translocation and utilization mechanisms of leaf intracellular water in *O. violaceus* and *B. napus* by analyzing the leaf anatomical and physical traits. Rapid water loss led to the timely shrinkage of *O. violaceus* leaves and mesophyll cells due to the better cell elasticity compared with *B. napus*, and the increased intracellular water translocation resistance helped to retain the intracellular water and maintain the turgor pressure. The water within the sponge parenchyma could also be translocated into the palisade parenchyma. Consequently, the PSII reaction center and photochemical efficiency were kept stable. Photosynthetic activity of *O. violaceus* was clearly inhibited after three hours from the onset of dehydration. Palisade parenchyma in *B. napus* leaves increased quickly to improve the intercellular water translocation due to the strong cell stiffness. Gradually increasing intracellular water translocation resistance, and the recovery of cell elasticity and thick upper epidermis, helped to slow down the leaf water loss, which, however, could not timely stop the damage on the PSII reaction center and photochemical efficiency. The photochemical efficiency was influenced by dehydration and was seriously inhibited until 4 and 5 h, which was attributed by the slow water loss in *B. napus* leaves. The response mechanism of intracellular water to dehydration can be investigated with the help of leaf electrophysiological traits, thereby providing a basis for improving the evaluating efficiency of plant drought resistance. However, the rapid and direct determination of plant drought resistance by using leaf electrophysiological information can still not be realized at present and needs further research.

Author Contributions: Conceptualization, D.X. and Y.W.; methodology, W.W., D.X. and Y.W.; validation, X.C.; resources, X.Q. and R.Y.; data curation, W.W. and X.C.; writing—original draft preparation, W.W.; writing—review and editing, D.X. and M.L.; funding acquisition, Y.W. All authors have read and agreed to the published version of the manuscript.

Funding: This research was funded by the Support Plan Projects of Science and Technology of Guizhou Province [No. (2021) YB453], the National Key Research and Development Program of China [No. 2021YFD1100300], the Priority Academic Program Development (PAPD) of Jiangsu Higher Education Institutions, and the Graduate Innovative Projects of Jiangsu Province [2014(KYLX_1061)].

Institutional Review Board Statement: Not applicable.

Informed Consent Statement: Not applicable.

Data Availability Statement: The datasets during or analyzed during the current study are available from the corresponding author upon reasonable request.

Conflicts of Interest: The authors declare no conflict of interest.

References

1. Wu, Y.Y.; Xing, D.K.; Hang, H.T.; Zhao, K. *Principles and Technology of Determination on Plant's Adaptation to Karst Environment*; Science Press: Beijing, China, 2018.
2. Wu, Y.Y.; Xu, W.X. Effect of plant growth regulators on the growth of *Orychophragmus violaceus* plantlets in vitro. *Planta Med.* **2011**, *77*, 1292–1293. [CrossRef]
3. Wang, R.; Wu, Y.Y.; Hang, H.T.; Liu, Y.; Xie, T.X.; Zhang, K.Y.; Li, H.T. *Orychophragmus violaceus* L., a marginal land-based plant for biodiesel feedstock: Heterogeneous catalysis, fuel properties, and potential. *Energy Convers. Manag.* **2014**, *84*, 497–502. [CrossRef]
4. Xing, D.k.; Chen, X.L.; Wu, Y.Y.; Xu, X.J.; Chen, Q.; Li, L.; Zhang, C. Rapid prediction of the re-watering time point of *Orychophragmus violaceus* L. based on the online monitoring of electrophysiological indexes. *Sci. Hortic.* **2019**, *256*, 108642. [CrossRef]
5. Lu, Y.; Wu, Y.Y.; Zhang, K.Y. Dose bicarbonate affect the nitrate utilization and photosynthesis of *Orychophragmus violaceus*? *Acta Geochim.* **2018**, *6*, 875–885. [CrossRef]
6. Bhardwaj, A.R.; Joshi, G.; Kukreja, B.; Malik, V.; Arora, P.; Pandey, R.; Shukla, R.N.; Bankar, K.G.; Katiyar-Agarwal, S.; Goel, S.; et al. Global insights into high temperature and drought stress regulated genes by RNA-Seq in economically important oilseed crop *Brassica juncea*. *BMC Plant Biol.* **2015**, *15*, 9. [CrossRef]
7. Vargas-Ortiz, E.; Ramírez-Tobias, H.M.; González-Escobar, J.L.; Gutiérrez-García, A.K.; Bojorquez-Velazquez, E.; Espitia-Rangel, E.; de la Rosa, A.P.B. Biomass, chlorophyll fluorescence, and osmoregulation traits let differentiation of wild and cultivated *Amaranthus* under water stress. *J. Photochem. Photobiol. B Biol.* **2021**, *220*, 112210–112220. [CrossRef]
8. Jagiełło-Kubiec, K.; Nowakowska, K.; Łukaszewska, A.J.; Pacholczak, A. Acclimation to ex vitro conditions in ninebark. *Agronomy* **2021**, *11*, 612. [CrossRef]
9. Das, K.; Roychoudhury, A. Reactive oxygen species (ROS) and response of antioxidants as ROS-scavengers during environmental stress in plants. *Front. Environ. Sci.* **2014**, *2*, 53. [CrossRef]
10. Xing, D.K.; Chen, X.L.; Wu, Y.Y.; Li, Z.Y.; Shanjida, K. Changes in elastic modulus, leaf tensity and leaf density during dehydration of detached leaves in two plant species of Moraceae. *Chil. J. Agric. Res.* **2021**, *81*, 434–447. [CrossRef]
11. Zhang, Z.; Huang, M.; Zhao, X.; Wu, L. Adjustments of leaf traits and whole plant leaf area for balancing water supply and demand in Robinia pseudoacacia under different precipitation conditions on the Loess Plateau. *Agric. For. Meteorol.* **2019**, *279*, 107733. [CrossRef]
12. Matschi, S.; Vasquez, M.F.; Bourgault, R.; Steinbach, P.; Chamness, J.; Kaczmar, N.; Gore, M.A.; Molina, I.; Smith, L.G. Structure-function analysis of the maize bulliform cell cuticle and its potential role in dehydration and leaf rolling. *Plant Direct* **2020**, *10*, e00282. [CrossRef]
13. Vastag, E.; Cocozza, C.; Orlović, S.; Kesić, L.; Kresoja, M.; Stojnić, S. Half-sib lines of pedunculate oak (*Quercus robur* L.) respond differently to drought through biometrical, anatomical and physiological traits. *Forests* **2020**, *11*, 153. [CrossRef]

14. García-Orellana, Y.; Ortuño, M.F.; Conejero, W.; Ruiz-Sánchez, M.C. Diurnal variations in water relations of deficit irrigated lemon trees during fruit growth period. *Span. J. Agric. Res.* **2013**, *11*, 137–145. [CrossRef]
15. Zhu, S.D.; Chen, Y.J.; Ye, Q.; He, P.C.; Liu, H.; Li, R.H.; Fu, P.L.; Jiang, G.F.; Cao, K.F. Leaf turgor loss point is correlated with drought tolerance and leaf carbon economics traits. *Tree Physiol.* **2018**, *5*, 658–663. [CrossRef]
16. Xing, D.K.; Chen, L.; Wu, Y.Y.; Janusz, J.Z. Leaf physiological impedance and elasticity modulus in *Orychophragmus violaceus* seedlings subjected to repeated osmotic stress. *Sci. Hortic.* **2021**, *276*, 109763. [CrossRef]
17. Bochicchio, A.; Vazzana, C.; Puliga, S.; Alberti, A.; Cinganelli, S.; Vernieri, P. Moisture content of the dried leaf is critical to desiccation tolerance in detached leaves of the resurrection plant *Boea hygroscopica*. *Plant Growth Regul.* **1998**, *24*, 163–170. [CrossRef]
18. Wang, Z.Q.; Huang, H.; Wang, H.; Peñuelas, J.; Sardans, J.; Niinemets, Ü.; Niklas, K.J.; Li, Y.; Xie, J.B.; Wright, I.J. Leaf water content contributes to global leaf trait relationships. *Nat. Commun.* **2021**, *13*, 5525. [CrossRef]
19. Li, Z.L. *Plant Section Technique*; Science Press: Beijing, China, 1978.
20. Xing, D.K.; Chen, X.L.; Wu, Y.Y.; Chen, Q.; Li, L.; Fu, W.G.; Shu, Y. Leaf stiffness of two *Moraceae* species based on leaf tensity determined by compressing different external gripping forces under dehydration stress. *J. Plant Interact.* **2019**, *14*, 610–616. [CrossRef]
21. Zhang, M.; Wu, Y.; Xing, D.; Zhao, K.; Yu, R. Rapid measurement of drought resistance in plants based electrophysiological properties. *Trans. ASABE* **2015**, *58*, 1441–1446. [CrossRef]
22. Xing, D.K.; Mao, R.L.; Li, Z.Y.; Wu, Y.Y.; Qin, X.J.; Fu, W.G. Leaf intracellular water transport rate based on physiological impedance: A possible role of leaf internal retained water in photosynthesis and growth of tomatoes. *Front. Plant Sci.* **2022**, *13*, 845628. [CrossRef]
23. Zhang, C.; Wu, Y.; Su, Y.; Xing, D.; Dai, Y.; Wu, Y.; Fang, L. A plant's electrical parameters indicate its physiological state: A study of intracellular water metabolism. *Plants* **2020**, *9*, 1256. [CrossRef] [PubMed]
24. Petrov, P.; Petrova, A.; Dimitrov, I.; Tashev, T.; Olsovska, K.; Brestic, M.; Misheva, S. Relationships between leaf morpho-anatomy, water status and cell membrane stability in leaves of wheat seedlings subjected to severe soil drought. *J. Agron. Crop Sci.* **2018**, *204*, 219–227. [CrossRef]
25. Vitalevna, P.O.; Dmitrievna, G.O.; Dmitrievich, K.S.; Vitalevna, K.O. Physiological Features of Red Currant Adaptation to Drought and High Air Temperatures. In *Drought-Detection and Solutions*; Ondrasek, G., Ed.; IntechOpen: London, UK, 2019; pp. 1–13.
26. Oliveira, I.; Meyer, A.; Afonso, S.; Gonçalves, B. Compared leaf anatomy and water relations of commercial and traditional *Prunus dulcis* (Mill.) cultivars under rain-fed conditions. *Sci. Hortic.* **2018**, *229* (Suppl. C), 226–232. [CrossRef]
27. Xing, D.K.; Xu, X.J.; Wu, Y.Y.; Liu, Y.J.; Wu, Y.S.; Ni, J.H.; Azeem, A. Leaf tensity: A method for rapid determination of water requirement information in *Brassica napus* L. *J. Plant Interact.* **2018**, *13*, 380–387. [CrossRef]
28. Li, Z.L.; Li, R.A. Anatomical observation of assimilating branches of nine xerophytes in Gansu. *Acta Bot. Sin.* **1981**, *23*, 181–185.
29. Luan, Z.H.; Shao, D.K.; Qi, Q.G.; Zhang, Q.C.; Gao, X.; Luan, J.E.; Lin, M.F.; Jiang, W.Q. Variation of leaf traits with altitude in *Lonicera caerulea* var. *edulis* (Caprifoliaceae) from Northeastern China. *Pak. J. Bot.* **2021**, *53*, 949–957. [CrossRef]
30. Chen, J.; Shen, Z.J.; Lu, W.Z.; Liu, X.; Wu, F.H.; Gao, G.F.; Liu, Y.L.; Wu, C.S.; Yan, C.L.; Fan, H.Q. Leaf miner-induced morphological, physiological and molecular changes in mangrove plant *Avicennia marina* (Forsk.) Vierh. *Tree Physiol.* **2017**, *37*, 82–97. [CrossRef]

horticulturae

MDPI

Article

Mitigation of Drought Damages by Exogenous Chitosan and Yeast Extract with Modulating the Photosynthetic Pigments, Antioxidant Defense System and Improving the Productivity of Garlic Plants

Khaled Abdelaal [1,*], Kotb A. Attia [2,*], Gniewko Niedbała [3], Tomasz Wojciechowski [3], Yaser Hafez [1], Salman Alamery [2], Talal K. Alateeq [4] and Sally A. Arafa [5]

1 Plant Pathology and Biotechnology Lab, Department of Agricultural Botany, EPCRS Excellence, Faculty of Agriculture, Kafrelsheikh University, Kafr El-Sheikh 33516, Egypt; yaserhafez@gmail.com
2 Department of Biochemistry, College of Science, King Saud University, P.O. Box 2455, Riyadh 11451, Saudi Arabia; salmanalamery@gmail.com
3 Department of Biosystems Engineering, Faculty of Environmental and Mechanical Engineering, Poznań University of Life Sciences, Wojska Polskiego 50, 60-627 Poznań, Poland; gniewko.niedbala@up.poznan.pl (G.N.); tomasz.wojciechowski@up.poznan.pl (T.W.)
4 Department of Plant Production, College of Food Science and Agriculture, King Saud University, P.O. Box 2455, Riyadh 11451, Saudi Arabia; talalalateq@gmail.com
5 Department of Agricultural Botany, College of Agriculture, Mansoura University, Mansoura 35516, Egypt; sallyarafa@gmail.com
* Correspondence: khaled.elhaies@gmail.com (K.A.); kattia@ksu.edu.sa (K.A.A.)

Citation: Abdelaal, K.; Attia, K.A.; Niedbała, G.; Wojciechowski, T.; Hafez, Y.; Alamery, S.; Alateeq, T.K.; Arafa, S.A. Mitigation of Drought Damages by Exogenous Chitosan and Yeast Extract with Modulating the Photosynthetic Pigments, Antioxidant Defense System and Improving the Productivity of Garlic Plants. *Horticulturae* **2021**, 7, 510. https://doi.org/10.3390/horticulturae7110510

Academic Editor: Yanyou Wu

Received: 12 October 2021
Accepted: 18 November 2021
Published: 19 November 2021

Abstract: Garlic is an important vegetable in terms of its economic value and also as a medicinal plant. In this study, chitosan (300 mM) and yeast extract (8 g/L) were used individually or in combination to improve the yields of garlic plants under drought conditions (i.e., 75% and 50% of the water they would normally receive from irrigation) for two seasons. Significant decreases in numbers of leaves per plant and plant height, plant dry weight, relative water content, and chlorophyll a and b concentrations were found in stressed garlic plants in both seasons. The greatest reductions in these characters were recorded in plants that received only 50% of the normal irrigation in both seasons. Levels of hydrogen peroxide, products of lipid peroxidation such as malondialdehyde, and superoxide, as well as percentages of electrolyte leakage, were elevated considerably and were signals of oxidative damage. The application of the yeast extract (8 g/L) or chitosan (300 mM) individually or in combination led to a remarkable increase in the most studied characters of the stressed garlic plants. The combination of yeast extract (8 g/L) plus chitosan (300 mM) led to increase plant height (44%), ascorbic acid levels (30.2%), and relative water content (36.8%), as well as the chlorophyll a (50.7%) and b concentrations (79%), regulated the proline content and levels of antioxidant enzymes in stressed garlic plants that received 75% of the normal irrigation, and this decreased the signs of oxidative stress (i.e., percentage of electrolyte leakage and levels of malondialdehyde, hydrogen peroxide, and superoxide).

Keywords: garlic; drought stress; chitosan; yeast; antioxidant system; reactive oxygen species

1. Introduction

Garlic (*Allium sativum* L.) is an important vegetable crop in Egypt, where the annual production during the 2018 season was 286,213 tons obtained from 315.85 ha [1]. Garlic is the second most important species of the *Allium* genus and has several constituents including phenolic compounds, saponins, organosulfur compounds, and polysaccharides [2]. Garlic bulbs also contain numerous bioactive compounds, such as alliin, allicin, diallyl disulfide, and S-allylcysteine [3]. Shang et al. [3] found that these valuable bioactive compounds are very important and play significant roles as antioxidant, antimicrobial, anti-inflammatory, and anticancer compounds. Furthermore, the volatile oil of garlic can be used as a herbicide and insecticide to improve yield production [4,5]. During germination and development plants are exposed to many stresses, such as biotic [6–9] and abiotic stresses [10–15].

Drought is a very detrimental abiotic factor that obstructs the growth and decrease the yield of many plants. It causes a decrease in morphological features such as leaf number, leaf area, and plant height [16–19]. Physiological features such as relative water content (RWC) and chlorophyll concentrations [20–22] are also significantly reduced. Drought has negative effects on biochemical characters such as enzyme activity, the production of hydrogen peroxide and superoxide, and lipid peroxidation [23–25]; and decreases the yield [22–27]. Under drought conditions, biochemical and physiological features such as proline content, levels of malondialdehyde (MDA), percentages of electrolyte leakage (EL%), levels of superoxide and hydrogen peroxide, and enzyme activities were adversely affected in plants [27]. A decrease in chlorophyll and the level of photosynthesis is a very important signal in drought stress [28,29]. It is associated with a decrease in carbon dioxide uptake, closed stomatal pores, and a reduction in the activity of enzymes of the Calvin cycle pathway [30]. Reactive oxygen species (ROS), EL%, and levels of MDA are important indicators of various stress factors [31,32]. In drought, ROS, especially superoxide and hydrogen peroxide, have accumulated in numerous species [28,29]. The extreme occurrence of these parameters could be due to damage to membranes in many plants, such as sugar beets and barley. Oxidative damage can be controlled with the up-regulation of antioxidant components, which enhances plant tolerance of stress conditions and can scavenge ROS [33]. This mechanism depends on nonenzymatic and enzymatic compounds such as ascorbic acid, catalase, and peroxidase, which decrease the damage to membranes, proteins, and DNA [18].

Chitosan is an important polysaccharide and plays a pivotal role in human life because of its biological activity and its safety in agricultural processes [27]. Application of chitosan can increase leaf numbers, plant height, and chlorophyll content during stress (predominantly drought) by enhancing the nutrient status and antioxidant system of plants [34,35]. The foliar application of chitosan led to increased yields, nutrient uptakes, and chlorophyll concentrations [36]. Ahmed [37] reported that chitosan (4 and 6 mL L^{-1}) led to improve productivity and storability of garlic plants. Additionally, Bistgani et al. [38] reported that chitosan at 400 μL L^{-1} led to improved dry weights of thyme plants that were stressed by drought.

Yeast is a natural, safe source of biofertilizer and can improve growth characters and plant yields. It contains many essential components, such as cytokines, riboflavin, thiamine, pyridoxine, and vitamins B1, B2, and B12, as well as other nutrients [39]. The valuable components of yeast, such as cytokinins, can stimulate cell division and elongation, as well as the synthesis of proteins and nucleic acids, and can increase mineral nutrients [40,41]. Application of yeast in drought conditions led to increased vegetative and yield characters of wheat plants, such as grain yields and 100-grain weights [26]. Abdelaal et al. [29] found that application of yeast led to improved plant fresh and dry weights, plant heights, and chlorophyll concentrations, as well as yields of wheat under water stress. Moreover, the application of yeast helped calendula plants to tolerate salt stress by improving their morphological, physiological, and anatomical features [14]. In a study by Haider et al. [42], yeast treatment at 6g/L resulted in maximum spike lengths, spike numbers, total phenols, prolines, and carbohydrates in wheat under drought conditions. Additionally, yeast application led to enhanced plant growth and differentiation and resulted in a remarkable increase in the numbers of stems per plant, plant height, leaf area, and chlorophyll content in potatoes [43]. Shalaby and El-Ramady [44] found that yeast extract led to improve yield, components and storability of garlic. Also, Ali [45] reported that application of yeast extract led to increase yield plant^{-1} and total yield of garlic plants.

Little information is available on the effect of chitosan and yeast on the physiological, morphological, and biochemical characters of garlic plants in drought conditions. Hence, the aim of our study is to assess the impact of chitosan and yeast individually or in combination as an environmentally friendly strategy for improving the yield of garlic plants associated with their biochemical, morphological, and physiological characters in drought conditions.

2. Materials and Methods

2.1. Experimental Site, Plant Materials, and Cultural Practices

Two field trials were conducted at a private farm in the Gharbia governorate during 2019/2020 and 2020/2021 to study the influence of chitosan (300 mM) and yeast extract (8 g/L) applied in a foliar spray on the vegetative, physiological, biochemical, and yield characters of garlic plants (*Allium sativum* L.) cv. Sids40 in drought conditions. The physio-biochemical studies were carried out at the EPCRS Excellence Center, Faculty of Agriculture, Kafrelsheikh University, Egypt. The experimental unit area was 14 m^2 and consisted of six rows, and the planting date was 25 September during both seasons. The experiment was planned in a complete randomized block design with four replicates and each replicate contain 20 plants. During soil preparation, 48 m^3 ha^{-1} of organic manure, 110 kg P_2O_5 ha^{-1}, and 150 kg sulphur ha^{-1} were added to and mixed with the soil. Garlic cloves of uniform size were sown on both sides of each row 7 cm apart. Fertilization (240 kg nitrogen and 135 kg potassium ha^{-1}) was done three times, the first time at 30 days from planting, the second time at 60 days from planting, and the third time at 90 days from planting. Chitosan was purchased from Sigma (Sigma-Aldrich, St. Louis, MO, USA), chitosan solution was prepared by dissolving 0.3 g in 0.1 N HCl and diluting with distilled water with pH adjusted at 6.5 by 0.1 NaOH. Yeast extract was prepared by inoculating 8 g of active dry yeast with 1 L of nutrient broth and incubated for 48 h. The foliar spray application of yeast extract and chitosan was done twice, the first time at 40 days from planting and the second time at 80 days from planting. The fertilization rates and other cultural practices were carried out as recommended by the Egyptian Ministry of Agriculture. Experimental soil was taken to study the physical and chemical characters according to AOAC [46] as follow: electrical conductivity, 1.7 dS m^{-1}; available nitrogen, 32.6 ppm; available potassium, 284 ppm; available phosphorus, 10.8 ppm; sand, 17.4%; silt, 34.6%; clay, 46.9%.

The foliar spray application of yeast extract and chitosan (1000 L ha^{-1}) was done twice with an apparatus from the Jining Bafang Mining Machinery Co., Ltd. (Jining Yanzhou, China). The treatments were as follows:

- The plants (control) were irrigated eight times to simulate normal conditions of rainfall (100% irrigation).
- The plants were irrigated six times to simulate 75%, or moderate, drought.
- The plants received 75% irrigation and were sprayed with yeast 8 g/L.
- The plants received 75% irrigation and were sprayed with chitosan 300 mM.
- Some plants received 75% irrigation and were sprayed with yeast 8 g/L plus chitosan 300 mM.
- The plants received four irrigations to simulate 50%, or severe, drought.
- The plants received 50% irrigation and were sprayed with yeast 8 g/L.
- The plants received 50% irrigation and were sprayed with chitosan 300 mM.
- The plants received 50% irrigation and were sprayed with yeast 8 g/L plus chitosan 300 mM.

The harvest dates were 16 April 2020, and 19 April 2021 after 200 days from sowing.

2.2. Morphological Characters

The samples were taken for morphological studies at 150 days from transplanting; Plant height (cm), leaves number $plant^{-1}$, and dry weight of plant (g) were recorded.

2.3. Physiological and Biochemical Studies

Physiological and biochemical studies were recorded in the fifth leaf at 150 days from sowing as follow:

2.3.1. Determination of Chlorophyll A, B Concentration and RWC

According to Lichtenthaler [47], samples of garlic fresh leaves were kept in solution of 80% acetone and 95% ethanol in the refrigerator. The chlorophylls concentrations were measured in extract at 663, 645 and 470 nm using a spectrophotometer. Relative water content (RWC%) was calculated as follows: RWC= (FW − DW)/(TW − DW) × 100, where FW is fresh weight, DW is dry weight and TW is turgid weight [48].

2.3.2. Determination of Proline Content

Garlic leaves (0.5 g) were placed in 3% sulphosalicylic acid and centrifuged for 20 min at 3000× *g*. 2 mL supernatant from extract was added to 2 mL ninhydrin reagent and 2 mL of glacial acetic acid. Proline was determined as mg g^{-1} FW at 520 nm using a spectrophotometer [49].

2.3.3. Assay of Electrolyte Leakage (EL%)

Ten discs of garlic leaves were placed in 25 mL deionized water and shaken for 20 h, then initial electrical conductivity was recorded. The discs were heated in a water bath at 80 °C for 1 h and shaken at 21 °C, then the final conductivity was determined. EL% was calculated as follow: initial/final conductivity × 100 [50].

2.3.4. Determination of Ascorbic Acid (AsA)

Garlic fresh leaves (500 mg) were taken to determine AsA, the samples was extracted in 10 mL trichloroacetic acid 6% (TCA) and centrifuged for 20 min at 1000× *g*, then 4 mL of the extract was mixed with 2 mL of dinitrophenyl hydrazine, then 1 drop of thiourea was added to the mixture and boiled for 15 min. The mixture was cooled to room temperature, 5 mL of H_2SO_4 80% were added to the mixture. AsA was determined in supernatant by a spectrophotometer at 530 nm as mg g^{-1} FW [51].

2.3.5. Assay of H_2O_2, O_2^- and MDA

For detecting O_2^-, garlic leaf samples were extracted in 50 mM phosphate buffer (pH 7.5) at a ratio of 1:8 (*w*/*v*) and centrifuged twice at 18,000× *g*. for 20 min. The reaction mixture for detecting O_2^- consisted of 4 mM epinephrine as an electron acceptor in 100 mM Tris-HCl buffer (pH 7.8) in the presence or absence of 2100 U/mL CuZn-SOD [52]. Absorbance was measured at 480 nm by employing Asys Expert 96 microplate spectrophotometer (Shanghai, China) supported by Kim software.

The H_2O_2 were assayed according to the method described by Yu et al. [53]. Samples of garlic leaf were extracted by homogenizing 0.5 g of garlic leaf with 3 mL of 50 mM K-phosphate buffer (pH 6.5) at 4 °C. The samples were centrifuged for 15 min at 11,500× *g*. 3 mL of supernatant was mixed with 1 mL of 0.1% TiCl4 in 20% H_2SO_4 (*v/v*), then the mixture was centrifuged at room temperature for 12 min at 11,500× *g*. The absorption of the supernatant was measured spectrophotometrically at 410 nm to determine the H_2O_2 content and expressed as arbitrary units (nmol g^{-1} fresh weight).

The lipid peroxidation was measured as malondialdehyde (MDA), a decomposed product of the peroxidized polyunsaturated fatty acid component of the membrane lipid, using thiobarbituric acid (TBA) as the reactive material following the method of Heath and Packer [54]. Garlic fresh leaves (500 mg) were homogenized in 3 mL 5% (*w*/*v*) trichloroacetic acid (TCA) and the homogenate was centrifuged at 11,500× *g* for 10 min. 1 mL supernatant was mixed with 4 mL of TBA reagent (0.5% of TBA in 20% TCA). The reaction mixture was heated at 95 °C for 30 min in a water bath and then quickly cooled in an ice bath and centrifuged at 11,500× *g* for 15 min. The absorbance of the supernatant was measured at 532 nm and was corrected for non-specific absorbance at 600 nm. The concentration of MDA was calculated as μ mol g^{-1} fresh weight.

2.3.6. Determination of Enzymes Activity

Garlic fresh leaves (500 mg) were homogenized and centrifuged (12,000× *g*) for 20 min at 4 °C, then the supernatant was used to measure the activities of total soluble enzyme using spectrophotometer. Activity of catalase (CAT) was determined at 240 nm in the supernatant based on the consumption rate of H_2O_2 as μmol min^{-1} mg protein^{-1} [55]. The reaction mixture contained 50 mM K-phosphate buffer (pH 7.0), 15 mM hydrogen peroxide and enzyme solution in a final volume of 700 μL. The reaction was initiated with enzyme extract and the activity was calculated using the extinction coefficient of 39.4 M^{-1} cm^{-1}.

SOD activity was determined as μmol min^{-1} mg protein^{-1} at 560 nm. The activity was assayed based on the competition between SOD and NBT for the production of superoxide from xanthine and xanthine oxidase interaction following Spitz and Oberley [56].

POX activity was measured as μmol min^{-1} mg protein^{-1} as described by Castillo et al. [57]. The reaction mixture contained 10 mM phosphate buffer at pH 6.1, 12 mM H_2O_2, 96 mM guaiacol and enzyme extract. The blank contained a complete reaction mixture without H_2O_2. Absorbance was recorded after 1 min at 470 nm and the activity was measured using the extinction coefficient of 26.6 mM^{-1} cm^{-1}.

2.4. *Yield Characteristics*

At harvest date (200 days), the plants were harvested for each replicate to determine total yield ha^{-1} while, total cured yield (ton ha^{-1}) was calculated after curing for 7 days. A random sample of twenty bulbs was taken from each replicate to determine bulb diameter (cm).

2.5. *Statistical Analysis*

The obtained data were statistically analyzed using ANOVA procedures using the MSTAT-C statistical software package [58]. Duncan's test was used to compare the means between treatments [59] when the difference was deemed significant ($p \leq 0.05$).

3. Results

3.1. *Effect of Yeast Extract and Chitosan on Plant Height, Number of Leaves per Plant, and Dry Weight of Plant for Garlic Plants in Drought Conditions*

Drought stress significantly ($p \leq 0.05$) decreased the number of leaves per plant, plant height, and plant dry weight of the garlic plants compared with the control plants that received normal irrigation during both seasons (Figure 1). These decreases were more pronounced in stressed garlic plants that received 50% of the normal irrigation. The application of yeast extract or chitosan individually or in combination caused a remarkable increase in the dry weight and height of plants and in the number of leaves per plant of stressed garlic plants that received 75% or 50% of the normal irrigation compared with stressed untreated garlic plants. The number of leaves per plant, plant height, and plant dry weight of garlic plants were greatly augmented by the application of yeast extract plus chitosan under drought conditions. The best results were obtained in the plants that received 75% irrigation followed by chitosan treatment during both seasons.

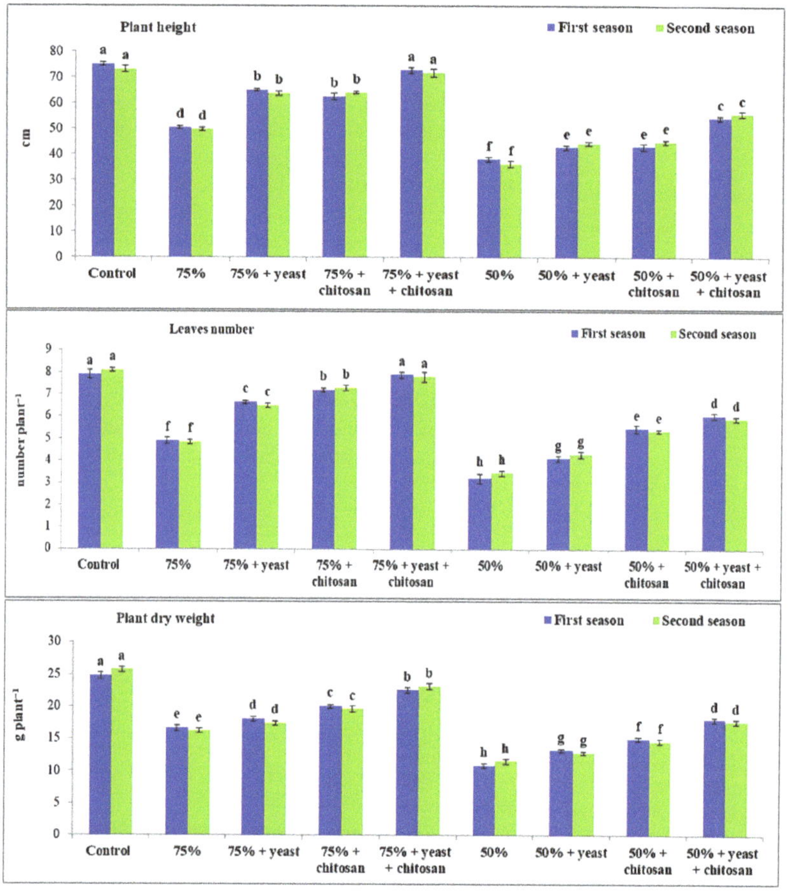

Figure 1. Effect of yeast and chitosan on plant height, leaves number, plant dry weight of garlic plants under drought during 2019/2020 and 2020/2021 seasons. The letters on the columns show significant differences between the treatments according to ANOVA, Duncan's multiple range test at 0.05 level. Data is the mean (±SE) of four replicates.

3.2. *Effect of Yeast Extract or Chitosan on Concentrations of Chlorophyll A and B and on RWC in Stressed Garlic Plants*

A significant decrease ($p \leq 0.05$) in concentrations of chlorophyll and in the RWC was recorded in stressed garlic plants (75% and 50% of normal irrigation). The lowest concentrations of chlorophyll a and b and lowest RWC were observed in stressed plants that received 50% of normal irrigation compared with the control plants and the plants that received 75% of normal irrigation (Figure 2). Spraying the stressed plants (75% and 50% of normal irrigation) with yeast extract or chitosan caused a remarkable increase ($p \leq 0.05$) in concentrations of chlorophyll a and b and in the RWC compared with stressed untreated garlic plants. The combination of yeast extract plus chitosan caused a remarkable increase in concentrations of chlorophyll a and b and in the RWC in the stressed garlic plants (75% of normal irrigation) without any significant differences compared with the control plants.

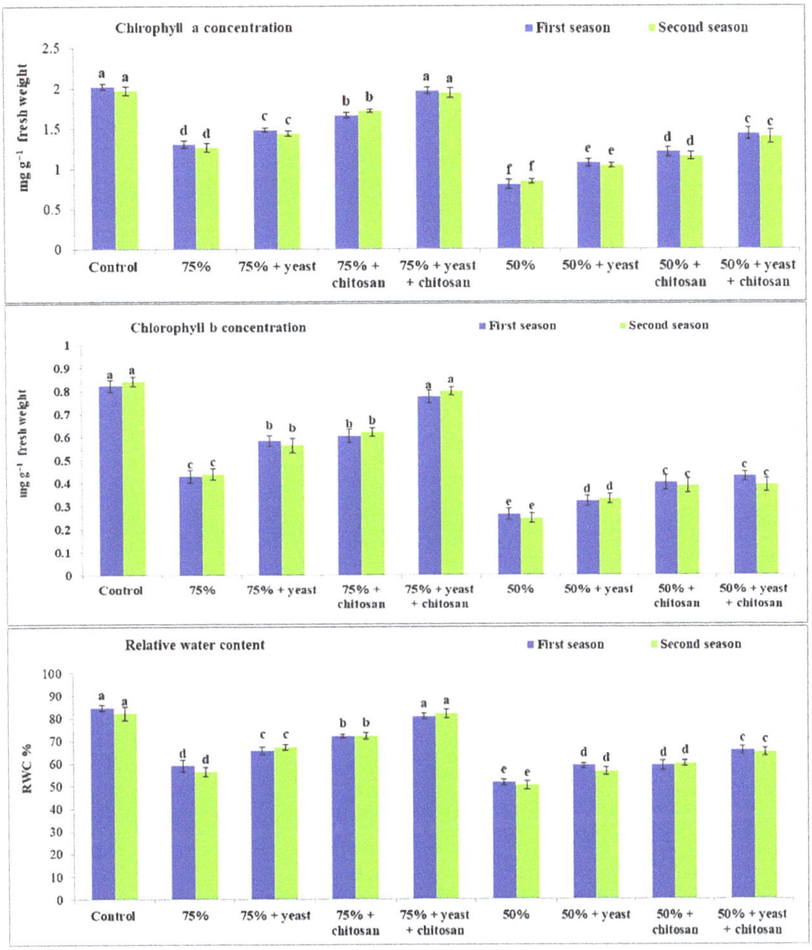

Figure 2. Effect of yeast and chitosan on chlorophyll a, chlorophyll b and relative water content of garlic plants under drought during 2019/2020 and 2020/2021 seasons. The letters on the columns show significant differences between the treatments according to ANOVA, Duncan's multiple range test at 0.05 level. Data is the mean (±SE) of four replicates.

3.3. *Effect of Yeast Extract or Chitosan on Proline Levels, Percentage of Electrolyte Leakage and Concentration of Ascorbic Acid in Garlic Plants in Drought Conditions*

Figure 3 shows that the proline content and EL% were considerably increased ($p \leq 0.05$) in garlic plants in drought conditions (75% and 50% of normal irrigation) compared with the control plants. The garlic plants that received 50% of normal irrigation had high values for proline and EL% compared with the control plants and plants that received 75% of normal irrigation. Similarly, ascorbic acid was considerably improved in stressed garlic plants in drought conditions, especially in plants that received 50% of normal irrigation in both seasons, compared with controls. The foliar application of yeast or chitosan or the combination of yeast plus chitosan led to a notable reduction in EL% in stressed garlic plants in both seasons. The application of yeast plus chitosan regulated the levels of proline in stressed garlic plants. The best results were observed with yeast plus chitosan in plants that had received 75% irrigation. Likewise, the maximum value of ascorbic acid was recorded with 50% irrigation plus yeast plus chitosan in both seasons.

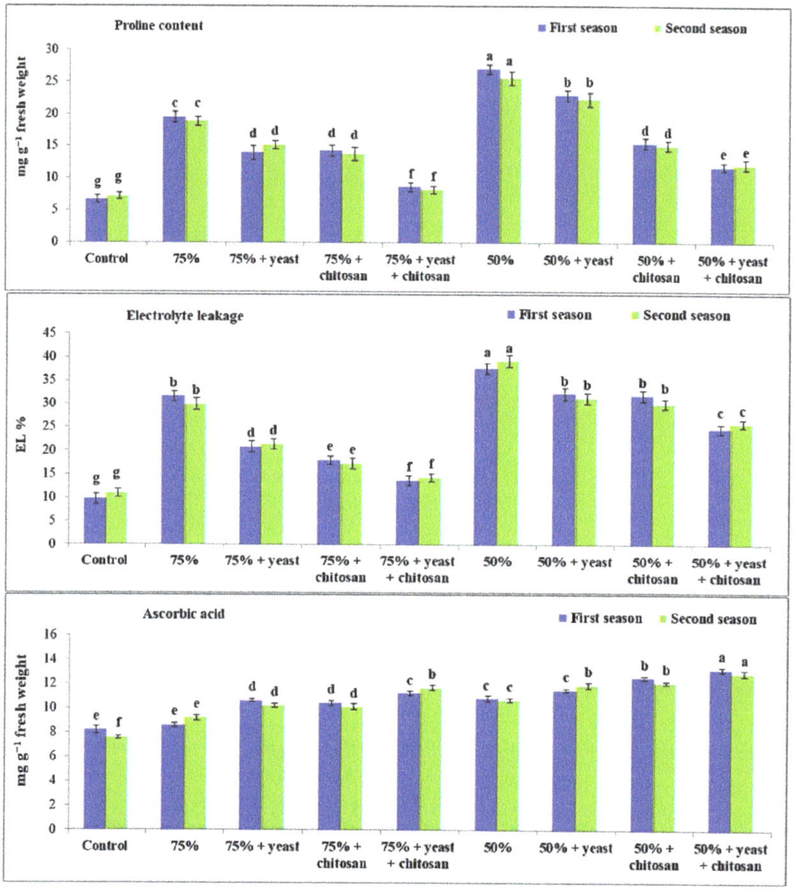

Figure 3. Effect of yeast and chitosan on proline content, electrolyte leakage and ascorbic acid of garlic plants under drought during 2019/2020 and 2020/2021 seasons. The letters on the columns show significant differences between the treatments according to ANOVA, Duncan's multiple range test at 0.05 level. Data is the mean (±SE) of four replicates.

3.4. Effect of Yeast Extract or Chitosan on Levels of Hydrogen Peroxide, Superoxide, and MDA of Garlic Plants in Drought Conditions

Hydrogen peroxide, superoxide, and MDA are very important components of garlic plants in drought conditions (Figure 4). Drought stress had a significant effect ($p \leq 0.05$) on the content of each of these components. These studied characters were significantly augmented in garlic plants in drought conditions compared with control plants in both seasons. The highest levels of all three components were found in the plants with 50% of normal irrigation during both seasons, followed by plants with 75% of normal irrigation, compared with the controls. Superoxide, hydrogen peroxide, and MDA were decreased significantly in stressed garlic plants following the application of yeast or chitosan individually or in combination. The best effects were recorded in the stressed plants with 75% irrigation plus yeast plus chitosan, without any significant differences compared with the controls.

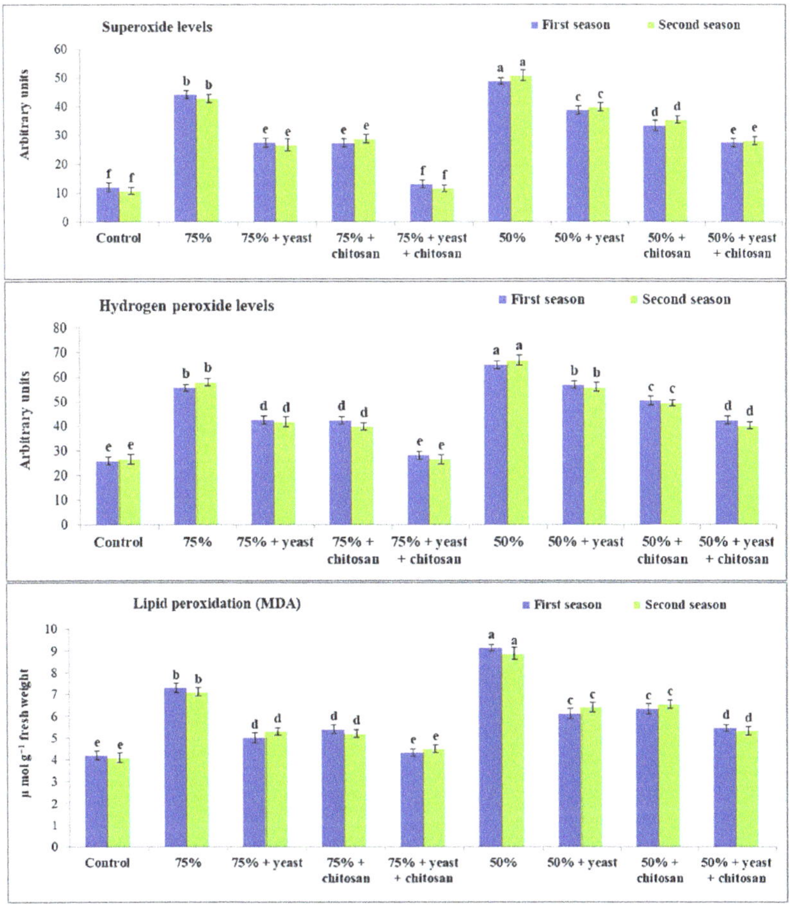

Figure 4. Effect of yeast and chitosan on hydrogen peroxide level, superoxide level and lipid peroxidation of garlic under drought during 2019/2020 and 2020/2021 seasons. The letters on the columns show significant differences between the treatments according to ANOVA, Duncan's multiple range test at 0.05 level. Data is the mean (±SE) of four replicates.

3.5. Effect of Yeast Extract or Chitosan on Catalase, Peroxidase, and Superoxide Dismutase Activity of Garlic in Drought Conditions

Drought stress induced the up-regulation of enzyme activity in garlic plants during both seasons; catalase, superoxide dismutase, and peroxidase activity significantly increased ($p \leq 0.05$) in garlic plants that received two drought treatments compared with controls (Figure 5). The maximum activities of the three substances catalase, peroxidase, and superoxide dismutase were observed in the stressed garlic plants that received 50% of normal irrigation during both seasons, followed by plants that received 75% of normal irrigation, compared with the controls. However, yeast or chitosan applied individually or in combination was effective in regulating the catalase, superoxide dismutase, and peroxidase activity for both the 75% and 50% irrigations. Of all treatments, the greatest results of catalase and peroxidase were observed during both seasons for the 75% irrigation plus yeast plus chitosan, without any significant differences with controls.

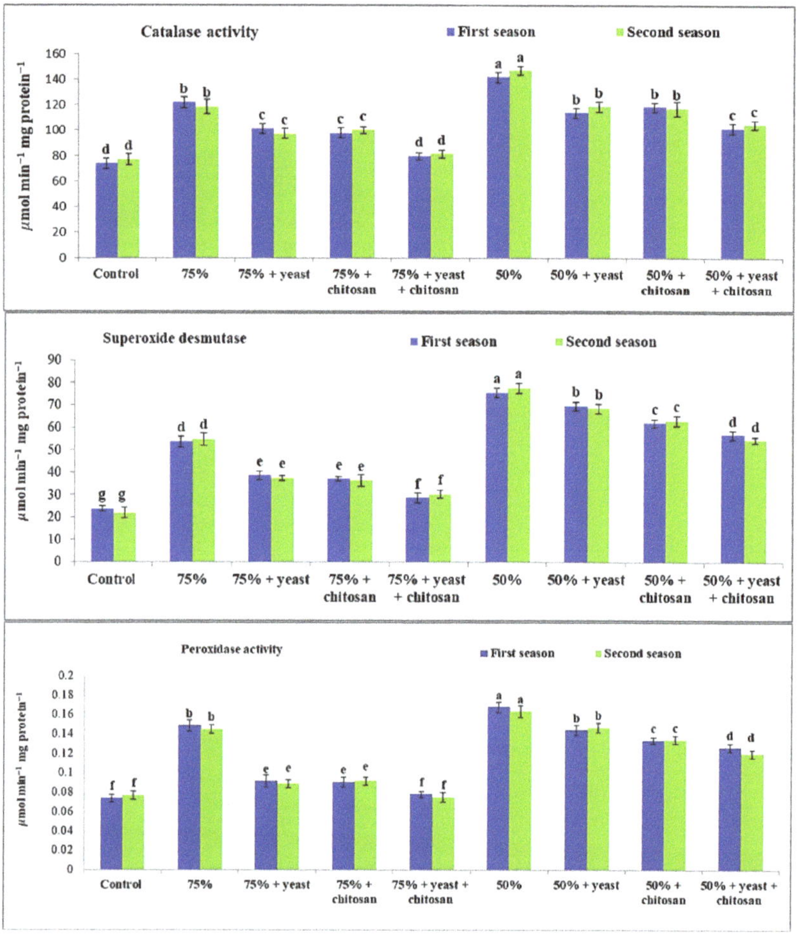

Figure 5. Effect of yeast and chitosan on catalase (CAT), superoxide dismutase (SOD) and peroxidase activity (POX) of garlic under drought during 2019/2020 and 2020/2021 seasons. The letters on the columns show significant differences between the treatments according to ANOVA, Duncan's multiple range test at 0.05 level. Data is the mean (±SE) of four replicates.

3.6. Effect of Yeast Extract or Chitosan on Bulb Diameter, Total Yield, and Total Cured Yield of Garlic Plants in Drought Conditions

Bulb diameter (cm), total yield (ton ha^{-1}), and total cured yield (ton ha^{-1}) were decreased significantly ($p \leq 0.05$) in garlic plants that received 75% or 50% of normal irrigation during both seasons compared with controls (Figure 6). The lowest values of bulb diameter, total yield, and total cured yield were recorded in the stressed garlic plants that received 50% of normal irrigation. A significant increase was noted in these yield characters when plants received foliar treatment with yeast or chitosan separately or in combination compared with stressed untreated plants during both seasons. The treatment 75% irrigation plus yeast plus chitosan had the best values for bulb diameter, total yield, and total cured yield in comparison with other treatments and without any significant differences with controls.

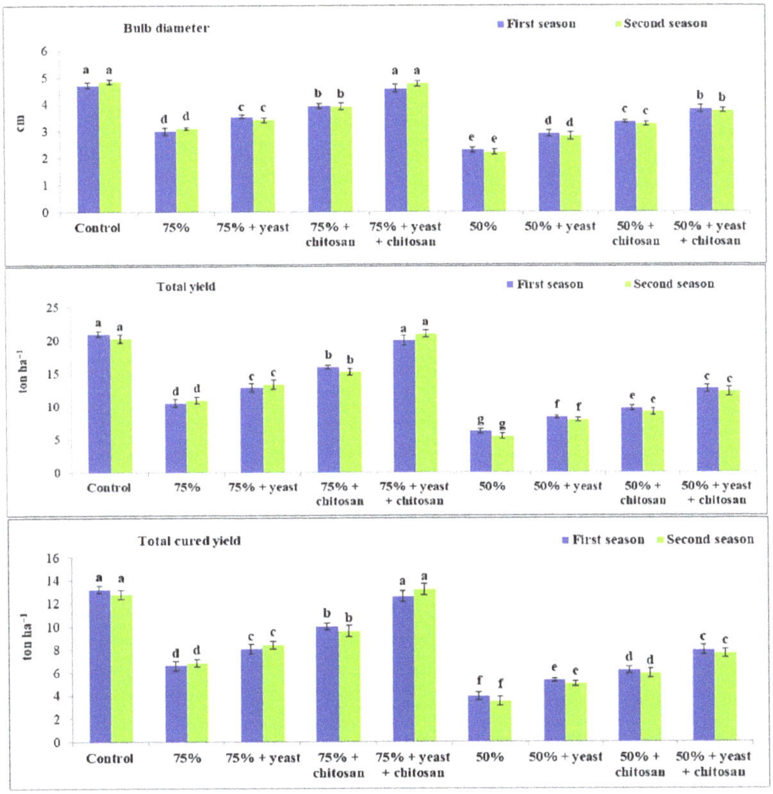

Figure 6. Effect of yeast and chitosan on bulb diameter (cm), total yield (ton ha^{-1}) and total cured yield (ton ha^{-1}) of garlic under drought during 2019/2020 and 2020/2021 seasons. The letters on the columns show significant differences between the treatments according to ANOVA, Duncan's multiple range test at 0.05 level. Data is the mean (±SE) of four replicates.

4. Discussion

It is well recognized that drought is a main factor that can harm plant production worldwide [26]. The stressed garlic plants in our study (subjected to 75% and 50% of normal irrigation) displayed a remarkable reduction in number of leaves, plant height, and plant dry weight during both seasons (Figure 1). The deleterious impact of drought on garlic plants could be due to the diminution in water absorption from soil to leaves, increased dehydration and reduced viscosity in the cells, and decreased cell division, all of which can negatively affect vegetative growth characters, especially plant height, number of leaves, and plant dry weight. The plants could adapt to drought conditions by producing fewer leaves and stomata and staying smaller in size as well as increasing the concentration of stress hormone such as ABA and salicylic acid [60]. A similar effect of drought or a water deficit was also seen in other plants such as maize [29], barley [27,28], and faba beans [33]. The application of yeast extract or chitosan individually or in combination considerably augmented the plant dry weight, plant height, and number of leaves in garlic plants in drought conditions. This increase may have been due to the synergistic role of yeast and chitosan in stimulating growth and increasing leaf numbers, root length, and plant dry weight compared with untreated plants in drought conditions. This could also be due to the fact that yeast is a biofertilizer and an essential source of many active compounds, such as vitamins, amino acids, and hormones, which induce plant growth [42,43]. These findings are in harmony with those recorded by El-Shawa et al. [14] and Abdelaal et al. [26]. Also, chitosan is a strong inducer of many secondary metabolites such as phenolic compounds in plants under stress [27,38,61]. Jasmonic acid biosynthesis plays a pivotal role in the regulation of water uptake in stressed plants. Chitosan is an anti-transpiration agent and can affect stomatal movement [62].

In the treatments in our study, concentrations of chlorophyll a and b were considerably reduced in garlic plants in both seasons (Figure 2). The reduction in chlorophylls could have been because drought injures chlorophyll pigment and causes destruction of light-harvesting protein complexes, decreases carbon dioxide fixation, and reduces $NADP^+$ production through the Calvin cycle pathway. Drought causes oxidative damage to lipids, proteins, and pigments in chloroplasts [63,64]. Our findings agreed with those of Shinde and Thakur [65], who found that drought significantly decreased chlorophyll a and b concentrations in chickpea plants [65] and barley plants [23,27,28]. Also, Gedam et al. [66] reported that, membrane stability index (MSI), RWC, total chlorophyll content and antioxidant enzyme activity as well as bulb yield were negatively affected in onion plants under drought stress.

In the current research, RWC was considerably decreased in stressed garlic plants in both seasons, and this reduction might be attributed to the detrimental impact on water absorption, conductivity, and availability in the plants. These results are in agreement with the findings of previous studies on cotton [67], *Zea mays* L. [68], and *Pisum sativum* L. [69]. The application of yeast and chitosan overcame the negative influences of drought and improved chlorophyll concentrations because yeast is a rich source of many vital components such as amino acids, which increase the chlorophyll content in garlic in drought conditions. Moreover, the positive impact of chitosan may have been due to the improvement in chloroplast numbers and chlorophyll synthesis because of an increase in potassium and nitrogen, which are essential for growth and good yields [70]. Farouk and Amany [63] and Khan et al. [71] found that chitosan can mitigate the negative effects of drought and improve chlorophyll concentrations and total carbohydrate and photosynthesis processes in maize and cowpea plants. Our findings showed a remarkable increase in EL% and levels of proline and ascorbic acid in stressed garlic plants in drought conditions compared with controls (Figure 3). This increase could be attributed to the oxidative stress experienced by plant cells in drought, which negatively affects plasma membranes and permeability. The resulting increase in proline, EL%, and ascorbic acid signals that oxidative damage is occurring. The MDA, hydrogen peroxide, and superoxide dismutase were significantly augmented in stressed garlic plants, as well, and this also signaled that oxidative damage was occurring in plants affected by drought compared with controls (Figure 4). These high levels were seen in several plants in numerous conditions of abiotic stress [32,72,73] and biotic stress [74–76]. These results are in harmony with those recorded by Hafez et al. [27], who found that the MDA levels, EL%, and ROS increased considerably in drought-stressed barley plants because of damage to plasma membranes and the cytoplasm. Abdelaal et al. [23] reported that the levels of superoxide dismutase and hydrogen peroxide, EL%, and MDA levels were considerably elevated in barley plants as a response to drought. Interestingly enough, yeast extract and chitosan treatments helped garlic plants to recover from drought stress and led to the regulation of the proline content, increased the ascorbic acids, and reduced the EL% and superoxide dismutase, hydrogen peroxide, and MDA levels compared with stressed untreated garlic plants. The helpful effect of chitosan on stressed plants could be due to its role in increasing and regulating proline as an osmolyte and very importantly, in stabilizing the plasma membrane and protein levels and scavenging of ROS under stress [23,77,78]. Chitosan is also a significant regulator of osmosis in drought stress, so the application of chitosan led to increased membrane stability and decreased lipid peroxidation in many plants [79,80]. Proline plays a central role in regulating the function of mitochondria, protecting the chloroplasts against oxidative damage, and activating the gene expression that helps plants to recover from stresses. Proline application significantly increased onion growth characters compared to untreated plants, this effect may be due to improve cell membrane stability and RWC as well as photosynthetic efficiency [81]. Also, Srmida et al. [81] stated that application of proline could mitigate drought effect by increasing sugar content and via improving plant self-defense system of onion plants.

The valuable impact of chitosan could be attributed to its role in enhancing membrane stability and decreasing the levels of superoxide dismutase, hydrogen peroxide, and MDA because of the existence of specific chitosan-like amino groups that react with ROS and produce nontoxic radicals [82]. The activities of antioxidant enzymes such as catalase, superoxide dismutase, and peroxidase were considerably augmented in drought-stressed garlic plants compared with controls in both seasons (Figure 5). These antioxidant enzymes play an important role in stressed plants to help them grow well and mitigate the oxidative damage that can occur. Because of the increase in ROS levels, these findings were recorded in many plants with different stresses [8,11,28,33]. The application of yeast or chitosan individually or in combination led to adjustments in the catalase, superoxide dismutase, and peroxidase activity, and this protected cells in drought stress compared with stressed untreated garlic plants. The best result was achieved with 75% irrigation plus yeast plus chitosan. This useful effect of chitosan could be due to the fact that chitosan decreases the transpiration rate and stimulates stomatal closure, as well as regulates the antioxidant enzymes, consequently mitigating the damaging impact of drought [83]. Also, chitosan can improve the production of important amino acids, such as aspartic acid, proline, serine, threonine, lysine, and phenylalanine, in drought [84]. Similarly, the use of yeast extract helped stressed garlic plants to recover their enzyme activity during both seasons compared with stressed untreated garlic plants. This adjustment with yeast treatments was recorded in many plants in stressful conditions [26,29,41,42]. This impact of yeast might be attributed to its role as a biostimulant and its ability to increase the hormonal activity in plants [85] and act as a nutritional increment factor, which can increase the growth and development of plants [86]. Bulb diameter, total yield, and total cured yield were considerably reduced in stressed garlic plants (Figure 6). This decrease could be due to the adverse effect of drought on morphophysiological features such as plant height, plant fresh weight, RWC, number of leaves, and chlorophyll content. These results are in agreement with the results recorded in several plants [23,30]. The application of yeast or chitosan or the combination of the two considerably increased the bulb diameter, total yield, and total cured yield in the stressed garlic plants. The effect of the yeast in this outcome could be explained by the production of many important compounds, such as amino acids, alkaloids, vitamins, and enzymes, as well as essential elements that increase the photosynthetic rate. Also, this result might have been due to the supportive effect of chitosan in inducing gene overexpression, which is involved in photosynthesis and protein and hormone metabolism, consequently improving the yield. These results are in agreement with those of Landi et al. [87] for strawberries, El-Shawa et al. [14] for calendula, Abdelaal et al. [26] for wheat, and Pongprayoon et al. [34] for rice plants. In general, the utilization of yeast extract and chitosan for increasing the yield production of garlic plants in drought has multiple advantages because these natural compounds are nontoxic, inexpensive, and environmentally friendly. The application of yeast plus chitosan significantly increased the vegetative growth and bulb yield characters, as well as improved the physiobiochemical characters of garlic in drought conditions.

5. Conclusions

Generally, we revealed that the application of yeast extract (8 g/L) plus chitosan (300 mM) individually or in combination significantly increased the growth and bulb yield of garlic plants exposed to drought conditions (75% or 50% of normal irrigation). These treatments alleviated the adverse impacts of drought, increased the number of leaves per plant, plant height, plant dry weight, chlorophyll a and b concentrations, and RWC; decreased the oxidative stress signals such as EL% and levels of superoxide, MDA, and hydrogen peroxide; as well as adjusted the production of proline, ascorbic acid, and antioxidant enzymes such as peroxidase, catalase, and superoxide dismutase.

Our findings revealed that yeast extract plus chitosan could be used as an inexpensive and nontoxic technique that is safe for the environment compared with synthetic compounds for improving the yield production of garlic plants in conditions of drought.

Author Contributions: Conceptualization, K.A., K.A.A., Y.H. and S.A.A.; methodology, K.A.A., K.A.A., Y.H. and S.A.A.; software, K.A., G.N., T.W., K.A.A., Y.H.; validation, K.A.; formal analysis, K.A., K.A.A., Y.H.; investigation, K.A., G.N., T.W.; resources, K.A., G.N., T.W. and S.A.A.; data curation, K.A., K.A.A., Y.H. and S.A.A.; writing—original draft preparation, K.A., Y.H. and S.A.A.; writing—review and editing, K.A., G.N., T.W., K.A.A., Y.H. and S.A.A.; visualization, K.A., G.N., T.W. and S.A.A. supervision, K.A., G.N., T.W. and S.A.A.; funding acquisition, K.A., K.A.A., Y.H., S.A., T.K.A. and S.A.A. All authors have read and agreed to the published version of the manuscript.

Funding: This research was funded by Researchers Supporting Project number (RSP-2021/241), King Saud University, Riyadh, Saudi Arabia.

Institutional Review Board Statement: Not applicable.

Informed Consent Statement: Not applicable.

Acknowledgments: The authors extend their appreciation to Researchers Supporting Project number (RSP-2021/241), King Saud University, Riyadh, Saudi Arabia. Also many thanks to all members of Plant Pathology and Biotechnology Lab. and EPCRS Excellence Centre (Certified according to ISO/17025, ISO/9001, ISO/14001 and OHSAS/18001), Dept. of Agric. Botany, Fac. of Agric.; Kafrelsheikh University, Kafr-Elsheikh, Egypt.

Conflicts of Interest: The authors declare no conflict of interest.

References

1. FAO. *Retrieved December 2020 from the FAOSTAT on the World Wide*; Food and Agriculture Organization of the United Nations: Rome, Italy, 2018; Available online: http://www.fao.org/faostat/en/#data/QC (accessed on 25 August 2021).
2. Abdelrahman, M.; Hirata, S.; Mukae, T.; Yamada, T.; Sawada, Y.; El-Syaed, M.; Yutaka, Y.; Sato, M.; Hirai, M.; Shigyo, M. Comprehensive Metabolite Profiling in Genetic Resources of Garlic (*Allium sativum* L.) Collected from Different Geographical Regions. *Molecules* **2021**, *26*, 1415. [CrossRef]
3. Shang, A.; Cao, S.Y.; Xu, X.Y.; Gan, R.Y.; Tang, G.Y.; Corke, H.; Li, H.B. Bioactive compounds and biological functions of garlic (*Allium sativum* L.). *Foods* **2019**, *8*, 246. [CrossRef]
4. Ismail, M.S.M.; Ghallab, M.M.A.; Soliman, M.F.M.; AboGhalia, A.H. Acaricidal activities of some essential and fixed oils on the two-spotted spider mite, *Tetranychus urticae*. *Egypt. Acad. J. Biol. Sci.* **2011**, *3*, 41–48. [CrossRef]
5. Sharaby, A.; Montasser, S.A.; Mahmoud, Y.A.; Ibrahim, S.A. Natural plant essential oils for controlling the grasshopper (*Heteracris littoralis*) and their pathological effects on the alimentary canal. *Ecol. Balk.* **2012**, *4*, 39–52.
6. Omara, R.I.; El-Kot, G.A.; Fadel, F.M.; Abdelaal, K.A.; Saleh, E.M. Efficacy of Certain Bioagents on Patho-Physiological Characters of Wheat Plants under Wheat Leaf Rust Stress. *Physiol. Mol. Plant Pathol.* **2019**, *106*, 102–108. [CrossRef]
7. Hafez, Y.M.; Abdelaal, K.A. Investigation of susceptibility and resistance mechanisms of some Egyptian wheat cultivars (*Triticum aestivum* L.) inoculated with *Blumeria graminis* f.sp. *tritici* using certain biochemical, molecular characterization and SEM. *J. Plant Prot. Pathol. Mansoura Univ.* **2015**, *6*, 431–454. [CrossRef]
8. Abdelaal, K.A.; Hafez, Y.M.; Badr, M.M.; Youseef, W.A.; Esmaeil, S.M. Biochemical, histological and molecular changes in susceptible and resistant wheat cultivars inoculated with stripe rust fungus *Puccinia striiformis* f. sp. *tritici*. *Egypt J. Biol. Pest. Control* **2019**, *24*, 421–429.
9. Abdelaal, K.A.A.; Essawy, M.; Quraytam, A.; Abdallah, F.; Mostafa, H.; Shoueir, K.; Fouad, H.; Fahmy, A.S.H.; Hafez, Y.M. Toxicity of Essential Oils Nanoemulsion against Aphis Craccivora and Their Inhibitory Activity on Insect Enzymes. *Processes* **2021**, *9*, 624. [CrossRef]
10. El-Flaah, R.F.; El-Said, R.A.R.; Nassar, M.A.; Hassan, M.; Abdelaal, K.A.A. Effect of rhizobium, nano silica and ascorbic acid on morpho-physiological characters and gene expression of POD and PPO in faba bean (*Vicia faba* L.) under salinity stress conditions. *Fresenius Environ. Bull.* **2021**, *30*, 5751–5764.
11. Abdelaal, K.A.A.; EL-Maghraby, L.M.; Elansary, H.; Hafez, Y.M.; Ibrahim, E.I.; El-Banna, M.; El-Esawi, M.; Elkelish, A. Treatment of Sweet Pepper with Stress Tolerance-Inducing Compounds Alleviates Salinity Stress Oxidative Damage by Mediating the Physio-Biochemical Activities and Antioxidant Systems. *Agronomy* **2020**, *10*, 26. [CrossRef]
12. El-Banna, M.F.; Abdelaal, K.A.A. Response of Strawberry Plants Grown in the Hydroponic System to Pretreatment with H2O2 Before Exposure to Salinity Stress. *J. Plant Prod. Mansoura Univ.* **2018**, *9*, 989–1001. [CrossRef]
13. Zhou, X.; Condori-Apfata, J.A.; Liu, X.; Condori-Pacsi, S.J.; Valencia, M.V.; Zhang, C. Transcriptomic Changes Induced by Drought Stress in Hardneck Garlic during the Bolting/Bulbing Stage. *Agronomy* **2021**, *11*, 246. [CrossRef]

14. El-Shawa, G.M.R.; Rashwan, E.M.; Abdelaal, K.A.A. Mitigating salt stress effects by exogenous application of proline and yeast extract on morphophysiological, biochemical and anatomical characters of calendula plants. *Sci. J. Flowers Ornam. Plants* **2020**, *7*, 461–482. [CrossRef]
15. Hasan, M.K.; El Sabagh, A.; Sikdar, M.S.I.; Alam, M.J.; Ratnasekera, D.; Barutcular, C.; Abdelaal, K.A.A.; Islam, M.S. Comparative adaptable agronomic traits of Blackgram and mungbean for saline lands. *Plant Arch.* **2017**, *17*, 589–593.
16. ALKahtani, M.D.F.; Attia, K.A.; Hafez, Y.M.; Khan, N.; Eid, A.M.; Ali, M.A.M.; Abdelaal, K.A.A. Chlorophyll Fluorescence Parameters and Antioxidant Defense System Can Display Salt Tolerance of Salt Acclimated Sweet Pepper Plants Treated with Chitosan and Plant Growth Promoting Rhizobacteria. *Agronomy* **2020**, *10*, 1180. [CrossRef]
17. Hafez, Y.; Elkohby, W.; Mazrou, Y.S.A.; Ghazy, M.; Elgamal, A.; Abdelaal, K.A. Alleviating the detrimental impacts of salt stress on morpho-hpysiological and yield characters of rice plants (*Oryza sativa* L.) using actosol, Nano-Zn and Nano-Si. *Fresenius Environ. Bull.* **2020**, *29*, 6882–6897.
18. Abdelaal, K.A.A.; AlKahtani, M.D.F.; Attia, K.; Hafez, Y.; Király, L.; Künstler, A. The pivotal role of plant growth promoting bacteria in alleviating the adverse effects of drought and facilitating sustainable agriculture. *Biology* **2021**, *10*, 520. [CrossRef]
19. Marostica, T.F.; Cazarolli1, L.H.; Moura, G.S.; Luz, V.C.D.; Guimarães, E.A.C.M.; Cargnelutti, D. Does *Allium sativum* L. tolerate water deficit? *Sci. Elec. Arch.* **2019**, *12*, 43–51. [CrossRef]
20. Ragab, A.Y.; Geries, L.S.M.; Abdelaal, K.A.A.; Hanna, S.A. Growth and productivity of onion plant (*Allium cepa* L.) as affected by transplanting method and NPK fertilization. *Fresenius Environ. Bull.* **2019**, *28*, 7777–7786.
21. AlKahtani, M.D.F.; Hafez, Y.M.; Attia, K.; Rashwan, E.; Husnain, L.A.; AlGwaiz, H.I.M.; Abdelaal, K.A.A. Evaluation of Silicon and Proline Application on the Oxidative Machinery in Drought-Stressed Sugar Beet. *Antioxidants* **2021**, *10*, 398. [CrossRef]
22. Rashwan, E.; Alsohim, A.S.; El-Gammaal, A.; Hafez, Y.; Abdelaal, K.A.A. Foliar application of nano zink-oxide can alleviate the harmful effects of water deficit on some flax cultivars under drought conditions. *Fresenius Environ. Bull.* **2020**, *29*, 8889–8904.
23. Abdelaal, K.A.A.; Attia, K.A.; Alamery, S.F.; El-Afry, M.M.; Ghazy, A.I.; Tantawy, D.S.; Al-Doss, A.A.; El-Shawy, E.S.E.; Abu-Elsaoud, A.M.; Hafez, Y.M. Exogenous Application of Proline and Salicylic Acid can Mitigate the Injurious Impacts of Drought Stress on Barley Plants Associated with Physiological and Histological Characters. *Sustainability* **2020**, *12*, 1736. [CrossRef]
24. Zahedi, S.M.; Moharrami, F.; Sarikhani, S.; Padervand, M. Selenium and silica nanostructure-based recovery of strawberry plants subjected to drought stress. *Sci. Rep.* **2020**, *10*, 17672. [CrossRef] [PubMed]
25. Abdelaal, K.A.A.; Rashed, S.H.; Ragab, A.; Hossian, A.; El Sabagh, A. Yield and quality of two sugar beet (*Beta vulgaris* L. ssp. *vulgaris* var. altissima Doll) cultivars are influenced by foliar application of salicylic Acid, irrigation timing, and planting density. *Acta Agric. Slov.* **2020**, *115*, 239–248. [CrossRef]
26. Abdelaal, K.; Elafry, M.; Abdel-Latif, I.; Elshamy, R.; Hassan, M.; Hafez, Y. Pivotal role of yeast and ascorbic acid in improvement the morpho-physiological characters of two wheat cultivars under water deficit stress in calcareous soil. *Fresenius Environ. Bull.* **2021**, *30*, 2554–2565.
27. Hafez, Y.M.; Attia, K.A.; Alamery, S.; Ghazy, A.; Al-Dosse, A.; Ibrahim, E.; Rashwan, E.; El-Maghraby, L.; Awad, A.; Abdelaal, K.A.A. Beneficial Effects of Biochar and Chitosan on Antioxidative Capacity, Osmolytes Accumulation, and Anatomical Characters of Water-Stressed Barley Plants. *Agronomy* **2020**, *10*, 630. [CrossRef]
28. Abdelaal, K.A.A.; Hafez, Y.M.; El-Afry, M.M.; Tantawy, D.S.; Alshaal, T. Effect of some osmoregulators on photosynthesis, lipid peroxidation, antioxidative capacity and productivity of barley (*Hordeum vulgare* L.) under water deficit stress. *Environ. Sci. Pollut. Res.* **2018**, *25*, 30199–30211. [CrossRef]
29. Abdelaal, K.A.A.; Hafez, Y.M.; El Sabagh, A.; Saneok, H. Ameliorative effects of Abscisic acid and yeast on morpho-physiological and yield characteristics of maize plant (*Zea mays* L.) under water deficit conditions. *Fresenius Environ. Bull.* **2017**, *26*, 7372–7383.
30. Farooq, M.; Wahid, A.; Kobayashi, N.; Fujita, D.; Basra, S.M.A. Plant drought stress: Effects, mechanisms and management. *Agron. Sustain. Dev.* **2009**, *29*, 185–212. [CrossRef]
31. Omara, R.I.; Abdelaal, K.A.A. Biochemical, histopathological and genetic analysis associated with leaf rust infection in wheat plants (*Triticum aestivum* L). *Physiol. Mol. Plant. Pathol.* **2018**, *104*, 48–57. [CrossRef]
32. Abdelaal, K.A.A.; Mazrou, Y.S.A.; Hafez, Y.M. Silicon Foliar Application Mitigates Salt Stress in Sweet Pepper Plants by Enhancing Water Status, Photosynthesis, Antioxidant Enzyme Activity and Fruit Yield. *Plants* **2020**, *9*, 733. [CrossRef]
33. Abdelaal, K.A.A.; El-Afry, M.; Metwaly, M.; Zidan, M.; Rashwan, E. Salt tolerance activation in faba bean plants using proline and salicylic acid associated with physio-biochemical and yield characters improvement. *Fresenius Environ. Bull.* **2021**, *30*, 3175–3186.
34. Pongprayoon, W.; Roytrakul, S.; Pichayagkura, R.; Chadchawan, S. The role of hydrogen peroxide in chitosan-induced resistance to osmotic stress in rice (*Oryza sativa* L.). *Plant Growth Regul.* **2013**, *70*, 159–173. [CrossRef]
35. Monirul, I.M.; Humayun, K.M.; Mamun, A.N.K.; Monirul, I.; Pronabananda, D. Studies on yield and yield attributes in tomato and chilli using foliar application of oligo-chitosan. *GSC Biol. Pharm. Sci.* **2018**, *3*, 20–28.
36. Ahmed, K.B.M.; Khan, M.M.A.; Siddiqui, H.; Jahan, A. Chitosan and its oligosaccharides, a promising option for sustainable crop production—A review. *Carbohydr. Polym.* **2020**, *227*, 115331. [CrossRef] [PubMed]
37. Zeng, D.; Luo, X.; Tu, R. Application of bioactive coatings based on chitosan for soybean seed protection. *Int. J. Carbohydr. Chem.* **2012**, *2012*, 104565. [CrossRef]
38. Bistgani, Z.E.; Siadat, S.A.; Bakhshandeh, A.; Pirbalouti, A.G.; Hashemi, M. Morpho-physiological and phytochemical traits of (*Thymus daenensis* Celak.) in response to deficit irrigation and chitosan application. *Acta Physiol. Plant.* **2017**, *39*, 231. [CrossRef]

39. Barnett, J.A.; Payne, R.W.; Yarrow, D. (Eds.) *Yeasts: Characterisation and Identification*, 3rd ed.; Cambridge University Press: Cambridge, MA, USA, 1990. [CrossRef]
40. Matter, F.M.A.; Abou-Sreea, A.I.B. Influence of application methods of bio-fertilization on morhological growth characters, seed yield and chemical composition of fenugreek plants. *Egypt. J. Hort.* **2016**, *43*, 19–33.
41. Nagodowithana, W.T. *Yeast Technology*; Van Nostrsnd Reinhold: New York, NY, USA, 1991; 273p.
42. Haider, I.; Raza, M.; Sammar, A.; Iqbal, R.; Ahmad, S.; Aslam, M.U.; Israr, M.; Riaz, U.; Sarfraz, M.; Abbas, N.; et al. Alleviating the Drought Stress in Wheat (*Triticum aestivum* L.) by Foliar Application of Amino Acid and Yeast. *Pak. J. Agric. Res.* **2021**, *34*, 239–246. [CrossRef]
43. Sarhan, T.; Abdullah, O.K. Effect of Azotobacter inoculation. Dry bread yeast suspension varying levels of urea on growth of potato Cv. Desiree. In Proceedings of the Tropentag World Food System—A Contribution from Europe, Zurich, Switzerland, 14–16 September 2010.
44. Shalaby, T.A.; El-Ramady, H. Effect of foliar application of bio-stimulants on growth, yield, components, and storability of garlic (*Allium sativum* L.). *Aust. J. Crop Sci.* **2014**, *8*, 271–275.
45. Ali, M.A.M. Effect of some Bio-stimulants on Growth, Yield and Bulb Quality of Garlic Grown in Newly Reclaimed Soil, New Valley-Egypt. *J. Plant Prod. Mansoura Univ.* **2017**, *8*, 1285–1294. [CrossRef]
46. Association of Official Analytical Chemists (A.O.A.C.). *Official Methods of Analysis*, 26th ed.; A.O.A.C. International: Washington, DC, USA, 2005.
47. Lichtenthaler, H.K. Chlorophylls and Carotenoids: Pigments of Photosynthetic Biomembranes. *Methods Enzymol.* **1987**, *148*, 350–382.
48. Sanchez, F.J.; de Andrés, E.F.; Tenorio, J.L.; Ayerbe, L. Growth of epicotyls, turgor maintenance and osmotic adjustment in pea plants (*Pisum sativum* L.) subjected to water stress. *Field Crop. Res.* **2004**, *86*, 81–90. [CrossRef]
49. Bates, L.S.; Waldren, R.P.; Teare, I.D. Rapid determination of free proline for water-stress studies. *Plant Soil* **1973**, *39*, 205–207. [CrossRef]
50. Szalai, G.; Janda, T.; Padi, E.; Szigeti, Z. Role of light in post-chilling symptoms in maize. *J. Plant Physiol.* **1996**, *148*, 378–383. [CrossRef]
51. Mukherjee, S.P.; Choudhuri, M.A. Implications of water stress-induced changes in the levels of endogenous ascorbic acid and hydrogen peroxide in *Vigna* seedlings. *Physiol. Plant* **1983**, *58*, 166–170. [CrossRef]
52. Yesbergenova, Z.; Yang, G.; Oron, E.; Soffer, D.; Fluhr, R.; Sagi, M. The plant Mo-hydroxylases aldehyde oxidase and xanthine dehy-drogenase have distinct reactive oxygen species signatures and are induced by drought and abscisic acid. *Plant J.* **2005**, *42*, 862–876. [CrossRef]
53. Yu, C.W.; Murphy, T.M.; Lin, C.H. Hydrogen peroxide-induces chilling tolerance in mung beans mediated through ABA independent glutathione accumulation. *Funct. Plant Biol.* **2003**, *30*, 955–963. [CrossRef] [PubMed]
54. Heath, R.L.; Packer, L. Photoperoxidation in isolated chloroplast. I. Kinetics and stoichiometry of fatty acid peroxidation. *Arch. Biochem. Biophys.* **1968**, *125*, 189–198. [CrossRef]
55. Havir, E.A.; McHale, N.A. Biochemical and developmental characterization of multiple forms of catalase in tobacco leaves. *Plant Physiol.* **1987**, *84*, 450–455. [CrossRef]
56. Spitz, D.R.; Oberley, L.W. An assay for superoxide dismutase activity in mammalian tissue homogenates. *Anal. Biochem.* **1989**, *179*, 8–18. [CrossRef]
57. Castillo, F.I.; Penel, I.; Greppin, H. Peroxidase release induced by ozone in sedum album leaves. *Plant Physiol.* **1984**, *74*, 846–851. [CrossRef]
58. Gomez, K.A.; Gomez, A.A. *Statistical Procedures for Agricultural Research*, 2nd ed.; Wiley Inter Science: New York, NY, USA, 1984; 690p.
59. Duncan, B.D. Multiple ranges and multiple F-test. *Biometrics* **1955**, *11*, 1–42. [CrossRef]
60. Liu, F.; Christian, R.; Shahanzari, J.A.; Andersen, M.N.; Jacobsen, E.E. ABA regulated stomata control and photosynthetic water use efficiency of potato (*Solanum tuberosum* L.) during progressive soil drying. *Plant Sci.* **2005**, *168*, 831–836. [CrossRef]
61. Zhao, J.; Davis, L.C.; Verpoorte, R. Elicitor signal transduction leading to production of plant secondary metabolites. *Biotechnol. Adv.* **2005**, *23*, 283–333. [CrossRef] [PubMed]
62. Abu-Muriefah, S.S. Effect of chitosan on common bean (*Phaseolus vulgaris* L.) plants grown under water stress conditions. *Int. Res. J. Agric. Sci. Soil Sci.* **2013**, *3*, 192–199.
63. Farouk, S.; Amany, A.R. Improving growth and yield of cowpea by foliar application of chitosan under water stress. *Egypt. J. Biol.* **2012**, *14*, 14–16. [CrossRef]
64. Lai, Q.; Zhi-yi, B.; Zhu-Jun, Z.; Qiong-Qiu, Q.; Bi-Zeng, M. Effects of osmotic stress on antioxidant enzymes activities in leaf discs of PSAG12-IPT modified gerbera. *J. Zhejiang Univ. Sci.* **2007**, *8*, 458–464. [CrossRef]
65. Shinde, B.P.; Thakur, J. Influence of *Arbuscular mycorrhizal* fungi on chlorophyll, proteins, proline and total carbohydrates content of the pea plant under water stress condition. *Int. J. Curr. Microbiol. Appl. Sci.* **2015**, *4*, 809–821.
66. Gedam, P.A.; Thangasamy, A.; Shirsa, D.V.; Ghosh, S.; Bhagat, K.P.; Sogam, O.A.; Gupta, A.J.; Mahajan, V.; Soumia, P.S.; Salunkhe, V.N.; et al. Screening of Onion (*Allium cepa* L.) Genotypes for Drought Tolerance Using Physiological and Yield Based Indices Through Multivariate Analysis. *Front. Plant Sci.* **2021**, *12*, 600371. [CrossRef]

67. Massacci, A.; Nabiev, S.M.; Pietrosanti, L.; Nematov, S.K.; Chernikova, T.N.; Thor, K.; Leipner, J. Response of the photosynthetic apparatus of cotton (*Gossypium hirsutum*) to the onset of drought stress under field conditions studied by gas-exchange analysis and chlorophyll fluorescence imaging. *Plant Physiol. Biochem.* **2008**, *46*, 189–195. [CrossRef]
68. Nayyar, H.; Gupta, D. Differential sensitivity of C3 and C4 plants to water deficit stress: Association with oxidative stress and antioxidants. *Environ. Exp. Bot.* **2006**, *58*, 106–113. [CrossRef]
69. AL-Quraan, N.A.; Al-Ajlouni, Z.I.; Qawasma, N.F. Physiological and Biochemical Characterization of the GABA Shunt Pathway in Pea (*Pisum sativum* L.) Seedlings under Drought Stress. *Horticulturae* **2021**, *7*, 125. [CrossRef]
70. Possingham, J.V. Plastid replication and development in the life cycle of higher plants. *Annu. Rev. Plant Physiol.* **1980**, *31*, 113–129. [CrossRef]
71. Khan, W.M.; Prithiviraj, B.; Smiyh, D.L. Effect of foliar application of chitin oligosaccharides on photosynthesis of maize and soybean. *Photosynthetica* **2002**, *40*, 87. [CrossRef]
72. Gupta, A.; Medina-Rico, A.; Delgado-Cano, A. The physiology of plant responses to drought. *Science* **2020**, *368*, 266–269. [CrossRef]
73. Ghaffaria, H.; Tadayona, M.R.; Bahadora, M.; Razmjoo, J. Investigation of the proline role in controlling traits related to sugar and root yield of sugar beet under water deficit conditions. *Agric. Water Manag.* **2021**, *243*, 106448. [CrossRef]
74. Omar, A.; Zayed, B.; Abdel Salam, A.; Hafez, Y.M.; Abdelaal, K.A.A. Folic acid as foliar application can improve growth and yield characters of rice plants under irrigation with drainage water. *Fresenius Environ. Bull.* **2020**, *29*, 9420–9428.
75. Shahin, A.; Esmaeil, R.A.; Badr, M.; Abdelaal, K.A.A.; Hassan, F.A.S.; Hafez, Y.M. Phenotypic characterization of race-specific and slow rusting resistance to stem rust disease in promising wheat genotypes. *Fresenius Environ. Bull.* **2021**, *30*, 6223–6236.
76. Esmail, S.M.; Omara, R.I.; Abdelaal, K.A.; Hafez, M. Histological and biochemical aspects of compatible and incompatible wheat-Puccinia striiformis interactions. *Physiol. Mol. Plant Pathol.* **2019**, *106*, 120–128. [CrossRef]
77. Rocychoudhury, A.; Banerjee, A. Endogenous glycine betaine accumulation mediates abiotic stress tolerance in plants. *Trop. Plant Res.* **2016**, *3*, 105–111.
78. Kheradmand, M.A.; Shahmoradzadeh, F.S.; Fatahi, E.; Raoofi, M.M. Effect of water stress on oil yield and some characteristics of *Brassica napus*. *Int. Res. J. Appl. Basic Sci.* **2014**, *8*, 1447–1453.
79. Yang, F.; Hu, J.; Li, J.; Wu, X.; Qian, Y. Chitosan enhances leaf membrane stability and antioxidant enzyme activities in apple seedlings under drought stress. *Plant Growth Regul.* **2009**, *58*, 131–136. [CrossRef]
80. Jiao, Z.; Li, Y.; Li, J.; Xu, X.; Li, H.; Lu, D.; Wang, J. Effects of exogenous chitosan on physiological characteristics of potato seedlings under drought stress and rehydration. *Potato Res.* **2012**, *55*, 293–301. [CrossRef]
81. Semida, W.M.; Abdelkhalik, A.; Rady, M.O.A.; Marey, R.A.; Abd El-Mageed, T.A. Exogenously applied proline enhances growth and productivity of drought stressed onion by improving photosynthetic efficiency, water use efficiency and up-regulating osmoprotectants. *Sci. Hortic.* **2020**, *272*, 109580. [CrossRef]
82. Sun, T.; Xie, W.M.; Xu, P.X. Superoxide anion scavenging activity of graft chitosan derivatives. *Carbohydr. Polym.* **2004**, *58*, 379–382. [CrossRef]
83. Bittelli, M.; Flury, M.; Campbell, G.S.; Nichols, E.J. Reduction of transpiration through foliar application of chitosan. *Agric. For. Meteorol.* **2001**, *107*, 167–175. [CrossRef]
84. Li, Z.; Zhang, Y.; Zhang, X.; Merewitz, E.; Peng, Y.; Ma, X.; Yan, Y. Metabolic pathways regulated by chitosan contributing to drought resistance in white clover. *J. Proteome Res.* **2017**, *16*, 3039–3052. [CrossRef]
85. Su, Y.; Xia, S.; Zhong, R.; Wang, L. Phytohormonal quantification based on biological principles. *Horm. Metab. Signal. Plants* **2017**, *13*, 431–470.
86. Vasconcelos, A.C.F.D.; Chaves, L.H.G. Biostimulants and their role in improving plant growth under abiotic stresses. In *Biostimulants in Plant Science*; Intech Open: London, UK, 2019. [CrossRef]
87. Landi, L.; De Miccoli, R.M.; Pollastro, S.; Feliziani, E.; Faretra, F.; Romanazzi, G. Global transcriptome analysis and identification of differentially expressed genes in strawberry after preharvest application of Benzothiadiazole and chitosan. *Front. Plant Sci.* **2017**, *8*, 235. [CrossRef]

horticulturae

MDPI

Article

Effects of Foliage Spraying with Sodium Bisulfite on the Photosynthesis of *Orychophragmus violaceus*

Zhongying Li [1], Yanyou Wu [2,*], Deke Xing [1], Kaiyan Zhang [3], Jinjin Xie [1], Rui Yu [1], Tian Chen [1] and Rongrong Duan [1]

[1] Key Laboratory of Modern Agricultural Equipment and Technology, Ministry of Education, Institute of Agricultural Engineering, Jiangsu University, Zhenjiang 212013, China; 2211816010@stmail.ujs.edu.cn (Z.L.); xingdeke@ujs.edu.cn (D.X.); 2211816002@stmail.ujs.edu.cn (J.X.); 2111316005@stmail.ujs.edu.cn (R.Y.); 2211916001@stmail.ujs.edu.cn (T.C.); 2211916002@stmail.ujs.edu.cn (R.D.)

[2] Research Center for Environmental Bio-Science and Technology, State Key Laboratory of Environmental Geochemistry, Institute of Geochemistry, Chinese Academy of Sciences, Guiyang 550081, China

[3] State Engineering Technology Institute for Karst Desertification Control, Guizhou Normal University, Guiyang 550001, China; kaiyanzhang@126.com

* Correspondence: wuyanyou@mail.gyig.ac.cn; Tel.: +86-851-84391746

Abstract: Sulphurous acid derived from sulfur dioxide (SO_2) emission leads to the pollution of irrigation water and the inhibition of plant growth. The safe concentration threshold of $NaHSO_3$ in plants should be clarified to promote agricultural production. In this study, *Orychophragmus violaceus* seedlings were used as experimental materials and five $NaHSO_3$ concentrations (i.e., 0, 1, 2, 5, 10 mmol·L^{-1}) were simultaneously sprayed on the leaf surface of different seedlings separately. Leaf physiology responses under different concentrations were analyzed. The $NaHSO_3$ did not promote photosynthesis in *O. violaceus* under the 1 and 2 mmol·L^{-1} treatments. It was conducive to the net photosynthetic rate (P_N), photorespiration rate (R_p), chlorophyll content, actual photochemical quantum yield (Y_{II}) and photochemical quenching (qP) under the 5 mmol·L^{-1} treatment. However, quantum yield of regulated energy dissipation (Y_{NPQ}) and nonphotochemical quenching (NPQ) were inhibited. Under the 10 mmol·L^{-1} treatment, P_N, chlorophyll content, Y_{II}, qP, dark respiration rate (R_d) and electron transport rate (ETR) showed significant decreases, while the photorespiration portion (S_p) significantly increased. Our results demonstrated that $NaHSO_3$ provided a sulfur source for plant growth and interfered with the redox reaction of the plant itself, and its role as a photorespiratory inhibitor might be masked.

Keywords: agricultural production; redox; photorespiration; chlorophyll fluorescence; dose effect

Citation: Li, Z.; Wu, Y.; Xing, D.; Zhang, K.; Xie, J.; Yu, R.; Chen, T.; Duan, R. Effects of Foliage Spraying with Sodium Bisulfite on the Photosynthesis of *Orychophragmus violaceus*. *Horticulturae* **2021**, *7*, 137. https://doi.org/10.3390/horticulturae7060137

Academic Editor: Alessandra Francini

Received: 14 May 2021
Accepted: 4 June 2021
Published: 6 June 2021

1. Introduction

Orychophragmus violaceus is a member of the family *Brassicaceae* that is widely used for beautifying the city and ecological restoration [1]. *O. violaceus* is also a healthy seasonal vegetable that can be eaten year round and is widely distributed, especially in Yunnan, Guizhou and other southern cities [2]. The plant species has high economic and ornamental value. Sulfur dioxide (SO_2) is a widely diffused air pollutant, which is easily dissolved in the water of rivers or lakes and which forms sulfite and sulfuric acid. If the water source polluted by SO_2 is used for irrigation or spraying on greening plants, it may not be conducive to the plants' growth. Studies have shown that the toxicity of SO_2 to plants was mainly attributed to the highly active intermediate bisulfite [3]. Katainen et al. has also reported that the treatment of sphagnum moss with 0.1 mmol·L^{-1} of H_2SO_3 increased the net photosynthetic rate [4]. Therefore, HSO_3^- may have a two-way effect on the photosynthesis and growth of plants when it is used for irrigation.

$NaHSO_3$ is one of the most commonly used sulfites, which can be used as a photosynthetic accelerator in agricultural production [5–8]. However, the effect of $NaHSO_3$ on the photosynthetic growth of plants depends on its concentration. Studies have shown that 0.5 mmol·L^{-1} of $NaHSO_3$ is the best concentration to promote the photosynthetic oxygen release of Anabaena, while 1 mmol·L^{-1} of $NaHSO_3$ can increase the net photosynthetic rate of *Satsuma mandarin* by approximately 15% [9,10]. In general, low concentrations of $NaHSO_3$ (<1 mmol·L^{-1}) can significantly improve the photosynthetic oxygen release rate and dry matter accumulation of algae and other lower plants [6,10,11], while most higher plants after low concentrations of $NaHSO_3$ (<8 mmol·L^{-1}) spraying can significantly enhance the photosynthetic carbon assimilation ability [7–9,12]. Bisulfite can represent a sulfur source for plants. *Botryococcus braunii* reportedly stopped growing after surviving for 12 days in a sulfur-free medium, but grew well under a bisulfite treatment of 0.1 or 0.8 mmol·L^{-1} [6]. However, the promotion of plant growth by the addition of low concentrations of $NaHSO_3$ is not just attributed to the supply of sulfur nutrients. At present, the effect of $NaHSO_3$ on the photorespiration of plants is still controversial. Kang et al. [7] demonstrated that 5 mmol·L^{-1} of $NaHSO_3$ inhibited the photorespiration rate of *Caragana korshinskii*, and the content of glyoxylic acid decreased significantly. However, Chen et al. [8] found that photosynthetic and photorespiration rates increased simultaneously after soybean leaves were treated with 5 mmol·L^{-1} of $NaHSO_3$. Under normal conditions, photorespiration consumes approximately a quarter of the total output of photosynthesis and the portion of photorespiration will increase when the atmospheric carbon dioxide significantly affects the stomata [13]. In recent years, studies on the effects of foliar sprays of $NaHSO_3$ on plants have mainly focused on the response of the photorespiration rate to $NaHSO_3$ [7,8], whereas the proportion of photorespiration in total photosynthesis has not yet been reported. Therefore, variations in the portion of photorespiration must be determined when studying the photosynthetic physiological mechanism of $NaHSO_3$ in plants. In addition, high concentrations of $NaHSO_3$ (>8 mmol·L^{-1}) can cause certain toxicity to the photosynthetic physiology of plants. Ten mmol·L^{-1} of $NaHSO_3$ significantly decreased the net photosynthetic rate of strawberry leaves [9]. The photosynthetic electron transport of pea leaves was inhibited by high concentrations of sulfite [14]. It is interesting to note that $NaHSO_3$ is a chemical compound with both oxidizing and reducing properties. Sulfite in plants can be reduced to sulfide by sulfite reductase or oxidized to sulfate by sulfite oxidase [15]. During photosynthesis, plants produce and accumulate different forms of reactive oxygen species (i.e., ROS) and reducing agents (i.e., ascorbic acid, thioredoxin and reduced glutathione), which are important regulators of photosynthesis-related gene expression [16]. Wei et al. showed that HSO_3^- could react with superoxide anion to form SO_4^{2-} [17]. However, it has also been reported that $NaHSO_3$ oxidation destroys the structure of algae cell membranes [18]. When $NaHSO_3$, which has both oxidation and reduction properties, enters the plant, the normal redox reaction will be disturbed and indirectly affect photosynthesis. However, few reports have focused on the regulation of plant redox by $NaHSO_3$.

O. violaceus was used as experimental material in this study, the mechanisms of different concentrations of $NaHSO_3$ on photosynthesis were investigated, the safe concentration threshold of $NaHSO_3$ in plant leaves was clarified and the theoretical basis for promoting agricultural production and reducing agricultural ecological environment pollution could be provided.

2. Materials and Methods

2.1. Plant Culture and Treatment

The experiment was carried out in the Key Laboratory of Modern Agricultural Equipment and Technology of the Ministry of Education, College of Agricultural Engineering, Jiangsu University (N 32°11′ and E 119°27′). The seeds of *O. violaceus* were placed on wet gauze and germinated in a light incubator with a light intensity of 40 μmol·m^{-2}·s^{-1}. Water was sprayed every day to keep the gauze moist. The seeds were seeded in a 12-hole seedling tray with perlite and exposed to white light. Seedlings were cultivated in the tray with a small amount of 1/4-strength Hoagland solution until the 2 leaf stage. The culture conditions were as follows: photoperiod of 12 h, CO_2 concentration of 390 ± 10 μmol·mol^{-1}, relative humidity of air of 60 ± 5%, day/night cycle temperature of 28 °C/20 °C and light intensity of 280 ± 20 μmol·m^{-2}·s^{-1}.

After 45 days of growth, the leaves of different seedlings were sprayed with 0 (CK), 1 (NS_1), 2 (NS_2), 5 (NS_3) and 10 (NS_4) mmol·L^{-1} of $NaHSO_3$ solutions. The spraying was conducted from 9:00 to 10:00 in the morning. The 50 mL $NaHSO_3$ solution was sprayed on plants in each pot every 5 days, and the seedlings were sprayed 5 cm from the top in all directions. During the treatment period, the leaves of the seedlings were sprayed every 5 days for a total of 5 times, and the experiment was carried out 25 days after the spray treatment.

2.2. Gas Exchange Measurements

The third fully expanded leaves from the top were chosen for the gas exchange measurement at 9:00–12:00 a.m. on a sunny day. A portable LI-6400XT photosynthesis measurement system (LI-COR Inc., Lincoln, NE, USA) was used. The flow rate was set to 500 μmol·s^{-1}, and the leaf temperature was 30 ± 2 °C. The net photosynthetic rate (P_N), stomatal conductance (g_s), intercellular carbon dioxide (C_i), transpiration rate (E) and other photosynthetic parameters were selected from the two response curves under a light intensity of 800 μmol·s^{-1} and a CO_2 concentration of 400 μmol·mol^{-1}. The P_N-PAR response curves were always fitted using the nonrectangular hyperbola equation [19], which is expressed as follows:

$$P_N = \frac{\alpha\, I + Amax - \sqrt{(\alpha\, I + Amax)^2 - 4\, k\, \alpha\, I\, Amax}}{2k} - R_d \tag{1}$$

where P_N is the net photosynthetic rate (μmol·m^{-2}·s^{-1}); I is the photosynthetically active radiation (μmol·m^{-2}·s^{-1}); α (apparent quantum efficiency) is the initial slope of the P_N-PAR curves (μmol·μmol^{-1}); Amax is the net photosynthetic rate at light saturation (μmol·m^{-2}·s^{-1}); k is the curve representing the degree of curvature of the curve angle, the value of which is [0,1]; and R_d is the dark respiration rate (μmol·m^{-2}·s^{-1}). The atmospheric CO_2 concentration during the measurement was 400 μmol·mol^{-1}. For every measurement, the PAR was set at 800, 600, 400, 300, 250, 200, 150, 100, and 50 μmol·m^{-2}·s^{-1}. After those photosynthetic parameters were acquired, the light saturation point (LSP) and light compensation point (LCP) for the photosynthetic capacity were obtained.

The P_N-C_i response curves were always fitted using the rectangular hyperbola equation [19], which is expressed as follows:

$$P_N = \frac{CE\ Bmax\ Ci}{CE\ Ci + Bmax} - R_t \tag{2}$$

where P_N is the net photosynthetic rate (μmol·m^{-2}·s^{-1}); CE (carboxylation efficiency) is the initial slope of the P_N-PAR curves (mol·m^{-2}·s^{-1}); C_i is the intercellular CO_2 concentration (μmol·mol^{-1}); Bmax is the net photosynthetic rate at CO_2 saturation (μmol·m^{-2}·s^{-1}); and R_t is the total respiratory rate (μmol·m^{-2}·s^{-1}). The photosynthetically active radiation during the measurement was 800 μmol·mol^{-1}. For every measurement, the CO_2 concentration was set at 1500, 1200, 1000, 800, 600, 400, 350, 300, 250, 200, 100, and 50 μmol·mol^{-1}. After those photosynthetic parameters were acquired, the CO_2 saturation point (CSP) and CO_2 compensation point (CCP) for the photosynthetic capacity were obtained.

The plant photorespiration portion was calculated as follows [20]:

$$R_p = R_t - R_d;\ P_t = P_N + R_t;\ \text{and}\ S_p = R_p/P_t \tag{3}$$

where the definitions of R_d and R_t are the same as those in Formulas (1) and (2); R_p was the photorespiration rate (μmol·m^{-2}·s^{-1}); P_N and P_t are the net photosynthesis rate (μmol·m^{-2}·s^{-1}) and total photosynthetic rate (μmol·m^{-2}·s^{-1}) under specific CO_2 concentrations and light intensities, respectively; and S_p is the photorespiratory portion.

2.3. Chlorophyll-A Fluorescence (ChlF) Measurement

The ChlF parameters were measured on the third fully expanded leaves from the top, which were the same leaves used for gas exchange measurements. Before the measurements, the leaves were dark-adapted for 30 min to ensure complete relaxation of all reaction centers. ChlF under dark adaptation was measured using a modulated chlorophyll fluorescence imaging system (IMAGING-PAM, Heinz Walz Gmbh) from 19:00 to 21:00. The minimum chlorophyll fluorescence (F_o) was determined using a measuring beam, whereas the maximum chlorophyll fluorescence (F_m) was recorded after a 0.8 s saturating light pulse (2800 $\mu mol \cdot m^{-2} \cdot s^{-1}$). Actinic light (340 $\mu mol \cdot m^{-2} \cdot s^{-1}$) was then applied for 3 min to drive photosynthesis. Maximum fluorescence in the light-saturated stage (F'_m), basic fluorescence after induction (F'_o) and fluorescence yield in the steady state (F_s) were determined. The actual photochemical quantum yield (Y_{II}) was calculated as $(F_m'\text{-}F)/F_m'$. The quantum yield of regulated energy dissipation (Y_{NPQ}) was calculated as $1\text{-}Y_{II}\text{-}1/(NPQ + 1 + qL(F_m/F_o - 1))$. The quantum yield of nonregulated energy dissipation (Y_{NO}) was calculated as $1/(NPQ + 1 + qL(F_m/F_o - 1))$. The photochemical quenching coefficient (qP) was calculated as $(F'_m - F_s)/(F'_m - F'_o)$, while the nonphotochemical quenching coefficient (NPQ) was calculated as $(F_m - F'_m)/F'_m = F_m/F'_m - 1$. Subsequently, the photosynthetic electron transport rate (ETR) was calculated as PAR $\times$ Y_{II} $\times$ 0.85 $\times$ 0.5, where 0.5 and 0.85 are the fractions of the excitation energy distributed to PSII and the fractional light absorbance, respectively, PAR is the photosynthetically active radiation, and PSII is photosystem II.

2.4. Chlorophyll and Carotene Content

The third fully expanded fresh leaves from the top were picked and immediately ground and extracted with 95% ethanol under dark conditions until the leaves turned white. The absorbance of chlorophyll a (Chl a), chlorophyll b (Chl b) and carotene was measured with a 7230 G spectrophotometer at 665 nm (OD_{665}), 649 nm (OD_{649}) and 470 nm (OD_{470}), respectively. The corresponding chlorophyll concentration was calculated from the measured optical density values, and the chlorophyll content was determined by using the following formula [21].

$$\text{chlorophyll content}\left(\text{mg}\cdot\text{g}^{-1}\ \text{FW}\right) = \frac{C \times V \times A}{W \times 1000} \tag{4}$$

where C is the chlorophyll concentration ($mg \cdot L^{-1}$); V is the the amount applied for the extraction (mL); A is the dilution ratio; W is the fresh weight of the sample (g).

2.5. Statistical Analysis

All measurements were based on 3 replicate plants. The statistical analysis included a 1-way analysis of variance (ANOVA), and significant differences between the means were tested using Duncan's multiple range test at 95% confidence.

3. Results

3.1. Effects of Foliage Spraying of $NaHSO_3$ on Gas Exchange of O. violaceus

The values of P_N, g_s and E in the NS_3 treatment were significantly higher than those in the CK (Figure 1A,B,D). However, the values of P_N, g_s and E in the NS_1, NS_2, NS_4 and CK treatments showed no significant difference. The values of P_N, g_s and E in the NS_4 treatment were significantly lower than those in the NS_3 treatment (Figure 1A,B).

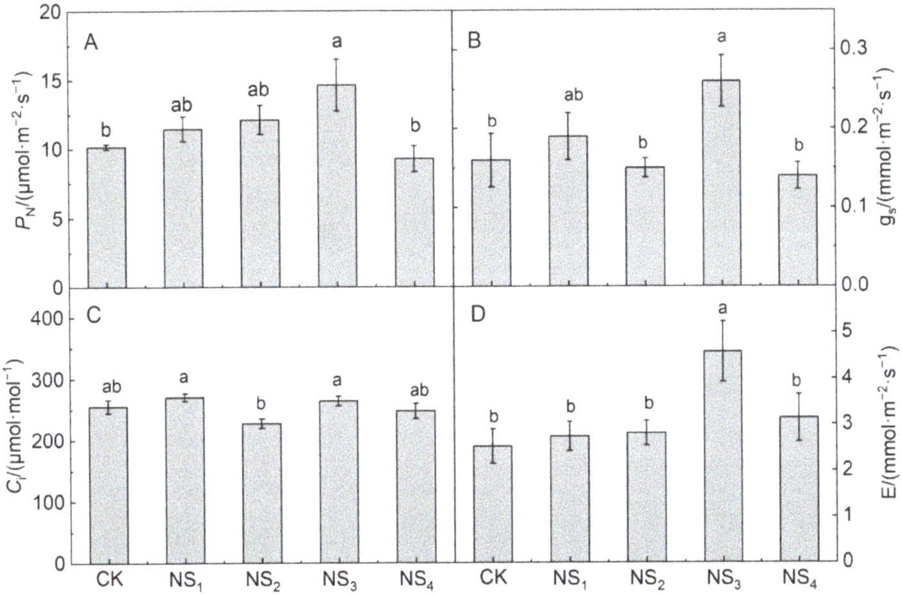

Figure 1. Effects of foliage spraying of $NaHSO_3$ on et photosynthetic rate (P_N) (**A**), stomatal conductance (g_s) (**B**), intercellular carbon dioxide concentration (C_i) (**C**), and transpiration rate(E) (**D**) of *O. violaceus*. Values are the means of five repetitions ± SE. Bars with different letters show significant differences at $p < 0.05$ (Duncan).

3.2. Responses of Net Photosynthetic Rate of O. violaceus to Photosynthetically Active Radiation (PAR) and Intracellular CO_2 Concentration (C_i) under Foliage Spraying of $NaHSO_3$

The correlation coefficients (R^2) of the P_N-PAR curve fitted by the nonrectangular hyperbolic model and the P_N-C_i curve fitted by the rectangular hyperbolic model were all higher than 0.98, which indicated that the two models fit the curves mentioned above well.

Different concentrations of $NaHSO_3$ affected the light response process differently. When the PAR was less than 200 μmol·m^{-2}·s^{-1}, the P_N increased rapidly as the PAR increased, but significant differences were not observed between the values of P_N in different treatments (Figure 2A). When the PAR was greater than 200 μmol·m^{-2}·s^{-1}, the P_N value in the NS_4 treatment increased more slowly than that in the CK as PAR increased. The P_N value was significantly lower than that in the CK when the PAR reached the light saturation point (LSP). The foliar application of 10 mmol·L^{-1} of $NaHSO_3$ decreased the LSP and inhibited the photosynthetic efficiency (Figure 2A). The P_N in the NS_3 treatment exhibited a clearer increase than that in the CK as the PAR increased. The foliar application of 5 mmol·L^{-1} of $NaHSO_3$ promoted the photosynthetic capacity of *O. violaceus* (Figure 2A).

Different concentrations of $NaHSO_3$ affected the CO_2 response process differently (Figure 2B). When the CO_2 concentration was less than 400 μmol·mol^{-1}, the values of P_N clearly increased as the CO_2 concentration increased but slowed down when the CO_2 concentration was greater than 600 μmol·mol^{-1} (Figure 2B). The values of P_N in the NS_2 and NS_3 treatments were higher than those in the CK, and the P_N value in the NS_3 treatment was the highest. The values of P_N in the NS_1 and NS_4 treatments exhibited no significant difference compared to those in the CK under different CO_2 concentrations (Figure 2B). The photosynthetic capacity of *O. violaceus* was the highest under the foliar application of 5 mmol·L^{-1} of $NaHSO_3$, which was the optimal concentration.

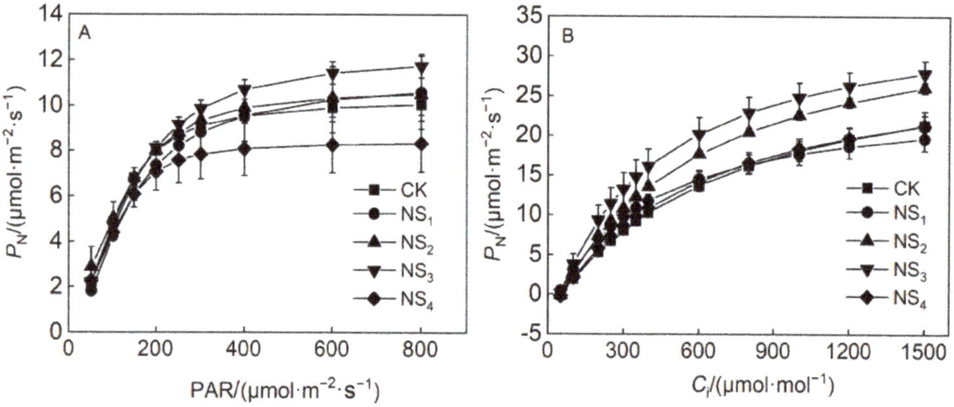

Figure 2. Net photosynthetic rate (P_N) −photosynthetically active radiation (PAR) curve (**A**) and net photosynthetic rate-intercellular carbon dioxide (C_i) curve (**B**) of *O. violaceus* under different concentrations of $NaHSO_3$. Values are the means ± SE. Symbols with different letters show significant differences at $p < 0.05$ (Duncan).

Significant differences were not observed between the values of Amax and Bmax in all treatments (Table 1). The apparent quantum efficiency (α) is an important index that reflects the light energy utilization rate of plants [22]. The light compensation point (LCP) reflects the ability of plants to overcome their own assimilation resistance. The lower the LCP, the less the consumption of photosynthetic products and the stronger the ability to use low light intensity [23]. In this study, the values of α and LCP in the NS_4 treatment decreased by 20.97% and 76.08% of those in the CK, respectively (Table 1). The *O. violaceus* treated with 10 mmol·L^{-1} of $NaHSO_3$ showed improvement in the ability to use weak light and lower consumption of photosynthetic products to resist the stress of high concentrations of sulfite. The initial carboxylation efficiency (CE) can reflect the activity of ribulose-1,5-bisphosphate carboxylase/oxygenase (Rubisco) and the ability of plants to utilize CO_2 [24]. In this study, the value of CE in the NS_3 treatment increased by 105.77% relative to that in the CK, and the values in the other treatments exhibited no significant difference compared to those in the CK (Table 1).

Table 1. Photosynthetic parameters under different concentrations of $NaHSO_3$.

	CK	NS_1	NS_2	NS_3	NS_4
Amax	11.33 ± 0.26 a	12.53 ± 2.27 a	11.66 ± 1.46 a	13.06 ± 0.57 a	8.61 ± 1.05 a
α	0.062 ± 0.00 a	0.055 ± 0.01 a	0.058 ± 0.01 b	0.058 ± 0.00 ab	0.049 ± 0.00 b
CE	0.052 ± 0.00 b	0.078 ± 0.00 ab	0.0745 ± 0.00 ab	0.107 ± 0.02 a	0.059 ± 0.01 b
Bmax	34.92 ± 4.03 ab	28.14 ± 2.35 b	40.19 ± 1.28 a	41.02 ± 0.51 a	33.31 ± 1.46 ab
LSP	314.28 ± 8.87 a	370.24 ± 35.24 a	327.22 ± 35.43 a	344.00 ± 7.09 a	287.31 ± 8.75 a
LCP	14.09 ± 0.59 ab	17.78 ± 2.74 a	14.12 ± 1.62 ab	9.88 ± 0.41 b	3.37 ± 0.54 c
CSP	1410.10 ± 75.77 a	919.69 ± 12.95 b	1203.71 ± 27.46 ab	1110.06 ± 249.37 ab	1292.54 ± 109.82 ab
CCP	52.97 ± 7.49 a	42.63 ± 0.23 a	51.64 ± 3.32 a	49.83 ± 8.29 a	54.78 ± 5.08 a

Note: Amax: maximum photosynthetic value of P_N-PAR curve; α: apparent quantum efficiency; CE: initial carboxylation efficiency; Bmax: maximum photosynthetic value of P_N-C_i curve; LSP: light saturation point; LCP: light compensation point; CSP: CO_2 saturation point; CCP: CO_2 compensation point; Values are the means ± SE. Bars with different letters show significant differences at $p < 0.05$ (Duncan).

As the $NaHSO_3$ concentration increased, the values of the total respiratory rate (R_t) initially increased and then decreased in the NS_4 treatment and the value of R_t in the NS_3 treatment increased by 80.63% compared with that in the CK (Figure 3B). The values of the photorespiration rate (R_p) and R_t in each treatment showed the same change trends as follows: $NS_3 > NS_2 > NS_4 > NS_1 >$ CK (Figure 3A,B). The values of the photorespiration rate (R_d) gradually decreased as the $NaHSO_3$ concentration increased, and the value in the NS_4 treatment decreased by 85.56% of that in the CK (Figure 3A). The values of the photorespiratory portion (S_p) gradually increased, and the value in the NS_4 treatment increased by 53.85% of that in the CK (Figure 3A).

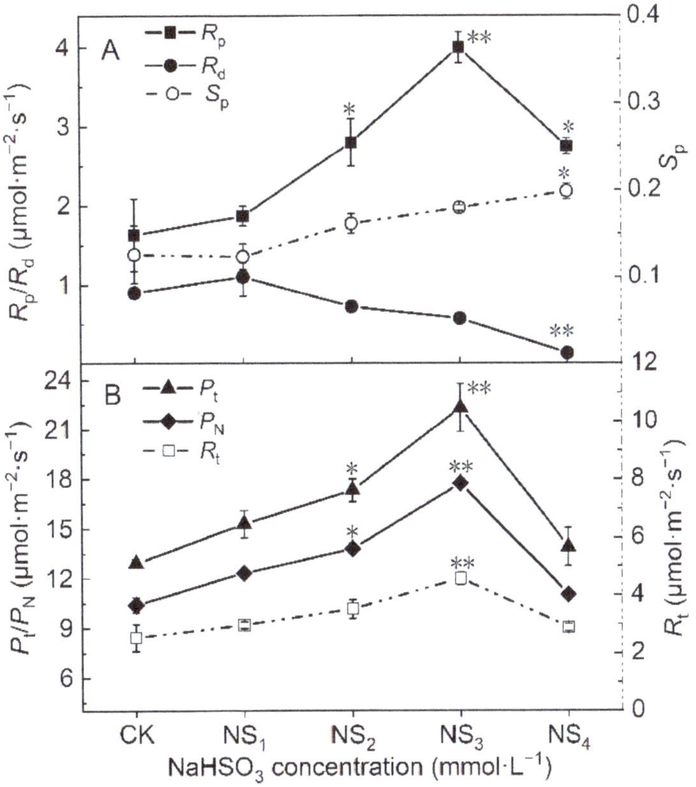

Figure 3. Photorespiration related parameters ((**A**): R_p, R_d and S_p; (**B**):P_t, P_N and R_t) under different concentrations of $NaHSO_3$. Significant differences between the control and treatment groups are indicated by asterisks (* $p < 0.05$, ** $p < 0.01$). R_d: the dark respiration rate; R_p: the photorespiration rate; S_p: the photorespiratory portion; P_N: the net photosynthetic rate; P_t: the total photosynthetic rate; R_t: the total respiratory rate.

3.3. Effects of Foliage Spraying of $NaHSO_3$ on Chlorophyll Content in Leaves of O. violaceus

Chlorophyll is a necessary molecule for the photosynthesis of plants. The NS_1, NS_2 and NS_3 treatments promoted the synthesis of chlorophyll a and b in *O. violaceus* (Table 2). The chlorophyll b contents in the NS_4 treatment had no significant difference compared with those in the CK (Table 2). The chlorophyll a, chlorophyll b and total chlorophyll contents in the NS_4 treatment were slightly lower than those in the NS_1, NS_2 and NS_3 treatments (Table 2). The chlorophyll a/b in each treatment had no significant difference compared with that in the CK (Table 2). The $NaHSO_3$ promoted the synthesis of chlorophyll in *O. violaceus* as a synchronous change of chlorophyll a and chlorophyll b. There was no significant difference in carotenoids between the CK and other treatments (Table 2).

Table 2. Effects of foliage spraying of $NaHSO_3$ on chlorophyll content in *O. violaceus* leaves.

Treatment	Chlorophyll a /(mg·g^{-1}FW)	Chlorophyll b /(mg·g^{-1}FW)	Chlorophyll(a + b)/(mg·g^{-1}FW)	Chlorophyll a/b	Carotenoid /(mg·g^{-1}FW)
CK	0.812 ± 0.026 b	0.323 ± 0.019 b	1.135 ± 0.041 b	2.525 ± 0.104 a	0.141 ± 0.008 a
NS_1	1.074 ± 0.046 a	0.445 ± 0.040 a	1.582 ± 0.074 a	2.459 ± 0.182 a	0.173 ± 0.018 a
NS_2	1.099 ± 0.069 a	0.440 ± 0.022 a	1.540 ± 0.120 a	2.526 ± 0.059 a	0.159 ± 0.019 a
NS_3	1.134 ± 0.076 a	0.440 ± 0.048 a	1.574 ± 0.261 a	2.648 ± 0.182 a	0.177 ± 0.014 a
NS_4	1.038 ± 0.049 a	0.394 ± 0.013 ab	1.432 ± 0.122 a	2.632 ± 0.061 a	0.157 ± 0.009 a

Note: Values are the means ± SE. Bars with different letters show significant differences at $p < 0.05$ (Duncan).

3.4. Effect of Foliage Spraying of $NaHSO_3$ on Chlorophyll a Fluorescence Parameters of *O. violaceus* Leaves

The light energy absorbed by the PSII reaction center is mainly distributed into three parts: photochemical pathway (Y_{II}), energy used for photoprotection mechanism (Y_{NPQ}) and other nonphotochemical energy (Y_{NO}) and $Y_{II} + Y_{NPQ} + Y_{NO} = 1$ [25]. With increasing $NaHSO_3$ concentrations, the value of Y_{II} gradually increased. The value in the NS_3 treatment increased by 17.95% of that in the CK, and thereafter, a decreasing trend was observed (Table 3). However, Y_{NPQ} is the opposite of Y_{II} and showed an initial decrease and then an increase. In the NS_3 treatment, Y_{NPQ} decreased by 16.28% relative to that in the CK, although the value of Y_{NO} in each treatment had no significant difference compared with that in the CK (Table 3). Photochemical quenching (qP) and nonphotochemical quenching (NPQ) are two forms of energy dissipation in chloroplasts [26]. qP is the part of light energy used for photochemical electron transfer, which reflects the utilization of light energy to a certain extent, while NPQ is the part where excess light energy is dissipated in the form of heat energy [26]. The values of NPQ in the NS_2 and NS_3 treatments decreased by 26.09% and 17.39% of those in the CK, respectively, while the values of qP in NS_2 and NS_3 increased by 18.57% and 14.29% of those in the CK, respectively (Table 3). The values of NPQ and qP in the NS_4 treatment exhibited no significant difference compared to those in the CK. The reduction in the photochemical reaction in the NS_4 treatment might be due to the excessive $NaHSO_3$ stress on *O. violaceus*, which would offset the appropriate amount of $NaHSO_3$ to promote the photochemical pathway. The apparent photosynthetic electron transport rate (ETR) mainly reflects the electron transport in the PS II reflection center [27]. The value of ETR in the NS_4 treatment decreased by 18.90% of that in the CK, while the values in other treatments showed no significant difference compared with those in the CK.

Table 3. Effects of foliage spraying of $NaHSO_3$ on chlorophyll *a* fluorescence parameters in *O. violaceus* leaves.

Treatment	Y_{II}	Y_{NPQ}	Y_{NO}	NPQ	qP	ETR
CK	0.39 ± 0.015 bc	0.43 ± 0.014 a	0.18 ± 0.021 a	0.46 ± 0.021 a	0.70 ± 0.020 bc	23.49 ± 1.064 ab
NS_1	0.39 ± 0.033 bc	0.43 ± 0.038 a	0.18 ± 0.198 a	0.45 ± 0.057 a	0.72 ± 0.037 abc	21.91 ± 1.844 bc
NS_2	0.45 ± 0.021 ab	0.36 ± 0.017 b	0.19 ± 0.020 a	0.34 ± 0.020 b	0.83 ± 0.061 a	25.31 ± 1.165 ab
NS_3	0.46 ± 0.06 a	0.36 ± 0.01 b	0.18 ± 0.016 a	0.38 ± 0.016 ab	0.80 ± 0.010 ab	26.09 ± 0.337 a
NS_4	0.34 ± 0.008 c	0.47 ± 0.004 a	0.19 ± 0.024 a	0.48 ± 0.024 a	0.63 ± 0.030 c	19.05 ± 0.447 c

Note: Y_{II}: actual photochemical quantum yield; Y_{NPQ}: quantum yield of regulated energy dissipation; Y_{NO}: quantum yield of nonregulated energy dissipation; NPQ: nonphotochemical quenching; ETR: electron transport efficiency; Values are the means ± SE. Bars with different letters show significant differences at $p < 0.05$ (Duncan).

4. Discussion

Sulfur is an essential mineral element for plants, and it is fourth in the list of major plant nutrients after nitrogen, phosphorus and potassium [28]. Higher plants mainly uptake inorganic sulfate from the soil by their roots, and they can also absorb the atmospheric SO_2 and exogenous HSO_3^-, SO_3^- and S^{2-} through leaf stomata. During the process of sulfur metabolism, exogenous sulfur is first converted into the form of sulfate (SO_4^{2-}), which can be absorbed by plants. After activation and reduction, sulfite (SO_3^{2-}) can be produced, which has potential cytotoxicity [29]. Many metabolic pathways of SO_3^{2-}

are observed in plants, and their metabolites are closely related to chlorophyll synthesis. First, SO_3^{2-} is reduced to sulfide (S^{2-}) under the action of sulfite reductase and S^{2-} reacts with acetylserine (OAS) to form cysteine (Cys) [30]. As the precursor of sulfur-containing amino acids, Cys is further synthesized into various sulfur-containing proteins, thus guaranteeing the early synthesis of chlorophyll; then, SO_3^{2-} in chloroplasts could enter the thiolipid reduction pathway to synthesize sulfoquinovosyldiacylglycerol (SQDG) through two consecutive steps. SQDG is a sulfur-containing nonphosphorus glycerolipid that participates in the formation of the granum lamellae of chloroplasts, and its content is positively correlated with the chlorophyll concentration in the process of chloroplast dedifferentiation and regeneration [31]. Although sulfur is not the main component of chlorophyll, it obviously affects the synthesis of chlorophyll. It is noteworthy that the variation of chlorophyll content will directly affect the absorption, transformation and utilization of light energy by plants [32]. Ribulose-1,5-bisphosphate (RuBP) carboxylase/oxygenase (Rubisco) catalyzes the first step of the reaction of CO_2 assimilation and photorespiration carbon oxidation in photosynthesis and is considered the main factor controlling the rate of photosynthesis. To ensure its catalytic ability, Rubisco must be activated by Rubisco activase (RCA). Studies have found that $NaHSO_3$ could promote the expression of RCA genes at the transcription and translation levels, thus enhancing the initial activity of Rubisco in plants [33]. RCA activity was sensitive to the ATP/ADP ratio observed in the chloroplast matrix, and the activation of RCA depended on the hydrolysis of ATP and was inhibited by ADP [34]. Wang et al. [35] reported that 1 $mmol \cdot L^{-1}$ of $NaHSO_3$ acted similarly as phenazine methyl sulfate (PMS), a cofactor that catalyzed cyclic photophosphorylation, by promoting photophosphorylation and increasing the ATP supply, thereby maintaining high levels of photosynthesis. Moreover, 5 $mmol \cdot L^{-1}$ of $NaHSO_3$, as a sulfur source absorbed and utilized by plants, may play an active role in *O. violaceus*. On the one hand, the increase in HSO_3^- in the leaves accelerated the metabolism of sulfate, and its metabolites directly or indirectly promoted the increase in chlorophyll content, which was conducive to the absorption of light energy by the treated leaves, which was consistent with the results of Li et al. [36]. Meanwhile, the ratio of light energy to the photochemical pathway and light protection mechanism was adjusted. As a result, more light energy was allocated to the photochemical pathway (Y_{II} increased significantly, while Y_{NPQ} decreased significantly), the light energy utilization rate increased and the final photosynthetic rate increased. On the other hand, $NaHSO_3$ may increase the expression of RCA genes by promoting photophosphorylation and increasing the supply of ATP, thus increasing the initial activity of Rubisco, which can catalyze the two reactions of RuBP carboxylation (photosynthesis) and oxidation (photorespiration) simultaneously, and the photosynthetic rate and photorespiration rate increase synchronously.

Photorespiration is a process in which plants fix oxygen and release CO_2 under light conditions. Photorespiration can alleviate photoinhibition, eliminate toxic intermediate products and provide raw materials for other metabolic activities, and it plays an active role in photosynthesis [37]. Studies have suggested that the activities of RuBP carboxylase and RuBP oxygenase in high-yield genotype wheat were higher than those in low-yield genotype wheat. High photosynthesis and photorespiration intensities are important preconditions for ensuring high wheat yield [38]. The electron transport rate of PSII and ATP production increased when a low concentration of $NaHSO_3$ was sprayed on citrus leaves, which decreased photoinhibition and thereby increased the net photosynthetic rate [12]. Foliage sprayed with an appropriate concentration of $NaHSO_3$ increased the total photosynthetic and photorespiration rates, which was consistent with the results reported in the studies mentioned above. Photorespiration consumed excess light energy and protected the photosynthetic apparatus when the consumption ratio of photorespiration to photosynthate was maintained, which indirectly maintained photosynthesis. However, photorespiration consumes photosynthetic products without producing ATP, and it has also been considered a negative factor in photosynthesis. Stomata are important channels for CO_2 and water exchange between plants and the environment [39]. In this study, the stomatal conductance of *O. violaceus* leaves in the NS_4 treatment decreased, which may be due to the stress caused by higher concentrations of $NaHSO_3$. To respond to the deficiency of water and CO_2 caused by the decrease in stomatal conductance, the gene expression of carbonic anhydrase (CA) in leaves is upregulated, which catalyzes the conversion of intracellular HCO_3^- into H_2O and CO_2 [40]. The contents of chlorophyll a and b, the electron transport rate and the photosynthetic rate of soybean all reportedly increased when the leaves were sprayed with an appropriate concentration of HCO_3^- [41]. This demonstrated that the effect of HCO_3^- on plants was similar to that of HSO_3^- in this study. One possible hypothesis was that competition may occur between HCO_3^- and HSO_3^- in the process of photosynthesis due to their similar structure. Under high-concentration $NaHSO_3$ treatment, excessive HSO_3^- accumulated in the cell sap to compete with HCO_3^- for the active site of CA, thereby hindering the combination of HCO_3^- and CA. Plants could not offset the deficiency of H_2O and CO_2 in their leaves by converting intracellular HCO_3^- when stomatal conductance decreased. The carboxylation of RuBP was inhibited, and the total photosynthetic rate decreased. In addition, *O. violaceus* would suffer from stress when they were sprayed with high concentrations of $NaHSO_3$ (the reason for stress will be explained later). A high photorespiration rate and S_p in plants had a protective effect against photosynthetic apparatus damage in response to stress conditions, while a high proportion of photorespiration would also consume photosynthates and therefore decrease the net photosynthetic rate.

Among sulfites, the valence of sulfur is +4, which is both reducing and oxidizing. Sulfite dissolved in water can not only obtain electrons to form sulfur precipitates but also lose electrons to form sulfates: $SO_3^{2-} + 3H_2O + 4e^- \rightleftharpoons S + 6OH^-$ $E = -0.66$; $SO_3^{2-} + H_2O - 2e^- \rightleftharpoons SO_4^{2-} + 2H^+$ $E = +0.2$. Sulfite in plants has both reduction and oxidation properties, and it has dual effects on plant photosynthesis due to its concentration, which is protective or inhibitory. As a nucleophilic substance, sulfite can attack diverse substrates by splitting the disulfide bonds into peptides and cause inactivation of these compounds, which is called sulfitolysis. Sulfitolysis can lead to chlorophyll destruction, photosynthesis suppression, necrotic damage and growth retardation [42]. Therefore, if sulfite accumulates in plants and cannot be metabolized rapidly, it will cause serious damage at the cellular and even the entire plant level [30]. Sulfate oxidase (SO) plays a vital role in relieving this toxicity, and it can serve as a 'safety valve' to detoxify excess amounts of sulfite and protect the cells from sulfitolysis [43]. Wei et al. [44] found that an appropriate amount of $NaHSO_3$ could react with the superoxide anion produced by the PSI receptor of *Chlamydomonas reinhardtii*; as a result, an anaerobic environment was established, hydrogenase (H_2ase) was activated and the hydrogen production capacity was significantly improved. Golan and Whitaker [45] also proved that $NaHSO_3$ could be used as a reducing agent to inhibit the activity of mushroom polyphenol oxidase (PPO), thereby playing a certain role in preventing browning. Therefore, $NaHSO_3$ had a certain degree of reducibility. At appropriate concentrations, it oxidized into sulfate to enter sulfate metabolism, and it detoxified or reacted with active oxygen to reduce the damage of strong oxidizing substances to cells. However, when the concentration of $NaHSO_3$ increased to a certain extent, its oxidation led to adverse impacts on plants. Lüttge et al. [18] indicated that a certain concentration of bisulfite compounds interfered with membrane proteins and lipids, which impaired membrane integrity and inhibited photosynthetic CO_2 fixation and ion transport processes. Lin et al. [46] reported that the active oxygen content in the leaves of rice seedlings increased significantly as the $NaHSO_3$ concentration increased. Chlorophyll *a* fluorescence technology is often used to study photosynthesis under adversity [47,48]. In this study, the ETR and Y_{II} in the NS_4 treatment decreased significantly compared to those in the NS_3 treatment, while the Y_{NPQ} increased. The results demonstrated that the leaves of *O. violaceus* suffered from mild stress when they were sprayed with 10 mmol·L^{-1} of $NaHSO_3$. Excessive HSO_3^- not only had oxidative properties but also induced the production of active oxygen. These strong oxidizing substances attacked the cell biofilm system of plants, injured the photosynthetic apparatus and even a variety of organelles and affected the processes of photosynthetic CO_2 absorption and ion transport, thereby inhibiting the photosynthetic carbon assimilation and reducing the efficiency of photosynthetic electron transport. To avoid further damage to plants caused by excess light energy, plants need to convert part of the captured light energy into heat energy through a heat dissipation mechanism. Physiological activities, such as protein synthesis, nutrient absorption and transport were affected under stress, which reduced the dark respiration rate.

5. Conclusions

The 5 mmol·L^{-1} of $NaHSO_3$ was the appropriate concentration, which promoted the photosynthetic capacity and increased production, while a concentration of 10 mmol·L^{-1} inhibited the photosynthesis and caused pollution of *O. violaceus*. Photorespiration had a certain protective effect on plants that suffered from stress, but an excessive photorespiration portion consumed photosynthates and decreased the net photosynthetic rate. 5 mmol·L^{-1} of $NaHSO_3$ absorbed by plants could be considered a sulfur source. The results helped to better understand the dose effect of HSO_3^- on plant photosynthetic physiology, which provided a theoretical basis for the reasonable utilization of $NaHSO_3$ and promotion of agricultural production.

Author Contributions: Conceptualization, Z.L. and Y.W.; methodology, Z.L., D.X. and Y.W.; validation, K.Z., J.X. and R.Y.; resources, Z.L.; data curation, T.C. and R.D.; writing—original draft preparation, Z.L.; writing—review and editing, D.X. and Y.W.; project administration, Z.L.; funding acquisition, Y.W. All authors have read and agreed to the published version of the manuscript.

Funding: This research was funded by by the project of the National Key Research and Development Program of China [2016YFC0502602], the National Natural Science Foundation of China [No. U1612441], Support Plan Projects of Science and Technology Department of Guizhou Province [No. (2021)YB453].

Institutional Review Board Statement: Not applicable.

Informed Consent Statement: Not applicable.

Data Availability Statement: The datasets during or analyzed during the current study available from the corresponding author on reasonable request.

Conflicts of Interest: The authors declare no conflict of interest. The funders had no role in the design of the study.

References

1. Zhang, Y.; Ji, H.B. Physiological responses and accumulation characteristics of turfgrasses exposed to potentially toxic elements. *J. Environ. Manag.* **2019**, *246*, 796–807. [CrossRef] [PubMed]
2. Xing, D.K.; Wu, Y.Y.; Fu, W.G.; Li, Q.L.; Wu, Y.S. Regulated deficit irrigation scheduling of *Orychophragmus violaceus* based on photosynthetic physiological response traits. *Trans. ASABE* **2016**, *59*, 1853–1860.
3. Bayat, L.; Askari, M.; Amini, F.; Zahedi, M. Effects of *Rhizobium* inoculation on *Trifolium resupinatum* antioxidant system under sulfur dioxide pollution. *Biol. J. Microb.* **2014**, *2*, 37–50.
4. Katainen, H.S.; Mäkinen, E.; Jokinen, J.; Kellomäki, S. Effects of SO_2 on the photosynthetic and respiration rates in scots pine seedlings. *Environ. Pollut.* **1987**, *46*, 241–251. [CrossRef]
5. Tombuloglu, H.; Ablazov, A.; Filiz, E. Genome-wide analysis of response to low sulfur (LSU) genes in grass species and expression profiling of model grass species *Brachypodium distachyon* under S deficiency. *Turk. J. Biol.* **2016**, *40*, 934–943. [CrossRef]
6. Yang, S.L.; Wang, J.; Cong, W.; Cai, Z.L.; Ouyang, F. Effects of bisulfite and sulfite on the microalga *Botryococcus braunii*. *Enzym. Microb. Technol.* **2004**, *35*, 46–50. [CrossRef]
7. Kang, T.; Wu, H.D.; Lu, B.Y.; Luo, X.J.; Gong, C.M.; Bai, J. Low concentrations of glycine inhibit photorespiration and enhance the net rate of photosynthesis in *Caragana korshinskii*. *Photosynthetica* **2018**, *56*, 512–519. [CrossRef]
8. Chen, G.K.; Wang, X.Y.; Kang, H.J.; Sun, J. Effect of different $NaHSO_3$ concentrations on gas exchange and fluorescence parameters in beans and maize. *J. Nucl. Agr. Sci.* **2017**, *31*, 379–385. (In Chinese)
9. Guo, Y.P.; Hu, M.J.; Zhou, H.F.; Zhang, L.C.; Su, J.H.; Wang, H.W.; Shen, Y.G. Different pathways are involved in the enhancement of photosynthetic rate by sodium bisulfite and benzyladenine, a case study with strawberry (*Fragaria* × *Ananassa Duch*) plants. *Plant Growth Regul.* **2006**, *48*, 65–72. [CrossRef]

10. Wang, L.; Ming, C.; Wei, L.; Gao, F.; Lv, Z.; Wang, Q.; Ma, W. Treatment with moderate concentrations of $NaHSO_3$ enhances photobiological H production in the cyanobacterium *Anabaena* sp. strain PCC 7120. *Int. J. Hydrogen Energy* **2010**, *35*, 12777–12783. [CrossRef]
11. Wang, H.; Mi, H.; Ye, J.; Deng, Y.; Shen, Y. Low concentrations of $NaHSO_3$ increase cyclic photophosphorylation and photosynthesis in cyanobacterium *Synechocystis* PCC 6803. *Photosynth. Res.* **2003**, *75*, 151–159. [CrossRef] [PubMed]
12. Guo, Y.P.; Hu, M.J.; Zhou, H.F.; Zhang, L.C.; Su, J.H.; Wang, H.W.; Shen, Y.G. Low concentrations of $NaHSO_3$ increase photosynthesis, biomass, and attenuate photoinhibition in Satsuma mandarin (*Citrus unshiu Marc.*) plants. *Photosynthetica* **2006**, *44*, 333–337. [CrossRef]
13. Busch, F.A. Photorespiration in the context of Rubisco biochemistry, CO_2 diffusion and metabolism. *Plant J.* **2020**, *101*, 919–939. [CrossRef] [PubMed]
14. Veeranjaneyulu, K.; Charlebois, D.; Soukpoé-Kossi, C.N.; Leblanc, R.M. Sulfite inhibition of photochemical activity of intact pea leaves. *Photosynth. Res.* **1992**, *34*, 271–278. [CrossRef]
15. Galina, B.; Dmiry, Y.; Albert, B.; Vladislav, G.; Lnna, G.K.; Aaron, F.; Rachel, A.; Robert, F.; Moshe, S. Sulfite oxidase activity is essential for normal sulfur, nitrogen and carbon metabolism in tomato leaves. *Plants* **2015**, *4*, 573–605.
16. Queval, G.; Foyer, C.H. Redox regulation of photosynthetic gene expression. *Philos. Trans. R. Soc. B* **2012**, *367*, 3475–3485. [CrossRef] [PubMed]
17. Wei, L.; Yi, J.; Wang, L.; Huang, T.; Gao, F.; Wang, Q.; Ma, W. Light intensity is important for hydrogen production in $NaHSO_3$ treated *Chlamydomonas reinhardtii*. *Plant Cell Physiol.* **2017**, *58*, 451–457.
18. Lüttge, U.; Osmond, C.B.; Ball, E.; Brinckmann, E.; Kinze, G. Bisulfite compounds as metabolic inhibitors: Nonspecific effects on membranes. *Plant Cell Physiol.* **1972**, *13*, 505–514.
19. Ye, Z.P. A review on modeling of responses of photosynthesis to light and CO_2. *Chin. J. Plant Ecol.* **2010**, *34*, 727–740. (In Chinese)
20. Wu, Y.Y.; Rao, S.; Zhang, K.Y.; Lu, Y.; Zhao, L.H.; Liang, Z. A Quantitative Method for Determining the Portion of Photorespiratory Pathway in Plants. China Patent 2016105277715, 13 February 2018.
21. Wang, J.; Lu, W.; Yu, T.; Yang, Q. Leaf morphology, photosynthetic performance, chlorophyll fluorescence, stomatal development of Lettuce (*Lactuca sativa* L.) exposed to different ratios of red light to blue light. *Front. Plant Sci.* **2016**, *7*, 250.
22. Herrmann, H.; Schwartz, J.M.; Johnson, G.N. From empirical to theoretical models of light response curves—Linking photosynthetic and metabolic acclimation. *Photosynth. Res.* **2020**, *145*, 5–14. [CrossRef]
23. Duan, M.; Yang, W.C.; Mao, X.M. Effects of water deficit on photosynthetic characteristics of spring wheat under plastic mulching and comparison of light response curve models. *Trans. Chin. Soc. Agri. Mach.* **2018**, *49*, 219–227. (In Chinese)
24. Ren, B.; Li, J.; Tong, X.J.; Mei, Y.M.; Meng, P.; Zhang, J.S. Simulation on photosynthetic-CO_2 response of quercus variabilis and *Robinia pseudoacacia* in the southern foot of the Taihang Mountain, China. *Chin. J. Appl. Ecol.* **2018**, *29*, 1–10. (In Chinese)
25. Kramer, D.M.; Johnson, G.; Kiirats, O.; Edwards, G.E. New fluorescence parameters for the determination of Q_A redox state and excitation energy fluxes. *Photosynth. Res.* **2004**, *79*, 209–218. [CrossRef] [PubMed]
26. Zai, X.M.; Zhu, S.N.; Qin, P.; Wang, X.Y.; Luo, F.X. Effect of *Glomus mosseae* on chlorophyll content, chlorophyll fluorescence parameters, and chloroplast ultrastructure of beach plum (*Prunus maritima*) under NaCl stress. *Photosynthetica* **2012**, *50*, 323–328. [CrossRef]
27. Hu, H.; Wang, L.H.; Wang, Q.Q.; Jiao, L.Y.; Hua, W.Q.; Zhou, Q.; Huang, X.H. Photosynthesis, chlorophyll fluorescence characteristics and chlorophyll content of soybean seedlings under combined stress of bisphenol A and cadmium. *Environ. Toxicol. Chem.* **2014**, *33*, 2455–2462. [CrossRef] [PubMed]
28. Anjum, N.A.; Gill, R.; Kaushik, M.; Hasanuzzaman, M.; Pereira, E.; Tuteja, N.; Gill, S.S. ATP-sulfurylase, sulfur-compounds and plant stress tolerance. *Front. Plant Sci.* **2015**, *6*, 210. [CrossRef]
29. Stanislav, K.; Mario, M.; Hideki, T. Sulfur nutrition: Impacts on plant development, metabolism, and stress responses. *J. Exp. Bot.* **2019**, *70*, 4069–4073.
30. Brychkova, G.; Grishkevich, V.; Fluhr, R.; Sagi, M. An essential role for tomato sulfite oxidase and enzymes of the sulfite network in maintaining leaf sulfite homeostasis. *Plant Physiol.* **2013**, *161*, 148–164. [CrossRef]
31. Krzysztof, Z. *Encyclopedia of Lipidomics*, 1st ed.; Springer: Dordrecht, The Netherlands, 2017; pp. 1–4.
32. Masuda, T. Recent overview of the Mg branch of the tetrapyrrole biosynthesis leading to chlorophylls. *Photosynth. Res.* **2008**, *96*, 21–143. [CrossRef]
33. Chen, Y.; Jin, J.H.; Jiang, Q.S.; Yu, C.L.; Chen, J.; Xu, L.G.; Jiang, D.A. Sodium bisulfite enhances photosynthesis in rice by inducing Rubisco activase gene expression. *Photosynthetica* **2014**, *52*, 475–478. [CrossRef]
34. Portis, A.R. Rubisco activase—Rubisco's catalytic chaperone. *Photosynth. Res.* **2003**, *75*, 11–27. [CrossRef]
35. Wang, H.W.; Wei, J.M.; Shen, Y.G. Spraying low concentration sodium bisulfite can promote the photosynthetic phosphorylation and photosynthesis of wheat leaves. *Sci. Bull.* **2000**, *45*, 394–398. (In Chinese) [CrossRef]
36. Li, J.; Liu, X.L.; Zhang, C.L.; Guan, C.Y.; Dai, L.L.; Zhang, Y.L.; Tan, L.T.; Ma, N.; Yuan, Z.J. Effects of $NaHSO_3$ on photosynthetic characteristics and nitrogen metabolism of rapeseed seedlings. *Chin. J. Oil Crop Sci.* **2014**, *36*, 761–769. (In Chinese)
37. Sunil, B.; Saini, D.; Bapatla, R.B.; Aswani, V.; Raghavendra, A.S. Photorespiration is complemented by cyclic electron flow and the alternative oxidase pathway to optimize photosynthesis and protect against abiotic stress. *Photosynth. Res.* **2019**, *139*, 67–69. [CrossRef]

38. Aliyev, J.A. Photosynthesis, photorespiration and productivity of wheat and soybean genotypes. *Physiol. Plant.* **2012**, *145*, 369–383. [CrossRef] [PubMed]
39. Matthew, H.; James, H.; Mcelwain, J.C. Differences in the response sensitivity of stomatal index to atmospheric CO_2 among four genera of Cupressaceae conifers. *Ann. Bot. Lond.* **2010**, *3*, 411–418.
40. Hu, H.; Boisson-Dernier, A.; Israelsson-Nordström, M.; Böhmer, M.; Xue, S.; Ries, A.; Godoski, J.; Kuhn, M.J.; Schroeder, I.J. Carbonic anhydrases are upstream regulators of CO_2 controlled stomatal movements in guard cells. *Nat. Cell Biol.* **2010**, *12*, 87–93. [CrossRef]
41. Hao, J.J.; Huang, C.H.; Lu, H.; Yu, Y. Influence of K^+, Na^+ and HCO_3^- on photosynthesis of soybean seedlings. *Soybean Sci.* **2012**, *31*, 436–439. (In Chinese)
42. Yarmolinsky, D.; Brychkova, G.; Fluhr, R.; Sagi, M. Sulfite reductase protects plants against sulfite toxicity. *Plant Physiol.* **2013**, *161*, 725–743. [CrossRef]
43. Hänsch, R.; Mendel, R.R. Sulfite oxidation in plant peroxisomes. *Photosynth. Res.* **2005**, *86*, 337–343. [CrossRef] [PubMed]
44. Wei, L.; Li, X.; Fan, B.; Ran, Z.; Ma, W. A stepwise $NaHSO_3$ addition mode greatly improves H_2 photoproduction in *Chlamydomonas reinhardtii*. *Front. Plant Sci.* **2018**, *9*, 1532. [CrossRef] [PubMed]
45. Golan, A.; Whitaker, J.R. Effect of ascorbic acid, sodium bisulfite, and thiol compounds on mushroom polyphenol oxidase. *J. Agr. Food Chem.* **1984**, *32*, 1003–1009. [CrossRef]
46. Lin, Z.F.; Liu, N.; Chen, S.W.; Lin, G.Z.; Mo, H. Bisulfite (HSO_3) hydroponics induced oxidative stress and its effect on nutrient element compositions in rice seedlings. *Bot. Stud.* **2011**, *52*, 173–181.
47. Liu, X.; Li, M.L.; Li, J.M.; Su, C.L.; Lian, S.; Zhang, H.B.; Li, Y.X.; Ge, K.; Li, L. AhGLK1 affects chlorophyll biosynthesis and photosynthesis in peanut leaves during recovery from drought. *Sci. Rep.* **2018**, *8*, 139–158. [CrossRef]
48. Ghassemi-Golezani, K.; Hosseinzadeh-Mahootchi, A.; Farhangi-Abriz, S. Chlorophyll a fluorescence of safflower affected by salt stress and hormonal treatments. *SN Appl. Sci.* **2020**, *2*, 121–158. [CrossRef]

horticulturae

Article

Effect of Salt Treatment on the Growth, Water Status, and Gas Exchange of *Pyrus pyraster* L. (Burgsd.) and *Tilia cordata* Mill. Seedlings

Viera Paganová [1,*], Marek Hus [2] and Helena Lichtnerová [1]

1 Faculty of Horticulture and Landscape Engineering, Slovak University of Agriculture, 949 76 Nitra, Slovakia; helena.lichtnerova@uniag.sk
2 AgroBioTech Research Centre, Slovak University of Agriculture, 949 76 Nitra, Slovakia; marek.hus@uniag.sk
* Correspondence: viera.paganova@uniag.sk

Abstract: Two-year-old seedlings of *T. cordata* and *P. pyraster* were exposed to salinity for 50 days, whereby each plant was subject to regular applications of a substrate solution containing 100 mM NaCl, amounting to a cumulative volume of 365 mL per plant. The adaptive reactions of the tree species in coping with salt stress were studied. The measured parameters were the growth and distribution of mass to organs, root to shoot mass ratio (R:S), content of assimilation pigments in the leaves, gas exchange parameters (g_s, E, A_n), and water use efficiency (WUE). The relative increase in biomass was reduced under salt treatment for both species. A significant decrease in the total FW and DW was observed only for *T. cordata*, which deposited 4.5 times more Na^+ ions in the plant tissues compared with *P. pyraster*. In *P. pyraster* seedlings, Na^+ ions mainly accumulated in the root (75%), and their distribution was limited to aboveground organs. Thus, a balanced content of the assimilation pigments in the leaves was maintained under salt treatment. In the initial (osmotic) phase of salt stress, *P. pyraster* reduced water consumption and maintained a steady rate of photosynthesis (A_n) per unit area. *T. cordata* responded to salinity by regulating stomatal conductance and increasing water use efficiency (WUE). *T. cordata* was not effective in blocking salt intake and transported Na^+ ions to the leaves. Due to the high cumulative salt content in the substrate, the water potential of the leaf tissues and the rate of photosynthesis significantly decreased in salt-treated *T. cordata* seedlings. The results document the important role of the root system in the resistance of woody plants and in ensuring their survival in conditions of excessive salinity. The investment in root growth improved the water supply of *P. pyraster* seedlings and enhanced the retention of salt ions in the root system, thus limiting their transfer to leaves.

Keywords: salinity; biomass; chlorophyll; woody plants; tolerance

Citation: Paganová, V.; Hus, M.; Lichtnerová, H. Effect of Salt Treatment on the Growth, Water Status, and Gas Exchange of *Pyrus pyraster* L. (Burgsd.) and *Tilia cordata* Mill. Seedlings. *Horticulturae* **2022**, *8*, 519. https://doi.org/10.3390/horticulturae8060519

Academic Editor: Yanyou Wu

Received: 29 April 2022
Accepted: 10 June 2022
Published: 14 June 2022

1. Introduction

Salinity affects trees due to direct ionic toxicity, osmotic effects, and interfering with nutrient uptake [1]. High amounts of salt in the soil limit root water uptake, and excessive concentrations of toxic salts in plants adversely impact plant functions. Salt stress induces a decrease in the aerial part of the plant associated with leaf abscission [2] and decreases the rate of leaf area expansion [3]. Salinity significantly lowers the dry and fresh weight of leaves, stems, and roots [4,5] and results in decreased leaf chlorophyll and carotenoid contents [6,7]. The degradation of photosynthetic pigments lowers the photo-reception efficiency of photosystems (PSI and PSII), which reduces the overall level of photosynthesis [8,9].

Street trees are exposed to salinity due to the application of de-icing salts, which lead to increased salt ion content in the surrounding soil structure, alkalinization, reduced permeability, and soil aeration. Salinity has a negative impact on the growth and vitality of street trees, which have relatively short lifespans and low species diversity. Only a few genera and species of trees dominate European cities [10]. The growth of tolerant tree species, which successfully cope with salt stress, can help to increase biodiversity in the urban environment.

In our study, we evaluated the growth and responses of two species of European flora, Tilia cordata Mill. and Pyrus pyraster L. Burgsd., to substrate salinity. *T. cordata* is a shade-tolerant tree [11–13], with broad ecological amplitude [14]. It is considered a drought-resistant tree species, showing optimal growth on deep loamy soils [15–17]. *T. cordata* is often used as an urban tree, since it is relatively resistant to adverse urban conditions and responds well to pruning [18]. *P. pyraster* is found in Europe over a large temperate climate zone [19,20], has relatively wide ecological amplitude [21], and grows on all soil types [22]. *P. pyraster* has quite a high demand for light and thus grows in rather extreme or marginal site conditions (very dry or wet) without canopy growth competition [19,22,23]. It is not a typical urban tree but has strong seasonal dynamics, favorable aesthetic properties, and responds well to pruning, so it can be a promising tree for planting in urban conditions. However, the tolerance of *P. pyraster* to urban conditions has not yet been investigated in detail. There are little data in the literature on the responses of *P. pyraster*, more generally, to environmental stressors in addition to a lack of comparison with urban trees. Both of the studied species are tolerant to fluctuations in the soil water content [22,24]. The salinity tolerance in plants varies among different species and is strongly influenced by the environmental conditions and plant growth stage [3]. Woody plants are sensitive to salinity in the early seedling stage, while they become more tolerant with increasing age [25]. Therefore, in the present study, the responses to salinity were investigated in seedlings. The possible interspecific differences of the studied taxa in response to salinity would be of wide interest because *T. cordata* is often used as an alley tree in urban settlements and considered to be tolerant to the urban environment [18]. Similar to *P. pyraster*, most fruit species [26] can be sensitive to salinity, which can fundamentally affect the possibilities for their utilization in urban areas.

The aim of this study was to investigate the impact of salinity on woody plants in the seedling stage, particularly how severe salinity affects the growth, water regime, and physiological performance of tree species during early growth. The specific goals were to find out how seedlings of T. cordata and *P. pyraster* cope with salinity in terms of (I) how salinity affects their growth (mass accumulation in plant organs); (II) how salinity affects their physiological performance represented by the parameters of the stomatal conductance (g_s), net photosynthetic rate (A_n), transpiration rate (E), relative water content (RWC), and the water potential of leaf tissues (Ψ_{wl}); and (III) what mechanisms these tree species apply when coping with salt stress.

2. Materials and Methods

2.1. Plant Material

The experimental plants were grown from seeds collected on original stands with *P. pyraster* and *T. cordata* in Slovakia. The climatic conditions of these stands are fairly similar (Table 1). The conditions in these stands are optimal for the studied taxa (submontane altitudinal zone, an average January temperature of −3.5 °C, and an average July temperature of 16–18 °C) within their natural area of distribution in Central Europe.

Upon the initiation of bud swelling, the two-year-old seedlings were placed in plastic pots (volume of 0.5 L) containing a fertilized peat-based growth substrate (20% black peat and 80% white peat moss, 0–5 mm fraction, pH of 5.5–6.5, enriched with nutrients at 1.0 kg/m^3 NPK 14:16:18). The potted plants were placed in the plastic bag to avoid uncontrolled water and salt leakage.

Table 1. The source stand conditions are given according to the climatic characteristics [27] of the regions in Slovakia in which they are found.

Taxon	Location	Exposure	Altitude (m)	TI.(°C)	TVII. (°C)	Precipitation (mm)	Type
P. pyraster	Kremnica hills (Tŕnie)	S	540	−3	18	750	MW
T. cordata	Dubeň (Žilina)	S	410	−3.5	16	800	MW

TI.: the average temperature in January; TVII.: the average temperature in July; S: south exposure; MW: moderately warm region.

2.2. Experimental Design

The salt-treated plants were regularly saturated with 7.5 mL of the salt solution (100 mM NaCl with electric conductivity 10.1 $dS \cdot m^{-1}$) per plant per day. Control plants were saturated with water. The total amount of saline solution applied in the experiment was 365 mL per plant. There were 20 replications for salt-treated plants, as well as the corresponding replication for control plants, and the proposed sample size allowed sufficient homogeneity of variance to be maintained (verified by Bartlett's test at a significance level of $\alpha = 0.05$).

The water content in the growth substrate was calculated based on wet weight [28] and maintained at 80% water as per the weight of the fully saturated substrate.

$$M_n = \frac{(W_w - W_d)}{W_w} \times 100$$

M_n = moisture content (%) of the material n;
W_w = wet weight of the sample;
W_d = weight of the sample after drying.

The water regime was imposed and maintained using a gravimetric approach, and the pots were regularly weighed on a precision industrial scale (Kern & Sohn GmbH, Balingen, Germany) with laboratory accuracy (max = 8000 g, standard deviation = 0.05 g) at 2-day intervals.

Experimental plants were placed in a PolEko KK1450 (POL-EKO-APARATURA sp.j., Wodzisław Śląski, Poland) growth chamber in which a regulated environment was maintained with a 14/10 h photoperiod, the 250 $\mu mol\ m^{-2} \cdot s^{-1}$ irradiation density, and 65% air humidity. A temperature of 24 °C was maintained during the light period and 14 °C during the dark period. After 17 days of acclimatization, the plants were daily treated with a NaCl solution for 50 days from June to August (Figure 1).

The distribution of fresh and dry mass in the plant organs and the measurements of the morphometric traits were performed at the beginning and end of the experiment.

The leaf gas exchange parameters (g_s, A_n, E, WUE) were measured at the 20th, 30th, 40th, and 50th days of the experiment after a total applied dose of 155, 225, 295, and 365 mL NaCl solution, respectively. The earlier measurements were not performed due to the insufficient development of leaf area in the initial phenological stages.

The measurements of the water potential of the leaf tissue and relative water content were performed at the 1st, 20th, 30th, 40th, and 50th days of the experiment after a total applied dose of 0, 155, 225, 295 and 365 mL NaCl solution, respectively.

Figure 1. Two-year-old tree seedlings of (**A**,**B**) *T. cordata* and (**C**,**D**) *P. pyraster;* Control plants grown in the growth chamber for 50 days (**A**,**C**); and plants grown under the salt treatment (100 mM NaCl) (**B**,**D**).

2.3. Measurement and Analysis of Plant Parameters

The total fresh mass of the experimental plants (20 samples for each species and variant) was determined at 10 a.m., before seedlings were planted in the pots, and again at the end of the experiment. The plant roots were extracted from the growth substrate by hand and gently washed to minimize the fine root loss.

The root parameters, including the root length (RL), root surface area (RSA), root volume (RV), average root diameter (ARD), and the number of root tips (NRT), were measured using the WinRhizo REG 2009 system (Regent Instruments, Québec, QC, Canada, SK0410192). The total leaf area (LA) was determined by scanning fresh leaves using ImageJ software at the end of the experiment. The dry weight of the plant organs was determined after the plant material was dried at 105 °C until a constant weight was reached. Then, we calculated the leaf water content (LWC), specific root length (SRL), and root to shoot ratio (R:S). The specific leaf area (SLA) [29] was calculated as the ratio of the leaf area to the leaf dry mass [30].

2.4. Leaf Gas Exchange

The measurements were performed beginning 20 days after the first application of 100 mM NaCl. The net photosynthetic rate (A_n), stomatal conductance (g_s), transpiration rate (E), and water use efficiency (WUE) were measured using gasometer CIRAS-3 (PP-Systems, Amesbury, MA, USA) and a PLC3 universal leaf cuvette, fitted with a 1.75 cm^2 measurement window, on the fully expanded leaf for each plant on the upper part of the seedling [31,32]. The measurements were performed between 8 a.m. and 11 a.m. The molar flow rate of air entering the leaf chamber was kept constant at 300 $cm^3 \cdot min^{-1}$. The conditions were maintained as an average leaf temperature of approximately 26 °C (±0.26 °C SD), vapor partial pressure deficit of 1.38 ± 0.25 kPa, photosynthetically active radiation (PAR) at 250 $\mu mol \cdot m^{-2} \cdot s^{-1}$, and CO_2 concentration of 400 $\mu mol \cdot mol^{-1}$. Measurements were taken following the full stabilization of A_n and g_s after clumping of the leaf in the cuvette, which took up to 5 min.

2.5. Leaf Water Potential and Relative Water Content

The water potential of the leaf tissues (Ψ_{wl}) was determined by psychrometric measurement on a Wescor (model PSYPRO, EliTech Inc., Logan, UT, USA) using a C-52 sample chamber at an ambient temperature of 21 °C from 7 a.m. to 3 p.m. The leaf samples were taken from four plants of each taxon in the salt treatment and the control groups.

The relative water content (RWC; %) was determined using a gravimetric method [33] with a 4 h saturation of leaf samples in water at 4 °C in the dark. The leaf samples were taken from four plants of each taxon in the salt treatment and the control groups.

The RWC was calculated as:

$$RWC = [(FW - DW)/(SW - DW)] \times 100$$

where FW, DW, and SW denote the fresh, dry, and fully saturated weights of the leaf samples, respectively.

2.6. Determination of Chlorophyll and Carotenoid Pigments

The assimilation pigments chlorophyll a (Chl*a*), chlorophyll b (Chl*b*), and carotenoids were extracted with 80% acetone and $MgCO_3$ powder using a mortar. Six circular leaf pieces were taken from twenty plants of each taxon in the salt treatment and the control groups. Samples were taken from the middle part of the seedlings. After complete extraction, the mixture was filtered, and the volume adjusted to 10 mL with cold acetone. The resulting extracts were immediately assayed spectrophotometrically. The absorbance of the extract was measured at 440, 645, and 663 nm using a spectrophotometer (Lange DR 3900, Hach, Loveland, CO, USA). The levels of each pigment were determined using the coefficients and equations determined by Lichtenthaler and Buschmann [34].

2.7. Ion Contents

The analysis of selected elements (K, Na) was determined using an inductively coupled plasma optical emission spectrometer (ICP-OES 720, Agilent Technologies, Mulgrave, VIC, Australia) in axial plasma configuration together with the SPS 3 autosampler (Agilent Technologies, Basel, Switzerland).

The samples were mineralized in the high-performance microwave digestion system Ethos UP (Milestone S.r.l., Sorisole, Italy) in a solution of 5 mL $HNO_3 \geq 69.0\%$ (TraceSELECT®, Honeywell Fluka, Morris Plains, NJ, USA), 1 mL $H_2O_2 \geq 30\%$ for trace analysis (Sigma-Aldrich, Saint Louis, MO, USA), and 2 mL of ultrapure water (18.2 $M\Omega\ cm^{-1}$; 25 °C, Synergy UV, Merck Millipore, Molsheim, France). In the experiment, multielement standard solution V for ICP (Sigma-Aldrich Production GmbH, Basel, Switzerland) was used. The legitimacy of the whole method was verified using a certified reference material (CRM–ERM CE278 K, Sigma-Aldrich Production GmbH, Basel, Switzerland) [35].

2.8. Statistical Analysis

The normality and homogeneity of variance for all the variables were determined based on Shapiro–Wilk's test (at a significance level of $\alpha = 0.001$) and Levene's test (at a significance level of $\alpha = 0.05$). Grubbs' test was used to detect and remove single outliers in the experimental dataset. Two-way ANOVA analysis comparing the effects of taxon, salt treatment, and the interactions between them was used to assess differences between *P. pyraster* and *T. cordata* seedlings grown under salt treatment. The multiple comparison of means was performed using the Tukey honest significant difference (HSD) test (at significance levels of $\alpha = 0.05$). Regression analysis was applied for assessment of the relationships between RWC, Ψ_{wl}, and the cumulative salt uptake.

Statgraphics Centurion XVII software (StatPoint Technologies, Warrenton, VA, USA, XVIII, license number: B480-E10A-00EA-P00S-60PO) was used for statistical data analysis.

3. Results

3.1. Effect of Salinity on Growth and Mass Accumulation of Plant Organs

T. cordata is more efficient in biomass production compared to *P. pyraster*, as per the documented data on the fresh weight (FW) and relative biomass increment (RBI) for the control plants (Table 2). Salinity (salt stress) had a strong but variable effect on the growth and mass accumulation of plant organs. RBI was significantly reduced for both species, but only for *T. cordata* was the decrease in both total FW and DW considered significant. Under salt treatment, the dry mass values of the aboveground organs (DW_S) (−26%) as well as of the root (DW_R) (−46%) were significantly reduced for *T. cordata* seedlings. The stem increment and leaf dry mass (DW_L) were not affected, but salinity reduced the radial growth of the stem manifested by the decrease in the stem dry mass (SDW) (−35%) compared to the control (Table 2). The root growth in length (−47%) and volume (−54%) were also reduced. Under salt treatment, DW_S and DW_R were not significantly reduced for *P. pyraster* seedlings. *P. pyraster* adapted to salinity by investing in root growth, as it increased the number of root tips (NRT) and maintained a balanced root length. The average root diameter (ARD) and root volume (RV) (−32%) of the salt-treated seedlings were reduced. Salinity did not significantly change the R:S for *P. pyraster* seedlings, but this was reduced for *T. cordata*. Under salt treatment, *T. cordata* has limited root growth and preferentially accumulated dry mass in the aboveground organs (leaves). The substrate salinity negatively affected the growth and development of the leaves of both studied species, which is documented by the reduction in leaf area (LA) (*T. cordata* −26%, *P. pyraster* −35%) compared to control plants. In both species, the specific leaf area (SLA) was also reduced in response to salinity (Table 2). However, leaf injury symptoms only appeared in *Tilia* seedlings. The first signs of damage appeared after 35 days of the experiment (when 265 mL of 100 mM NaCl had already been applied to the substrate) in the form of leaf yellowing (Figure 2A) and, later, leaf edge burning (Figure 2B) and early leaf dropping (Figure 2B,C). *P. pyraster* seedlings showed no symptoms of leaf damage during the experiment.

3.2. Leaf Water Status under Salt Treatment

The water potential of the leaf tissues was influenced by the taxon as well as by the cumulative amount of applied saline solution at a concentration of 100 mM NaCl, which was regularly applied to the substrate. At the beginning of the experiment, the water potential of the leaf tissues was significantly higher for *T. cordata* (−0.71 MPa) compared to *P. pyraster* (−1.46 MPa). From the beginning of the salt treatment, *Tilia* maintained balanced Ψ_{wl} values (-1.00 ± 0.36) for 30 days, when 225 mL of 100 mM NaCl per plant had been added to the substrate (Figure 3). However, Ψ_{wl} was significantly reduced at the 40th day (−2.45 MPa) and 50th day (−3.22 MPa) of the salt treatment. *P. pyraster* maintained a balanced Ψ_{wl} (-1.50 ± 0.22 MPa) throughout the whole experiment (Figure 3). A significant increase in the water potential was observed only at day 30 of the experiment.

Table 2. The growth parameters and biomass allocation of *P. pyraster* and *T. cordata* seedlings in the pot experiment after 50 days of salt treatment. The multiple comparison of means ($n = 20$) was performed using the 95% Tukey honest significant difference (HSD) test. Data are the mean values and standard deviations (±SD). Mean values followed by different letters are significantly different.

	p-Value			Control		100 mmol NaCl	
Parameter	**T**	**S**	**T × S**	*P. pyraster*	*T. cordata*	*P. pyraster*	*T. cordata*
Stem length (mm)	0.00	0.02	0.84	281.94 (±64.82) b	399.90 (±91.42) a	242.79 (±49.69) b	353.50 (±74.51) a
Stem increment (mm)	0.00	0.02	0.42	117.44 (±62.65) b	235.70 (±82.90) a	65.86 (±37.06) c	210.40 (±71.10) a
LA (mm^2)	0.00	0.00	0.02	12,843.20 (±4810.70) b	51,364.80 (±10,898.80) a	8340.84 (±2732.16) c	37,860.00 (±9278.79) d
SLA ($mm^2 \cdot mg^{-1}$)	0.00	0.00	0.18	17.76 (±2.23) b	33.01 (±5.06) a	15.53 (±2.22) c	28.19 (±5.04) d
DW_L (mg)	0.00	0.01	0.74	723.78 (±256.67) b	1594.65 (±423.24) a	544.57 (±183.86) c	1364.20 (±332.04) a
SDW (mg)	0.00	0.00	0.01	1134.83 (±469.02) b	2357.10 (±779.90) a	979.21 (±369.60) b	1542.45 (±397.21) c
DW_S (mg)	0.00	0.00	0.08	1858.61 (±710.76) b	3951.75 (±1185.07) a	1523.79 (±505.60) b	2906.65 (±709.21) c
RBI (%)	0.01	0.00	0.36	123.80 (±67.41) b	175.13 (±70.47) a	54.87 (±37.68) c	80.49 (±45.98) c
FW_S (mg)	0.00	0.00	0.00	4218.50 (±1628.89) a	11,942.20 (±3353.35) b	3201.21 (±1059.03) a	7386.70 (±1866.87) c
FW_R (mg)	0.00	0.00	0.00	3588.11 (±1241.37) a	8170.40 (±2945.84) c	3083.79 (±1127.77) a	4164.90 (±885.26) a
RL (mm)	0.00	0.00	0.00	7826.26 (±1925.41) a	19,607.50 (±4212.76) b	8439.39 (±3935.52) ac	10,326.20 (±1812.39) c
SRL ($mm \cdot mg^{-1}$)	0.00	0.91	0.30	5.81 (±1.58) b	8.10 (±2.54) a	6.45 (±3.10) ab	7.59 (±2.05) a
DW_R (mg)	0.00	0.00	0.00	1405.33 (±420.96) b	2650.30 (±928.65) a	1377.79 (±511.18) b	1429.55 (±331.48) b
RSA (mm^2)	0.00	0.00	0.00	11,224.10 (±2955.99) b	25,545.90 (±6564.05) a	8812.52 (±3580.33) c	13,677.80 (±2451.54) d
ARD (mm)	0.13	0.00	0.00	0.54 (±0.06) a	0.49 (±0.06) b	0.40 (±0.06) c	0.49 (±0.04) b
NRT	0.00	0.00	0.00	2044.78 (±710.42) b	7375.80 (±2502.22) a	2742.93 (±1178.07) c	3923.70 (±976.10) d
RV (mm^3)	0.00	0.00	0.00	4981.76 (±1804.35) b	12,760.70 (±4552.20) a	3386.21 (±1260.69) c	5919.14 (±1548.73) b
R:S	0.00	0.93	0.00	0.83 (±0.22) a	0.68 (±0.16) b	1.00 (±0.42) a	0.50 (±0.09) c

T—taxon; S—salt treatment; T × S—interaction between the taxon and salt treatment; LA—leaf area; SLA—specific leaf area; DW_L—dry mass of leaves; SDW—dry mass of the stem; DW_S—dry mass of shoots; RBI—relative biomass increment; FW_S—fresh mass of shoots; FW_R—fresh mass of roots; RL—root length; SRL—specific root length; DW_R—dry mass of roots; RSA—root surface area; ARD—average root diameter; NRT—number of root tips; RV—root volume; R:S—root to shoot mass ratio.

Figure 2. The leaf damage on *T. cordata* seedlings due to a high concentration of Na+ ions. The first symptoms appeared after 35 days of salt treatment in the form of (**A**) leaf yellowing; (**B**) leaf edge burning; and (**C**) early leaf fall after 50 days of salt treatment.

RWC values for *P. pyraster* and *T. cordata* seedlings demonstrated differences in the species responses to salinity (Figure 4). At the beginning of the experiment, the RWC values were in the range of 96.95 ± 1.34% for *T. cordata* and in the range of 96.19 ± 0.26% for *P. pyraster*. On day 30 of the salt treatment, a significant decrease in the RWC (84.04 ± 7.5%) was observed for *P. pyraster*, but the values of this parameter later increased and remained stable until the end of the experiment (89.48 ± 4.80%). The RWC values for *T. cordata* steadily declined from the beginning of the saline treatment (Figure 4), but a significant decrease (80.71 ± 6.49) was demonstrated on day 50 of the experiment after the cumulative NaCl intake per plant exceeded 300 mL.

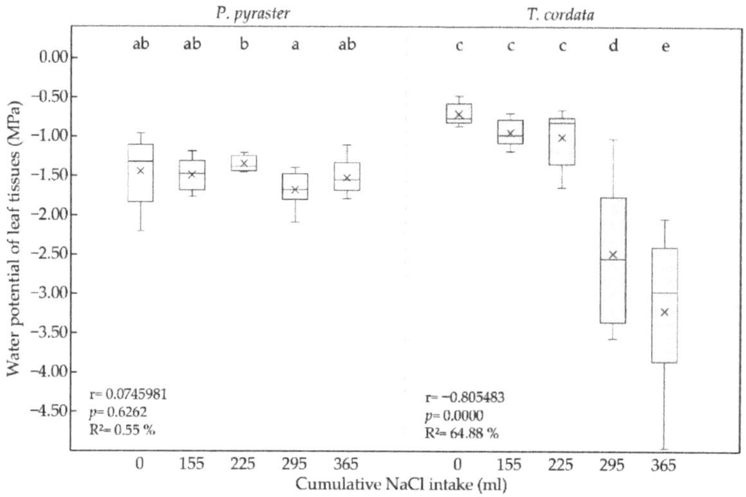

Figure 3. Box plot for the water potential of leaf tissues (Ψ_{wl}) of the seedlings of *P. pyraster* and *T. cordata* measured at the beginning of the experiment (0 mL) and during the experiment when treated with 100 mM NaCl solution. Significant differences between measurements ($p < 0.05$) are indicated by different letters.

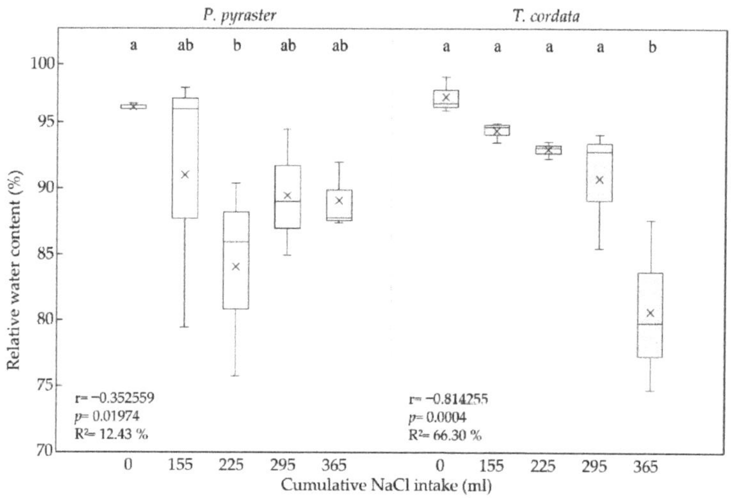

Figure 4. Box plot for the relative water content (RWC) for seedlings of *P. pyraster* and *T. cordata* measured at the beginning of the experiment (0 mL) and during the experiment when treated with 100 mM NaCl solution. Significant differences between measurements ($p < 0.05$) are indicated by different letters.

The obtained data document different reactions of the studied species to salinity and, therefore, varying mechanisms for regulating water uptake. The leaf tissues of *T. cordata* have a demonstrably higher water potential than *P. pyraster*; however, at a higher cumulative salinity level (above 300 mL per plant), the values of Ψ_{wl} decreased sharply, similar to the RWC. *P. pyraster* maintained a balanced water potential and relative water content in the leaves under salt treatment, even at a higher cumulative NaCl intake per plant.

3.3. Salt Ion Uptake and Distribution

Significant differences were observed in the studied species in terms of the uptake and distribution of salt ions to the plant organs. During the experiment, the same amount of salt solution was applied to the substrate per plant, but *T. cordata* seedlings absorbed significantly more Na^+ ions compared to *P. pyraster* (Figure 5). *P. pyraster* accumulated Na^+ ions mainly in the root (75%), and a low amount of Na^+ was distributed to the aboveground organs—stems (19%) and leaves (6%) (Figure 6). Similarly, *T. cordata* retained the majority of Na^+ ions in the root system (55%); however, a relatively high proportion was also distributed to the stems (34%) and leaves (11%). Compared to the untreated control, *T. cordata* significantly increased the uptake of K^+ ions, and preferentially distributed them to the stem and leaves under salt treatment. The K^+ content did not change in the leaves of *P. pyraster* compared with the control, but it was significantly reduced in the stem and root system (Figure 4). Increased NaCl uptake competed with K^+ intake and led to a significant change in the K^+/Na^+ ratio in the plant organs. The significant differences in the K^+/Na^+ ratio are influenced by the increased salt content in the substrate and are species-specific (Figure 5). In the roots of *P. pyraster*, the K^+/Na^+ ratio decreased (−81%) from 15.1 (control) to 2.8 (salt-treated seedlings). In *T. cordata* roots, the change in the K^+/Na^+ ratio was even more pronounced (−94%) and decreased from 17 (control) to 1 for salt-treated seedlings (Figure 5). A significant decrease (−96%) in the K^+/Na^+ ratio was detected in the leaves of *T. cordata* in comparison to both the control as well as salt-treated *P. pyraster* seedlings (−77%). Compared to *P. pyraster*, *T. cordata* distributed a six times higher amount of Na^+ to the leaves, which negatively affected the condition of the leaf apparatus and induced the development of necrosis (Figure 2). *P. pyraster* blocked salt uptake at the level of the root system, and thus protected its leaves from intoxication and damage.

3.4. Effect of Salinity on Photosynthetic Pigments

The content of photosynthetic pigments for the seedlings of the studied tree species is species-specific. Compared to *T. cordata*, *P. pyraster* has a significantly higher content of photosynthetic pigments according to the relevant parameters for the control and salt-treated plants (Table 3). A significant reduction in the chlorophyll content was observed in the leaves of *T. cordata* seedlings after the application of 100 mM NaCl in a cumulative volume of 365 mL NaCl per plant. The contents of total chlorophyll (Chl*ab*), Chl*a*, and Chl*b* were decreased by NaCl stress. The carotenoid content and the Chl *a/b* ratio did not change for the salt-treated seedlings compared to the control. *P. pyraster* seedlings maintained a balanced content of photosynthetic pigments in the leaves during 50 days of the continual salt treatment.

Table 3. The contents of photosynthetic pigments in the leaves of *P. pyraster* and *T. cordata* seedlings after 50 days of salt treatment. Mean values followed by different letters are significantly different.

Parameter	*p*-Value T	*p*-Value S	*p*-Value T × S	Control *P. pyraster*	Control *T. cordata*	100 mM NaCl *P. pyraster*	100 mM NaCl *T. cordata*
Chl *a* ($mg \cdot mm^{-2}$)	0.00	0.04	0.02	378.15 (±78.82) a	289.94 (±41.93) b	380.93 (±87.60) a	221.22 (±45.92) c
Chl *b* ($mg \cdot mm^{-2}$)	0.00	0.02	0.05	145.29 (±34.35) a	122.54 (±16.31) b	143.13 (±32.20) a	95.52 (±19.79) c
Chl *ab* ($mg \cdot mm^{-2}$)	0.00	0.03	0.03	523.31 (±112.26) a	412.37 (±57.72) b	523.92 (±117.65) a	316.67 (±65.38) c
Chl *a/b* ($mg \cdot mm^{-2}$)	0.00	0.96	0.25	2.62 (±0.14) a	2.37 (±0.10) b	2.66 (±0.29) a	2.32 (±0.11) b
Carotenoids ($mg \cdot mm^{-2}$)	0.00	0.62	0.22	109.79 (±19.55) a	83.78 (±11.82) b	112.87 (±23.30) a	76.63 (±14.00) b

T—taxon; S—salt treatment; T × S—interaction between the taxon and salt treatment.

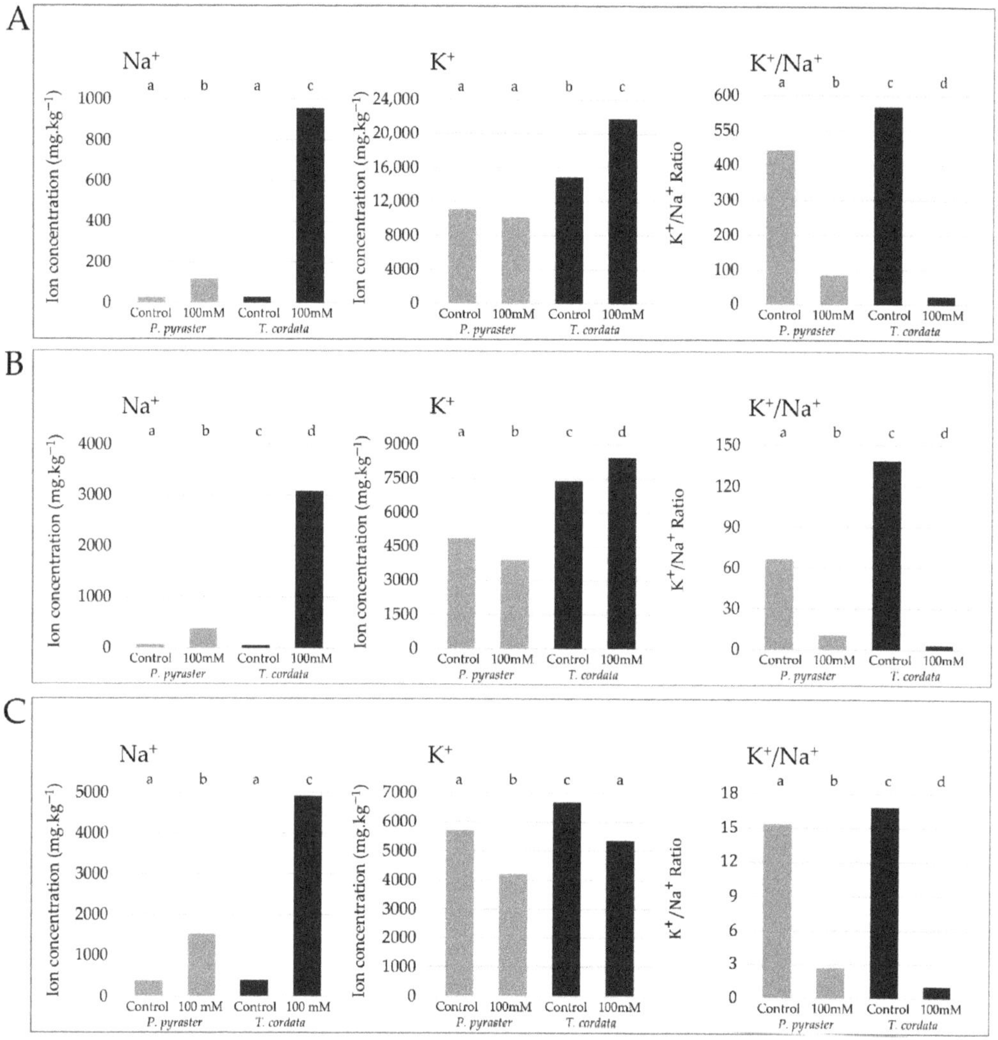

Figure 5. The salt ion content (Na^+, K^+, and K^+/Na^+) in the (**A**) leaves, (**B**) stems, and (**C**) roots of two-year-old seedlings of *P. pyraster* and *T. cordata* after 50 days of continuous 100 mM NaCl saline application and in the control. The bars indicate the mean values of the ion content in the plant organs from a sample of 10 plants. Significant differences ($p < 0.05$) are indicated by different letters.

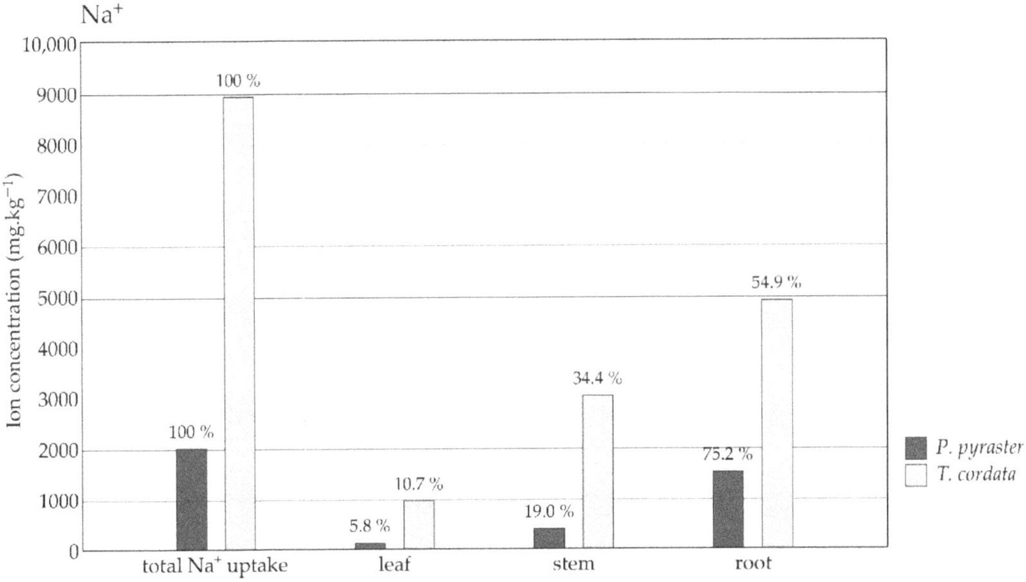

Figure 6. The content and distribution of Na+ ions in the plant tissues (leaves, stem, roots) of *P. pyraster* and *T. cordata* seedlings grown for 50 days under salt treatment with 100 mM NaCl solution. The values above the bars indicate distribution to the plant organs as a percentage of the total Na+ ion uptake.

3.5. Effect of Salinity on Leaf Gas Exchange

Species–specific differences were observed in all examined gas exchange parameters (g_s, E, A_n, WUE) (Table 4). The *P. pyraster* seedlings had significantly reduced g_s and E in response to salinity on day 20 of regular NaCl application to the substrate. The reductions in g_s and E were observed in response to the osmotic phase of the salinity stress. During the experiment, *P. pyraster* seedlings maintained balanced values of both parameters (g_s, E) compared to the control under increasing cumulative NaCl intake to the substrate. *P. pyraster* seedlings had balanced values of A_n and WUE without significant changes compared to the control throughout the whole experiment. In response to the salt treatment, the seedlings of *T. cordata* had reduced g_s and increased WUE on days 20 and 40 of the experiment, when the cumulative salt uptake was 155 and 295 mL per plant, respectively. The transpiration rate decreased significantly on day 40 of the experiment (0.27 ± 0.12 mmol H_2O $m^{-2} \cdot s^{-1}$) compared to the control (0.58 ± 0.19 mmol H_2O $m^{-2} \cdot s^{-1}$). A significant decrease in A_n was observed on day 50 of the experiment, which indicates that salt stress has a negative impacts on photosynthesis. The decreases in the stomatal conductance and transpiration rate were transient for *P. pyraster* seedlings, which maintained balanced photosynthesis even under conditions of higher cumulative salt content in the substrate.

Table 4. The parameters of leaf gas exchange in the leaves of *P. pyraster* and *T. cordata* seedlings after 50 days of salt treatment (n = 5). Mean values followed by different letters are significantly different.

CSI (mL Plant^{-1}) DOT	Parameter	*p*-Value			*P. pyraster*		*T. cordata*	
		T	S	T × S	Control	100 mM NaCl	Control	100 mM NaCl
155	g_s (mmol H_2O m^{-2} s^{-1})	0.06	0.00	0.53	83.00 (±19.65) a	46.80 (±14.60) bc	69.50 (±4.95) ac	21.50 (±0.71) b
20th	A_n (µmol CO_2 m^{-2} s^{-1})	0.85	0.34	0.13	4.66 (±1.17) a	4.36 (±0.57) a	4.00 (±0.44) a	5.20 (±0.42) a
	E (mmol H_2O m^{-2} s^{-1})	0.22	0.01	0.48	0.73 (±0.23) a	0.44 (±0.13) b	0.67 (±0.24) ab	0.23 (±0.01) b
	WUE (mmol CO_2 mol^{-1} H_2O)	0.00	0.00	0.00	7.11 (±3.15) a	8.61 (±1.23) a	6.61 (±2.55) a	22.60 (±0.46) b
225	g_s (mmol H_2O m^{-2} s^{-1})	0.00	0.00	0.38	144.60 (±24.11) a	96.40 (±22.50) b	59.25 (±15.35) c	28.00 (±11.63) d
30th	A_n (µmol CO_2 m^{-2} s^{-1})	0.00	0.13	0.61	7.44 (±0.80) a	6.98 (±0.39) a	5.46 (±1.02) b	4.56 (±1.31) b
	E (mmol H_2O m^{-2} s^{-1})	0.00	0.02	0.51	1.16 (±0.15) a	1.02 (±0.15) a	0.65 (±0.18) b	0.40 (±0.17) b
	WUE (mmol CO_2 mol^{-1} H_2O)	0.00	0.21	0.47	6.48 (±1.00) a	6.96 (±0.87) a	10.35 (±2.60) b	12.07 (±2.36) b
295	g_s (mmol H_2O m^{-2} s^{-1})	0.00	0.00	0.90	103.2 (±29.12) a	69.25 (±27.11) ab	61.89 (±23.18) b	25.40 (±14.72) c
40th	A_n (µmol CO_2 m^{-2} s^{-1})	0.00	0.64	0.36	7.62 (±0.70) a	7.06 (±1.18) a	5.71 (±1.03) b	4.13 (±2.05) b
	E (mmol H_2O m^{-2} s^{-1})	0.01	0.00	0.37	0.74 (±0.16) a	0.56 (±0.20) a	0.58 (±0.19) a	0.27 (±0.12) b
	WUE (mmol CO_2 mol^{-1} H_2O)	0.74	0.03	0.63	10.79 (±2.59) ab	13.35 (±4.48) ab	10.55 (±3.02) a	14.51 (±3.83) b
365	g_s (mmol H_2O m^{-2} s^{-1})	0.00	0.21	0.80	115.00 (±30.22) a	101.80 (±39.06) a	56.57 (±25.05) b	37.00 (±8.79) b
50th	A_n (µmol CO_2 m^{-2} s^{-1})	0.00	0.06	0.42	7.36 (±0.82) a	6.86 (±1.21) a	5.11 (±0.74) b	3.95 (±0.85) c
	E (mmol H_2O m^{-2} s^{-1})	0.00	0.27	0.93	0.87 (±0.17) a	0.79 (±0.22) ac	0.53 (±0.19) bc	0.43 (±0.13) b
	WUE (mmol CO_2 mol^{-1} H_2O)	0.35	0.75	0.96	8.47 (±1.62) a	8.91 (±0.91) a	9.66 (±3.05) a	9.98 (±3.47) a

DOT—day of treatment; CSI—cumulative NaCl intake per plant; T—taxon; S—salt treatment; T × S—interaction between the taxon and salt treatment; A_n—net assimilation rate; E—transpiration rate; g_s—stomatal conductance to water vapor; WUE—water use efficiency.

4. Discussion

Salinity causes severe physiological dysfunctions in plants in addition to indirect damage, even at low concentrations [25]. The growth reduction due to salinity has been documented for many plant species, but not halophytes [26]. Several authors particularly refer to the reduction in shoot growth [2,36–39]. In our study, salinity significantly reduced the fresh and dry matter of *T. cordata* seedlings (FW: 56%; DW: 34%). A significant reduction in shoot and root matter was not confirmed for *P. pyraster* seedlings under salt treatment, where the mass accumulation did not change compared to the control plants. *P. pyraster* invested in root growth, increased the number of root tips and maintained the balanced parameters RL and SRL under salt treatment. Our previous research shows that *P. pyraster* maintains a balanced ratio of the mass distribution between underground and aboveground organs (R:S) even under conditions of water scarcity [40,41]. The findings suggest the effective water uptake and retention of salt ions in the root system of *P. pyraster*. *T. cordata* demonstrably reduced the length of the root (−47%) and the number of root tips (−47%) under salt treatment, which reduced the exploratory capacity of the root system. The obtained data indicate both effective water uptake as well as retention of salt ions in the root system for *P. pyraster*. The continuous growth of roots allows for the continuous exploration of new soil for water and the partial alleviation of water stress through improvements in the water supply [42]. Increasing the proportion of root system mass in plants exposed to salinity has a positive effect through the increased retention of toxic salt ions in the root, thus limiting their transfer to aboveground organs (especially leaves) [2], which can be a typical mechanism of resistance or survival in condition of excessive salinity [43,44].

Under salt treatment, the seedlings of the studied tree species had reduced leaf area (*Pyrus*: 35%; *Tilia*: 26%), which is considered a typical response to salt stress. *P. pyraster* formed smaller leaves and showed significantly reduced shoot growth (−43%). According to Munns and Tester [1], a significant reduction in the size of individual leaves or the number of shoots is the main effect of salinity that manifests in dicotyledonous species. *T. cordata* seedlings maintained balanced stem growth and dry matter accumulation in the leaves even after salt treatment. The reduction in the leaf area was due to damage and early leaf fall. The first symptoms of leaf damage appeared on *T. cordata* seedlings on day 35 from the beginning of the salt treatment. The salt ions were accumulated in the leaves to the level of toxic concentration, which caused leaf damage and the decay of photosynthetic pigments. According to Munns and Tester [1], the "ion-specific" phase of plant response to salinity begins when a toxic level of salt concentration in leaf tissues is reached. According to these authors, the leaf development during long-term salinity is mainly influenced by the ability of plants to prevent the accumulation of salt ions in assimilation organs in addition to producing new leaves faster than salt-intoxicated leaves are dying.

The control of salt uptake by the roots and the regulation of salt distribution to the aboveground organs are important mechanisms in preventing the concentration of salt ions in plant leaves. It is considered to be a significant characteristic of the plant tolerance to salinity [45–49]. Both studied tree species accumulated salt ions mainly in the root (*Pyrus* 75% and *Tilia* 55%). However, when the same cumulative volume of the salt solution was applied to the substrate (365 mL) in each plant, *T. cordata* seedlings deposited 4.5 times more Na^+ ions in the plant tissues (Figure 5) and specifically distributed six times more Na^+ ions to the leaves compared to *P. pyraster*. A high salt uptake negatively affected the condition and functionality of the assimilation apparatus of *T. cordata*. *P. pyraster* blocked the salt uptake in the root system (75%) and deposited a portion of the Na^+ ions the stem (19%), thus protecting the leaf apparatus from intoxication and damage. The effective limitation of the salt uptake and the deposition of ingested salt ions in the root can be considered a manifestation of the tolerance of *P. pyraster* seedlings to salinity. The mechanisms for blocking salt ions in the stem and root system have been identified for *P. amygdaliformis* and *P. elaeagrifolia* [50]. These species showed no symptoms of leaf damage even after 30 days of 150 mM NaCl treatment nor any changes in the photosynthetic rates. A similar restriction and blocking of Na^+ ions in trunk wood is reported for *P. communis* by Boland et al. [51].

Salinity affects nutrient uptake and significantly reduces K^+ content in the plant tissues of many species [52–54]. The exclusion of Na^+ from plant tissues is considered a mechanism for the optimization of the K^+/Na^+ ratio [55,56]. The ability of plants to optimize the K^+/Na^+ ratio (through K^+ retention or the prevention of Na^+ accumulation in leaves) is considered to be a key feature of salt tolerance [57,58]. Under salt treatment, *T. cordata* significantly increased the uptake and distribution of K^+ to the leaves and stems, while *P. pyraster* maintained a balanced K^+ content in only the leaves, probably due to the effective restriction of Na^+ transfer and deposition in leaves (Figure 5A,B). The seedlings of the studied tree species each applied different strategies for the optimization of the K^+/Na^+ ratio in the plant tissues, i.e., an increased K^+ uptake (*T. cordata*) or restricted uptake and transfer of Na^+ (*P. pyraster*).

Salinity also affects the water–plant relationship, as the salts present in the substrate prevent the absorption of water by the root system due to the osmotic effect [59]. The accumulation of salts in the root zone causes a decrease in the osmotic potential, with a consequent decrease in the water potential [60–62]. Under salinity conditions, plants show symptoms of dehydration, indicating a lower water potential due to poor water uptake or less water availability in the substrate [63,64]. *P. pyraster* quickly reduced water uptake due to the higher concentration of the salt ions in the root zone. The decrease in the RWC (Figure 4) was the result of a high concentration of the external salt solution, which according to Greenway and Munns [65], causes osmotic stress and dehydration at the cellular level. *P. pyraster* was effective in the restriction of salt uptake and transport to the leaves. Salt ions were maintained in the root system and in the stem, which allowed the restoration of the water uptake through a slight increase in the water potential of the leaf tissues (Figure 3). The balanced values of the water potential were maintained for *P. pyraster* even under increasing cumulative salt content in the substrate. The decrease in the RWC and Ψ_{wl} under increasing salinity was observed for *T. cordata* seedlings. The significant decrease in Ψ_{wl} (−2.48 MPa) was accompanied by the severe symptoms of the leaf damage. *T. cordata* was not effective in blocking salt uptake by the root system and transported Na^+ ions to the leaves that were damaged due to salt accumulation. The *T. cordata* seedlings were particularly affected by ionic stress, which dominates only at high levels of salinity or in sensitive species that lack the ability to control Na^+ transport [1].

The data obtained in our study indicate the species-specific content of photosynthetic pigments in the leaves of *T. cordata* and *P. pyraster*. The contents of Chl*a*, Chl*b*, and Chl*ab* were significantly reduced only for *T. cordata* seedlings after the regular application of 100 mM NaCl solution in the total cumulative volume of 365 mL per plant. The chlorophyll parameters were not reduced for *P. pyraster* compared to the control plants. The balanced content of photosynthetic pigments indicates the tolerance of *P. pyraster* to salinity. Saline-tolerant plants have increased or unchanged chlorophyll content in the leaves when treated with salt, while chlorophyll content decreases in saline-sensitive plants [66,67]. Chlorophyll content in leaves is considered to be a biochemical indicator of the plant's tolerance to salinity [54]. Compared to *T. cordata*, *P. pyraster* seedlings have a significantly higher content of carotenoids, which, according to several studies, stabilize the photochemical processes of photosynthesis under stress conditions [68].

The seedlings of *P. pyraster* and *T. cordata* had reduced g_s in the early response to salt treatment (after 20 days). The decrease in the transpiration rate (E) was observed only for *P. pyraster*. Stomatal plant responses are induced by the osmotic effect of salinity in the root zone [1]. A reduction in g_s can prevent excessive water loss by transpiration, similarly to what has been observed for plants in desiccating soil [69]. *P. pyraster* limited water loss in the initial (osmotic) phase of salt stress and maintained a steady net photosynthetic rate (A_n) per unit area throughout the whole experiment. The rate of photosynthesis per leaf unit area often remains unchanged for salt-treated plants, even under reduced stomatal conductance [70]. It is associated with changes in the cell anatomy that give rise to smaller, thicker leaves and lead to higher chloroplast densities per unit leaf area [1]. As these authors state, when photosynthesis is expressed on the basis of a chlorophyll unit and not on the basis of leaf area, the reduction due to salinity can usually be measured. Therefore, the reduction in the leaf area due to salinity means that photosynthesis per plant is always reduced.

The ratio of stomatal opening and the level of photosynthesis (water use efficiency, WUE) are used as indicators of plant tolerance to osmotic stress [2,71]. *T. cordata* seedlings responded to salinity by increasing WUE. The increase in WUE was evident in two stages (on days 20 and 40) of the experiment, which is related to the increasing cumulative NaCl uptake per plant. Due to the high accumulation of Na^+ ions in the leaves (ionic stress), the abovementioned adaptive reactions decreased the water potential of the leaf tissues and also significantly decreased the net photosynthetic rate (A_n) on day 50 of the experiment. Despite the adaptive reactions of *T. cordata* seedlings to osmotic stress, Ψ_{wl} significantly decreased on day 40 of the experiment due to the high accumulation of Na^+ ions in the leaves (ionic stress). In the long term, a high salinity level had a negative impact on photosynthesis (decrease in A_n) in *T. cordata* seedlings. The accumulated salts in the leaves may inhibit enzymes involved in carbohydrate metabolism or exert a direct toxic effect on photosynthetic processes [1].

5. Conclusions

Differences in tolerance to salinity were identified for *P. pyraster* and *T. cordata* seedlings as well as different mechanisms when coping with salt stress. The results document the significance and role of the root system in the resistance and survival of woody plants under saline conditions. Balanced root growth improves water supply and has a positive effect via the retention of toxic salt ions, which restricts their transfer and accumulation in leaves.

Under salt treatment, *P. pyraster* maintained a balanced growth and biomass accumulation in the root, as observed in the balanced root parameters RL, SRL, FW_R, and DW_R, and the increased number of root tips. The *P. pyraster* seedlings reduced the size of the leaf area in response to osmotic stress, i.e., increased salinity in the root zone. The seedlings formed smaller leaves, effectively regulated stomatal conductance, and maintained balanced values of RWC and Ψ_{wl}. *P. pyraster* blocked salt uptake via the root system and in the stem, thus protecting the leaf apparatus from intoxication and damage. The rate of photosynthesis per leaf unit area (A_n) remained unchanged under increasing salinity throughout the whole experiment. The reduced water loss, effective limitation of the salt uptake, deposition of ingested salt ions in the stem and roots, balanced content of photosynthetic pigments, and protection of the leaf apparatus against intoxication observed in *P. pyraster* seedlings are indicative of their tolerance to salinity.

We consider the effective limitation of the salt uptake and deposition of the ingested salt ions in the root to be a manifestation of the tolerance of *P. pyraster* seedlings to salinity. This tree species reduced water loss in the initial (osmotic) phase of salt stress and maintained a steady rate of photosynthesis (A_n) per unit area throughout the whole experiment.

T. cordata seedlings regulated water loss (g_s) and maintained balanced photosynthesis (A_n) in the early response to osmotic stress. However, they were not effective in the restriction of salt uptake by the root system. Na^+ ions were transported to the leaves, which were damaged by the accumulation of salt ions. The seedlings were not able to cope with the ionic stress that dominates at high levels of salinity or in sensitive species that lack the ability to control Na^+ transport. Despite adaptive reactions (g_s regulation, growth restriction), a significant decrease in Ψ_{wl} and A_n was observed for salt-treated *T. cordata* seedlings. The obtained findings show weak tolerance to salinity for *T. cordata* in the juvenile stage of growth.

This comparison of the early stages of growth shows that *P. pyraster* can cope with salinity in urban conditions. However, a deeper study of the effects of different forms and levels of salinity on these tree species is necessary in order to further our understanding of the mechanisms of salinity tolerance.

Author Contributions: Conceptualization and design, V.P. and M.H.; investigation, M.H. and H.L.; methodology. V.P. and M.H.; resources, H.L.; formal analysis, V.P. and M.H.; validation, H.L.; writing the first draft, V.P. and M.H.; writing—review and editing, V.P. and M.H.; funding acquisition. V.P. All authors have read and agreed to the published version of the manuscript.

Funding: This research was funded by Kultúrna a Edukačná Grantová Agentúra MŠVVaŠ SR (Cultural and Educational Grant Agency of the Ministry of Education, Science, Research and Sport of the Slovak Republic) grant number 007SPU-4/2020. This publication was supported by the Operational Programme Integrated Infrastructure within the Sustainable Smart Farming Systems Taking into Account Future Challenges 313011W112 project. This study was co-financed by the European Regional Development Fund.

Institutional Review Board Statement: Not applicable.

Informed Consent Statement: Not applicable.

Data Availability Statement: The data presented in this study are available on request from the corresponding author.

Conflicts of Interest: The authors declare that they have no conflict of interest.

References

1. Munns, R.; Tester, M. Mechanisms of salinity tolerance. *Ann. Rev. Plant Biol.* **2008**, *59*, 651–681. [CrossRef] [PubMed]
2. Acosta-Motos, J.R.; Ortuño, M.F.; Bernal-Vicente, A.; Diaz-Vivancos, P.; Sanchez-Blanco, M.J.; Hernandez, J.A. Plant responses to salt stress: Adaptive mechanisms. *Agronomy* **2017**, *7*, 18. [CrossRef]
3. Kordrostami, M.; Rabiei, B. Salinity Stress Tolerance in Plants: Physiological, Molecular, and Biotechnological Approaches. In *Plant Abiotic Stress Tolerance*; Hasanuzzaman, M., Hakeem, K., Nahar, K., Alharby, H., Eds.; Springer: Cham, Switzerland, 2019.
4. Azza, M.A.M.; Fatma, A.M.; Farahat, M.M. Responses of ornamental plants and woody trees to salinity. *World J. Agric. Sci.* **2007**, *3*, 386–395.
5. Qados, A.M.A. Effect of salt stress on plant growth and metabolism of bean plant *Vicia faba* (L.). *J. Saudi Soc. Agric. Sci.* **2011**, *10*, 7–15.
6. Parida, A.K.; Das, A.B. Salt tolerance and salinity effects on plants: A review. *Ecotoxicol. Environ. Saf.* **2005**, *60*, 324–349. [CrossRef]
7. Ashraf, M.; Harris, P.J.C. Potential biochemical indicators of salinity tolerance in plants. *Plant Sci.* **2004**, *166*, 3–16. [CrossRef]
8. Geissler, N.; Hussin, S.; Koyro, H.W. Interactive effects of NaCl salinity and elevated atmospheric CO_2 concentration on growth, photosynthesis, water relations and chemical composition of the potential cash crop halophyte *Aster tripolium L.*. *Environ. Exp. Bot.* **2009**, *65*, 220–231. [CrossRef]
9. Zhang, Z.-H.; Liu, Q.; Song, H.-X.; Rong, X.-M.; Ismail, A.M. Responses of different rice (*Oryza sativa L.*) genotypes to salt stress and relation to carbohydrate metabolism and chlorophyll content. *Afr. J. Agric. Res.* **2012**, *7*, 19–27.
10. Pauleit, S.; Jones, N.; Garcia-Martin, G.; Garcia-Valdecantos, J.L.; Rivière, L.M.; Vidal-Beaudet, L.; Bodson, M.; Randrup, T.B. Tree establishment practice in towns and cities–Results from a European survey. *Urban For. Urban Green.* **2002**, *1*, 83–96. [CrossRef]
11. Pigott, C.D. *Tilia cordata* Miller. (*T. europaea* L. pro parte, *T. parvifolia* Ehrh. ex Hoffm., *T. sylvestris* Desf., *T. foemina folio minore* Bauhin). *J. Ecol.* **1991**, *79*, 1147–1207. [CrossRef]
12. Rameau, J.C.; Mansion, D.; Dumé, G.; Timbal, J.; Lecointe, A.; Dupont, P.; Keller, R. *Flore Forestière Française. Plaines et Collines*; IDF: Paris, France; ENGREF: Nancy, France, 1989.
13. Niinemets, Ü.; Valladares, F. Tolerance to shade, drought, and waterlogging of temperate Northern Hemisphere trees and shrubs. *Ecol. Monogr.* **2006**, *76*, 521–547. [CrossRef]
14. Pigott, D. *Lime-Trees and Basswoods: A Biological Monograph of the Genus Tilia*, 1st ed.; Cambridge University Press: New York, NY, USA, 2012; p. 405.
15. De Jaegere, T.D.; Hein, S.; Claessens, H. A Review of the Characteristics of Small-Leaved Lime (*Tilia cordata Mill.*) and Their Implications for Silviculture in a Changing Climate. *Forests* **2016**, *7*, 56. [CrossRef]
16. Moser, A.; Rötzer, T.; Pauleit, S.; Pretzsch, H. Structure and ecosystem services of small-leaved lime (*Tilia cordata Mill.*) and black locust (*Robinia pseudoacacia L.*) in urban environments. *Urban For. Urban Green.* **2015**, *14*, 1110–1121. [CrossRef]
17. Jensen, J.S. Lime (*Tilia cordata* and *Tilia platyphyllos*). In *EUFORGEN Technical Guidelines for Genetic Conservation and Use for Lime (Tilia spp.)*; International Plant Genetic Resources Institute: Rome, Italy, 2003.
18. Martynova, M.; Sultanova, R.; Odintsov, G.; Sazgutdinova, R.; Khanova, E. Growth of *Tilia cordata Mill.* in urban forests. *South East Eur. For. SEEFOR* **2020**, *11*, 51–59. [CrossRef]
19. Stephan, B.R.; Wagner, I.; Kleinschmit, J. *EUFORGEN Technical Guidelines for Genetic Conservation and Use for Wild Apple and Pear (Malus sylvestris and Pyrus pyraster)*; International Plant Genetic Resources Institute: Rome, Italy, 2003.
20. Wagner, I.; Büttner, R. Hybridization in wild pear (*Pyrus pyraster*) from various regions in Germany and from Luxembourg with respect to Pyrus × communis. In Proceedings of the III International Symposium on Horticulture in Europe-SHE2016, Chania, Greece, 17–21 October 2016; pp. 427–434.
21. Ellenberg, H. *Vegetation Mitteleuropas mit den Alpen in Ökologischer Sicht*, 3rd ed.; Ulmer Verlag: Stuttgart, Germany, 1983.
22. Paganová, V. Wild pear *Pyrus pyraster* (L.) Burgsd. requirements on environmental conditions. *Ekológia* **2003**, *22*, 225–241.
23. Milner, E. *Trees of Britain and Ireland*; Natural History Museum: London, UK, 2011.
24. Lawesson, J.E.; Oksanen, J. Niche characteristics of Danish woody species as derived from coenoclines. *J. Veg. Sci.* **2002**, *13*, 279–290. [CrossRef]
25. Shannon, M.C.; Grieve, C.M.; Francois, L.E. Whole-plant response to salinity. In *Plant—Environment Interactions*; Wilkinson, R.E., Ed.; Marcel Dekker: New York, NY, USA, 1994; pp. 199–244.
26. Kozlowski, T.T. Responses of woody plants to flooding and salinity. *Tree Physiol.* **1997**, *17*, 490. [CrossRef]
27. Lapin, M.; Faško, P.; Melo, M.; Šťastný, P.; Tomlain, J. Klimatické oblasti. In *Atlas Krajiny Slovenskej Republiky*; Ministerstvo životného prostredia: Banská Bystrica, Slovakia, 2002; p. 344.
28. Trautmann, N.; Richard, T. Moisture Content. Cornell Waste Management Institute. 1996. Available online: http://compost.css.cornell.edu/calc/moisture_content.html (accessed on 15 August 2021).
29. Ostonen, I.; Püttstep, U.; Biel, C.; Alberton, O.; Bakker, M.R.; Lõhmus, H.; Majdi, D.; Metcalfe, D.; Olsthoorn, A.F.M.; Pronk, A.; et al. Specific root length as an indicator of environmental changes. *Plant Biosyst.* **2007**, *141*, 3426–3442. [CrossRef]
30. Poorter, H.; Niklas, K.J.; Reich, P.B.; Oleksyn, J.; Poot, P.; Mommer, L. Biomass allocation to leaves, stems and roots: Meta-analyses of interspecific variation and environmental control. *New Phytol.* **2012**, *193*, 30–50. [CrossRef]
31. Parsons, R.; Weyers, J.D.B.; Lawson, T.; Godber, I.M. Rapid and straightforward estimates of photosynthetic characteristics using a portable gas exchange system. *Photosynthetica* **1998**, *34*, 265–279. [CrossRef]
32. Hunt, S. Measurements of photosynthesis and respiration in plants. *Physiol. Plant.* **2003**, *117*, 314–325. [CrossRef] [PubMed]

33. Barrs, H.D.; Weatherley, P.E. A re-examination of the relative turgidity technique for estimating water deficit in leaves. *Aust. J. Biol. Sci.* **1962**, *15*, 413–428. [CrossRef]
34. Lichtenthaler, H.K.; Buschmann, C. Extraction of photosynthetic tissues: Chlorophylls and carotenoids. *Curr. Protoc. Food Anal. Chem.* **2001**, *1*, F4.2.1–F4.2.6. [CrossRef]
35. Kovacik, A.; Tvrda, E.; Miskeje, M.; Arvay, J.; Tomka, M.; Zbynovska, K. Trace metals in the freshwaterfish *Cyprinus carpio*: Effect to serum biochemistry and oxidative status markers. *Biol. Trace Elem. Res.* **2018**, *188*, 494–507. [CrossRef]
36. Ziska, L.H.; Seeman, L.H.; DeJong, T.M. Salinity induced limitations on photosynthesis in *Prunus salicina*, a deciduous tree species. *Plant Physiol.* **1990**, *93*, 864–870. [CrossRef]
37. Zekri, M.; Parsons, L.R. Response of split-root sour orange seedlings to NaCl and polyethylene glycol stresses. *J. Exp. Bot.* **1990**, *41*, 35–40. [CrossRef]
38. Abassi, M.; Mguis, K.; Béjaoui, Z.; Albouchi, A. Morphogenetic responses of *Populus alba L.* under salt stress. *J. For. Res.* **2014**, *25*, 155–161. [CrossRef]
39. Chen, P.F.; Zuo, L.H.; Yu, X.Y.; Dong, Y.; Zhang, S.; Yang, M. Response mechanism in Populus× euramericana cv. '74/76′ revealed by RNA-seq under salt stress. *Acta Physiol. Plant* **2018**, *40*, 96. [CrossRef]
40. Paganová, V.; Jureková, Z.; Lichtnerová, H. The nature and way of root adaptation of juvenile woody plants Sorbus and Pyrus to drought. *Environ. Monit. Assess.* **2019**, *191*, 714. [CrossRef]
41. Paganová, V.; Hus, M.; Jureková, Z. Physiological performance of *Pyrus pyraster* L. (Burgsd.) and *Sorbus torminalis (L.)* crantz seedlings under drought treatment. *Plants* **2020**, *9*, 1496. [CrossRef]
42. Hsiao, T.C.; Xu, L.K. Sensitivity of growth of roots versus leaves to water stress: Biophysical analysis and relation to water transport. *J. Exp. Bot.* **2000**, *51*, 1595–1616. [CrossRef]
43. Cassaniti, C.; Leonardi, C.; Flowers, T.J. The effects of sodium chloride ornamental shrubs. *Sci. Hortic.* **2009**, *122*, 586–593. [CrossRef]
44. Cassaniti, C.; Romano, D.; Flowers, T.J. The response of ornamental plants to saline irrigation water. In *Irrigation Water Management. Pollution and Alternative Strategies*; Garcia-Garizabal, I., Ed.; InTech Europe: Rijeka, Croatia, 2012; pp. 132–158.
45. Boursier, P.; Läuchli, A. Growth responses and mineral nutrient relations of salt stressed sorghum. *Crop Sci.* **1990**, *30*, 1226–1233. [CrossRef]
46. Maathuis, F.J.M.; Amtmann, A. K+ nutrition and Na+ toxicity: The basis of cellular K+/Na+ ratios. *Ann. Bot.* **1999**, *84*, 123–133. [CrossRef]
47. Pérez-Alfocea, F.; Balibrea, M.E.; Alarçon, J.J.; Bolarín, M.C. Composition of xylemand phloem exudates in relation to the salt tolerance of domestic and wild tomato species. *J. Plant Physiol.* **2000**, *156*, 367–374. [CrossRef]
48. Colmer, T.D.; Munns, R.; Flowers, T.J. Improving salt tolerance of wheat and barley: Future prospects. *Aust. J. Exp. Agric.* **2005**, *45*, 1425–1443. [CrossRef]
49. Murillo-Amador, B.; Troyo-Diéguez, E.; García-Hernández, J.L.; López-Aguilar, R.; Ávila-Serrano, N.Y.; Zamora-Salgado, S.; Rueda-Puente, E.O.; Kaya, C. Effect of NaCl salinity in the genotypic variation of cowpea (*Vigna unguiculata*) during early vegetative growth. *Sci. Hortic.* **2006**, *108*, 423–441. [CrossRef]
50. Matsumoto, K.; Tamura, F.; Chun, J.; Tanabe, K. Native Mediterranean *Pyrus rootstock, P. amygdaliformis* and *P. elaeagrifolia* present higher tolerance to salinity stress compared with Asian natives. *J. Jpn. Soc. Hort. Sci.* **2006**, *75*, 450–457. [CrossRef]
51. Boland, A.M.; Jerie, P.; Maas, E. Long-term effects of salinity on fruit trees. *Acta Hortic.* **1997**, *449*, 599–606. [CrossRef]
52. Munns, R.; Termaat, A. Whole-plant responses to salinity. *Funct. Plant Biol.* **1986**, *13*, 143–160. [CrossRef]
53. Laffray, X.; Alaoui-Sehmer, L.; Bourioug, M.; Bourgeade, P.; Alaoui-Sossé, B.; Aleya, L. Effects of sodium chloride salinity on ecophysiological and biochemical parameters of oak seedlings (*Quercus robur L.*) from use of de-icing salts for winter road maintenance. *Environ. Monit Assess.* **2018**, *190*, 266. [CrossRef] [PubMed]
54. Rahneshan, Z.; Nasibi, F.; Moghadam, A.A. Effects of salinity stress on some growth, physiological, biochemical parameters and nutrients in two pistachio (*Pistacia vera L.*) rootstocks. *J. Plant Interact.* **2018**, *13*, 73–82. [CrossRef]
55. Munns, R. Comparative physiology of salt and water stress. *Plant Cell Environ.* **2002**, *25*, 239–250. [CrossRef] [PubMed]
56. Tester, M.; Davenport, R. Na+ Tolerance and Na+ Transport in Higher Plants. *Ann. Bot.* **2003**, *91*, 503–527. [CrossRef]
57. Shabala, S.; Cuin, T.A. Potassium transport and plant salt tolerance. *Physiol. Plant.* **2008**, *133*, 651–669. [CrossRef]
58. Zhang, X.; Liu, L.; Chen, B.; Qin, Z.; Xiao, Y.; Zhang, Y.; Yao, R.; Liu, H.; Yang, H. Progress in Understanding the Physiological and Molecular Responses of Populus to Salt Stress. *Int. J. Mol. Sci.* **2019**, *20*, 1312. [CrossRef]
59. Navarro, A.; Bañón, S.; Conejero, W.; Sánchez-Blanco, M.J. Ornamental characters. ion accumulation and water status in *Arbutus unedo* seedlings irrigated with saline water and subsequent relief and transplanting. *Environ. Exp. Bot.* **2008**, *62*, 364–370. [CrossRef]
60. Franco, J.A.; Bañón, S.; Vicente, M.J.; Miralles, J.; Martínez-Sánchez, J.J. Root development in horticultural plants grown under abiotic stress conditions—A review. *J. Hortic. Sci. Biotechnol.* **2011**, *86*, 543–556. [CrossRef]
61. Sánchez-Blanco, M.J.; Rodríguez, P.; Olmos, E.; Morales, M.A.; Torrecillas, A. Differences in the effects of Simulated Sea Aerosol on Water Relations, Salt Content, and Leaf Ultrastructure of Rock-Rose Plants. *J. Environ. Qual.* **2004**, *33*, 1369–1375. [CrossRef]
62. Slama, I.; Ghnaya, T.; Savouré, A.; Abdelly, C. Combined effects of long-term salinity and soil drying on growth, water relations, nutrient status and proline accumulation of *Sesuvium portulacastrum*. *C. R. Biologies* **2008**, *331*, 442–451. [CrossRef]

63. Álvarez, S.; Gómez-Bellot, M.J.; Castillo, M.; Bañón, S.; Sánchez-Blanco, M.J. Osmotic and saline effect on growth water relations and ion uptake and translocation in *Phlomis purpurea* plants. *Environ. Exp. Bot.* **2012**, *78*, 138–145. [CrossRef]
64. Sánchez-Blanco, M.J.; Rodríguez, P.; Morales, M.A.; Ortuño, M.F.; Torrecillas, A. Comparative growth and water relations of *Cistus albidus* and *Cistus monspeliensis* plants during water deficit conditions and recovery. *Plant Sci.* **2002**, *162*, 107–113. [CrossRef]
65. Greenway, H.; Munns, R. Mechanisms of Salt Tolerance in Non-Halophytes. *Annu. Rev. Plant Physiol. Plant Mol. Biol.* **1980**, *31*, 149–190. [CrossRef]
66. Stepien, P.; Johnson, G.N. Contrasting responses of photosynthesis to salt stress in the glycophyte Arabidopsis and the halophyte thellungiella: Role of the plastid terminal oxidase as an alternative electron sink. *Plant Physiol.* **2009**, *149*, 1154–1165. [CrossRef]
67. Ashraf, M.; Harris, P.J.C. Photosynthesis under stressful environments: An overview. *Photosynthetica* **2013**, *51*, 163–190. [CrossRef]
68. Demmig-Adams, B.; Adams, W.W. The role of xanthophylls cycle carotenoids in the protection of photosynthesis. *Trends. Plant Sci.* **1996**, *1*, 21–26. [CrossRef]
69. Davies, W.J.; Kudoyarova, G.; Hartung, W. Long-distance ABA signaling and its relation to other signaling pathways in the detection of soil drying and the mediation of the plant's response to drought. *J. Plant Growth Regul.* **2005**, *24*, 285–295. [CrossRef]
70. James, R.A.; Rivelli, A.R.; Munns, R.; Caemmerer, S.V. Factors affecting CO_2 assimilation.leaf injury and growth in salt-stressed durum wheat. *Funct. Plant Biol.* **2002**, *29*, 1393–1403. [CrossRef]
71. Chaves, M.M.; Osorio, J.; Pereira, J.S. Water use efficiency and photosynthesis. In *Water Use Efficiency in Plant Biology*; Bacon, M., Ed.; Blackwell Publishing: Oxford, UK, 2004; pp. 42–74.

horticulturae

MDPI

Article

Salt Spray and Surfactants Induced Morphological, Physiological, and Biochemical Responses in *Callistemon citrinus* (Curtis) Plants

Stefania Toscano, Giovanni La Fornara and Daniela Romano *

Department of Agriculture, Food and Environment (Di3A), Università degli Studi di Catania, Via Valdisavoia 5, 95123 Catania, Italy; stefania.toscano@unict.it (S.T.); giovannilafornara@alice.it (G.L.F.)
* Correspondence: dromano@unict.it

Abstract: The growth and aesthetic value of ornamental plant species used near coastlines are negatively influenced by salt spray. The presence of surfactants could enhance salt damage. To analyze the influences of salt spray and surfactants alone and in combination with each other, individual *Callistemon* plants were subjected to different treatments for 8 weeks: a solution simulating the composition of seawater (salt spray), a solution containing an anionic surfactant (surfactant), a solution with salt spray and anionic surfactant (salt plus surfactants), and deionized water (control). To study the influence of different climatic conditions, two growing periods, from January to March (I CP) and from May to July (II CP), were established. Salt spray, alone or with surfactant action, influences plants' growth and aesthetic features in different cycle periods. The percentage of leaf damage significantly increased with salt spray and salt plus surfactants during II CP (~27%). Additionally, the Na^+ and Cl^- contents were enhanced in the leaves in both CPs, but the contents in the roots were only enhanced in the II CP. The gas exchanges were significantly influenced by the treatments, especially during the II CP, when a reduction in net photosynthesis due to salt spray was observed starting from the second week of stress. At the end of the experiment, in both cycle periods, the leaf proline content increased in the salt spray and salt plus surfactants treatments. In both CPs, PCA revealed that the morphological and physiological parameters were directly associated with the control and surfactants treatments, whereas the mineral contents and biochemical parameters were directly correlated with the salt and salt plus surfactants treatments. The additive effect of surfactant stress, compared to salt stress, did not appear to be significant, with the exception of CP II, and for some parameters, the solubilization action of surfactants was favored by higher temperatures.

Keywords: coastal areas; ornamental plants; gas exchange; chlorophyll *a* fluorescence; enzyme activity; proline

Citation: Toscano, S.; La Fornara, G.; Romano, D. Salt Spray and Surfactants Induced Morphological, Physiological, and Biochemical Responses in *Callistemon citrinus* (Curtis) Plants. *Horticulturae* **2022**, *8*, 261. https://doi.org/10.3390/horticulturae8030261

Academic Editor: Yanyou Wu

Received: 18 February 2022
Accepted: 16 March 2022
Published: 17 March 2022

1. Introduction

In coastal gardens and ornamental green areas, one of the main problems observed is related to salt stress originating from sea spray, which damages plants [1,2]. In this landscape, salt spray may limit plant growth and survival [3], altering coastal species' composition [4] and thus influencing the ornamental and ecological value of plant communities [5,6].

The negative effects of salts on ornamental plants are observed above all on their aesthetic appearance, which is an essential trait [7]. The visual aspect, although a subjective parameter, is of fundamental importance when evaluating the salt tolerance of landscape plants [8]. Among the different species used in urban landscaping, salt tolerance differs considerably [9] according to the morphological and physiological characteristics of different genotypes. Most ornamental plant species are non-halophytes; therefore, the assessment of salt tolerance is necessary [10]. The species selected for inclusion in urban green areas must not only survive but must also maintain suitable aesthetic characteristics [11].

Although plant species typical of coastal areas are resistant to the presence of salt on their leaves [12], plant growth and reproduction can be negatively affected when they are exposed to salt spray, and when salt water penetrates into the groundwater [13]. When salt is applied to roots, the damage is less significant than that seen from salt sprays [14]. Salt spray can reduce plant growth because it determines water stress; leads to the disruption of membranes and enzyme systems; causes the inhibition of nutrient uptake, necrosis, and the loss of leaves, and can lead to mortality [15].

In modern societies, surfactants are often used in the form of detergents and pesticides. Given their wide usage, these substances cause pollution in aquatic environments. The foliar absorption of sea salt through stomatal and cuticular penetration can be increased by the presence of surfactants in seawater [16].

The wide uses of surfactants in chemical industries, such as household products, industrial cleaning, ink, pharmaceuticals, and personal care, can affect the environment and human health. In 2014, 15.93 million tons of surfactants were used, and in 2022 this figure is expected to be over 1/3 more (24.19 million tons of surfactants) [17]. The phenomenon is worrying for the coastal ecosystem because about 10% of this amount is released into the sea [18].

The damage to coastal flora is mainly caused by the synergic effects of marine salt and surfactants, but also by the direct actions of the surfactants [19] that are responsible for the solubilization of cell membranes [20], the increase in the permeability of cuticles [21], and the dissolution of epicuticular waxes [22], all of which are processes that facilitate the foliar absorption of salt, and its phytotoxic effects. The effect of surfactants is to degenerate epicuticular waxes and modify the photosynthetic processes by causing the leaves to absorb greater quantities of sea salt [23].

Bussotti et al. [24] observed that resistance to salt spray depends on different behaviors determined by leaf structure (sclerophyllia, cuticle thickness). Reductions in gas exchanges and photosystem II (PSII) efficiency have been observed in the presence of salt spray because, although the cuticle of the leaves protects against external agents, it cannot prevent the penetration of ions and the consequent osmotic and ionic stress [25]. Modifications in the chlorophyll content and consequently reductions in photosynthesis activity occur in the presence of saline stress [26]. Stomatal limitation reduces the diffusion of CO_2 in the mesophyll and can contribute to the generation of additional reactive oxygen species (ROS), which may cause irreversible damage [27].

Photosynthetic pigments can be damaged by ROS, altering the metabolism because they are highly reactive in the absence of protective mechanisms [28]. In cellular compartments, to counteract the effects of reactive oxygen species, plants have developed different mechanisms. Generally, plants activate various antioxidant enzymes (superoxide dismutase (SOD), catalase (CAT), and glutathione peroxidase (GPX)) [29]. In many species, a close relationship is clear between stress tolerance and antioxidant activity [30–32]. Catalase, guaiacol peroxidase, and L-ascorbate peroxidase are all involved in the plant's responses to biotic and abiotic stresses. In particular, these enzymes are involved in the scavenging of reactive oxygen species (ROS), partially reduced forms of atmospheric oxygen, that are highly reactive and capable of causing oxidative damage to the cell.

Among the ROS, the most dangerous is the superoxide anion; this is scavenged in plants by superoxide dismutase, which converts the superoxide anion to hydrogen peroxide [33]. Furthermore, in the context of the detoxification of toxic ROS, the superoxide anion is a fundamental component of the ascorbate-glutathione cycle [34].

In the presence of abiotic stress, plants accumulate metabolic components, which act as plant osmotic regulators, such as proline (Pro) and malondialdehyde (MDA) [35–37].

Few studies have been conducted on the response of plants to interactions between natural stresses in coastal areas; research has generally focused on single stress, leading to a suitable but incomplete understanding of the response mechanism of plants [38]. The effect of stress depends on both the intensity of the stress factor and on its duration and therefore makes it more difficult to understand the response mechanisms and interactions between other stresses [39]. In coastal areas, salt spray is often associated with pollutants, and the role of surfactants is critical in some cases of environmental damage to coastal vegetation [40]. Interactions with environmental conditions are often important as well; surfactant experiments carried out during winter produced damage lower down in the plants than experiments in warmer seasons [41].

To better understand the influence of some of the stronger factors affecting the ornamental plants used in the coastal landscape, we selected *Callistemon* of the Myrtaceae family; this entire genus is endemic of Australia and occurs in the form of shrubs or small trees. The *Callistemon* genus is characterized by a suitable tolerance to drought, salt, and root restriction stresses [42–44]. One of the most interesting ornamental species is *Callistemon citrinus* (Curtis) Skeels, characterized by fast growth and abundant blooming with bright colors and various shapes and volumes [45].

The research related to the effects of salt spray and its interactions with surfactants in ornamental shrubs is sparse. The aims of this study were to define (i) the influence of salt spray and surfactants and their interactions on *Callistemon* plants during two cycle periods with different climatic conditions; (ii) the morphological, physiological, and biochemical mechanisms involved in responding to salt spray and surfactants; (iii) the relationships among different parameters, in order to better understand the action mechanisms.

2. Materials and Methods

2.1. Plant Materials and Treatments

Both cycle periods were operated in an unheated greenhouse located near Catania, Italy (37°41′ N 15°11′ E 89 m a.s.l.). In 3 L pots (16 cm), filled with peat and perlite (2:1 *v*/*v*), rooted cutting (two months old) of *Callistemon* (*Callistemon citrinus* (Curtis) Skeels) were transplanted; 2 g L^{-1} of Osmocote Plus (14:13:13 N, P, K plus microelements) was added to each pot. A drip system with one 2 L h^{-1} emitter into each pot was used, and the plants were irrigated daily. The plants were subjected for 8 weeks to different treatments: (a) a solution simulating the composition of seawater-salt spray [46] with a concentration of 401.8 mM (salt spray); (b) a solution containing an anionic surfactant (sodium dodecylbenzene-sulfonate 82.52%, 50 mg L^{-1}) (surfactants) [3]; (c) a solution containing salt spray and anionic surfactant (salt plus surfactants); (d) control plants treated only with deionized water (control). The plants were sprayed twice a week with different aqueous solutions. To avoid salt deposition on the soil, plastic discs were placed over the soil surfaces of the potted plants. To investigate the effects of different climatic conditions, two growing periods were considered: January to March (I CP), with transplant in November, and May to July (II CP), with transplant in February.

2.2. Plant Materials and Treatments

At the ends of both cycle periods, three pots per replication were randomly chosen for the determination of shoots' and roots' fresh and dry biomass, total leaf area, leaf number, chlorophyll content, and leaf damage. The dry biomass (DW) of the shoots (leaves plus stems) and the whole plant (leaves, stems, and roots) was determined by drying the weighed samples in a thermo-ventilated oven at 70 °C until constant weights were obtained. A leaf area meter (Delta-T Devices Ltd., Cambridge, U.K.) was used for determining the total leaf area and percentage of leaf damage (necrotic area on total area). The protocol of Moran and Porath [47] was adopted to measure the chlorophyll content. Dry leaves and roots were ground in a Wiley Mill and then passed through a 20-mesh screen for the mineral analysis. Na^+ and Cl^- contents were determined by chromatography on a Dionex IC 25 (Dionex Corp., Sunnyvale, CA, USA). Ion concentrations were expressed in g kg^{-1} DW.

2.3. Chlorophyll Content, Gas Exchanges, Chlorophyll a Fluorescence and RWC Measurements

The gas exchanges were registered during both cycle periods, every two weeks (10:00 am and 2:00 pm). Net photosynthetic rate (A_N) and stomatal conductance (gs) were registered on fully expanded leaves through a CO_2/H_2O IRGA (LCi, ADC Bioscientific Ltd., Hoddesdon, UK). Simultaneously, the efficiency of PSII was registered using a modulated chlorophyll fluorimeter OS1-FL (Opti-Sciences Corporation, Tyngsboro, MA, USA). Using cuvettes, the leaves were darkened for 20 min before measurement. The chlorophyll *a* fluorescence was reported as the Fv/Fm ratio, which shows the maximal quantum yield of PSII photochemistry, where Fm = the maximal fluorescence and Fv = the variable fluorescence.

2.4. Estimation of Proline Content

The protocol of Bates et al. [48] was used for determining the proline content, using L-proline as the standard. In total, 1 g of fresh leaf samples was homogenized in 5 mL of 3% aqueous sulfosalicylic acid and centrifuged at 14,000× *g* for 15 min (Neya 10R, REMI, Mumbai, India). The supernatant (2 mL) was blended with an equal volume of acetic acid and acid ninhydrin, vortexed, and incubated for 1 h at 100 °C. An ice bath was used to stop the reaction, and the solution was extracted with 4 mL of toluene. The sample was read spectrophotometrically at 525 nm using toluene as the blank (7315 Spectrophotometer, Jenway, Staffordshire, UK).

2.5. Extraction and Assay of Antioxidant Enzymes

In total, 0.5 g fresh leaf sample was extracted with 4 mL of buffer (50 mM potassium phosphate, 1 mM EDTA, 1% PVP, 1 mM DTT and 1 mM PMSF, pH 7.8). The samples were centrifuged at 15,000× *g* for 30 min at 4 °C [49]. The supernatant was picked up and stored at −80 °C for the determination of enzyme activity. The catalase activity (CAT) was measured via the protocol of Aebi [50]. Then, 20 µL of supernatant was blended with 830 µL of potassium phosphate buffer (50 mM, pH 7). The reaction was initiated by the addition of 150 µL of H_2O_2, and we registered the decrease in absorbance (240 nm) for 2 min. The GPX activity was determined using the protocol described by Ruley et al. [34]. A reaction mixture consisting of suitable quantities of enzymatic extracts was employed, with equal amounts of 17 mM H_2O_2 and 2% guaiacol. The reaction was registered for 3 min at 510 nm. The Giannopolitis and Ries [51] protocol was used for determining the SOD activity (560 nm).

2.6. Climate Conditions

The climate conditions (air temperatures (°C) and relative humidity (RH %)) in both cycle periods were registered using a CR 1000 data logger (Campbell Scientific Ltd., Loughborough, UK). The mean air temperatures during I CP and II CP were 13.4 °C and 21.4 °C, respectively, and the mean relative humidity values were 86% and 76.0%, respectively (Figure S1).

2.7. Statistical Analysis

Both experimental trials were conducted via a randomized complete design; three replicates in each cycle period (CP) and for each treatment (T) were used. Statistical analysis of the data was performed using CoStat release 6.311 (CoHort Software, Monterey, CA, USA). The data were subjected to two- and one-way analysis of variance (ANOVA), and the means were compared via a Tukey's test at $p < 0.05$. The data presented in the figures are the means ± standard error (SE).

3. Results

Table 1 shows the main effects and the interaction effects (over cycle periods and treatments) of morphometric parameters and mineral contents. The two-way ANOVA showed that the parameters were influenced by cycle and treatments (Table 1). The total and shoot dry biomass were statistically different in different cycle periods. Significant differences were found in the root/shoot ratio depending on the two factors and their interactions. Leaf number and total leaf area were only statistically significant for the cycle periods. There were no significant differences in chlorophyll content for any factors. The mineral contents in the leaves and in the roots were significantly different between cycle periods and treatments, and their interactions had an effect too. Leaf damage was only significantly different between treatments. To better understand the effects of the cycle periods on the treatments, the two cycle periods were separately analyzed.

Table 1. Summary of the main effects and interaction effects of cycle periods (I CP and II CP) and treatments (control, salt spray, surfactants, and salt plus surfactants) on total and epigeous dry biomass, root/shoot ratio, leaf number, total leaf area, chlorophyll content, Na^+ and Cl^- in leaves and roots, and leaf damage of *Callistemon* plants with the corresponding significance of the F-values.

	Total Dry Biomass	Shoot Dry Biomass	R/S Ratio	Leaf Number	Total Leaf Area	Chlorophyll Content	Na^+ Leaves	Cl^- Leaves	Na^+ Root	Cl^- Root	Leaf Damage
Main Effects											
Cycle periods (CP)	F 12.21 **	F 24.42 ***	F 178.26 ***	F 21.50 ***	F 25.59 ***	F 0.68 ns	F 19.74 ***	F 852.36 ***	F 674.85 ***	F 98.51 ***	F 0.84 ns
Treatments (T)	F 1.06 ns	F 1.21 ns	F 3.60 *	F 1.33 ns	F 1.60 ns	F 0.12 ns	F 123.46 ***	F 522.64 ***	F 346.21 ***	F 12.27 ***	F 440.64 ***
Interaction											
CP × T	F 0.50 ns	F 0.66 ns	F 3.50 *	F 0.67 ns	F 0.57 ns	F 0.11 ns	F 7.37 **	F 184.15 ***	F 164.62 ***	F 13.94 ***	F 9.23 ns

Significance of differences in parameters: ns = not significant; * $p < 0.05$; ** $p < 0.01$; *** $p < 0.001$ with the corresponding significance of the F-values.

Salt spray, alone or in combination with the surfactant, significantly influenced the biomass accumulation in both growing periods (Table 2). The total dry biomass was reduced in I CP under all treatments by 30% in relation to the control plants (Figure 1a). During II CP, the total dry biomass was only reduced in the treatment with salt by 33% (Table 2). Shoot dry biomass showed a similar trend in both cycle periods (Table 2).

All treatments slightly modified root dry biomass (data not shown), and in I CP, enhanced the root/shoot ratio (Table 2).

The effects of treatments on leaf numbers and total leaf area were observed in both cycle periods (Table 2). Leaf number and total leaf area were significantly reduced during I CP in all treatments, with a reduction of 64% in relation to the control plants (Table 2). A similar trend was registered during II CP, but with only a minor reduction for both parameters (by 20% and 14% for leaf number and total leaf area, respectively).

The chlorophyll contents in both cycle periods did not change according to the treatments (Table 2).

At the end of I CP, a seven-fold higher leaf Na^+ content was registered in plants treated with salt spray and salt spray plus surfactants in relation to the control plants and those treated with surfactants only. In II CP, this increase was amplified; 8- and 11-fold increases were noted in plants treated with salt spray and salt spray plus surfactants, respectively (Table 2).

A similar trend was registered for the leaf Cl^- content; at the end of the I CP, a five-fold higher Cl^- content was noted in the plants treated with salt spray and with salt spray plus surfactants in relation to the control plants and those treated with surfactants only. In CP II, we noticed a six-fold increase in plants treated with salt spray and salt spray plus surfactants (Table 2).

No significant differences were observed in the root mineral content during I CP. In II CP, on the other hand, an increase in the Na^+ content by 33% was observed in plants treated with salt spray and by 44% in those treated with salt plus surfactants. The Cl^- content increased by 50% in plants treated with salt spray and salt spray plus surfactants (Table 2).

Table 2. Effects of the treatments on *Callistemon* grown in winter (I CP) and summer (II CP) on total and epigeous dry biomass, root/shoot ratio, leaf number, total leaf area, chlorophyll content, Na^+ and Cl^- in leaves and roots, and leaf damage.

	Treatments	Total Dry Weight (g Plant^{-1})	Shoot Dry Weight (g Plant^{-1})	R/S Ratio (g g^{-1})	Leaf (n·Plant^{-1})	Total Leaf Area (cm^2 Plant^{-1})	Chlorophyll Content (µg cm^{-2})	Na^+ Leaves (g kg^{-1} DW)	Cl^- Leaves (g kg^{-1} DW)	Na^+ Root (g kg^{-1} DW)	Cl^- Root (g kg^{-1} DW)	Leaf Damage (%)
Winter (I CP)	Control	11.3 ± 0.5a	7.6 ± 0.4a	0.5 ± 0.0b	146.0 ± 6.3a	478.3 ± 31.7a	39.1 ± 2.8	2.6 ± 0.3b	2.9 ± 0.4b	3.4 ± 0.1	2.9 ± 0.6	0.0 ± 0.0b
	Salt spray	8.5 ± 0.8b	4.8 ± 0.6b	0.8 ± 0.1a	56.5 ± 7.9b	181.2 ± 29.7b	27.8 ± 1.9	17.9 ± 1.1a	14.8 ± 1.3a	4.0 ± 0.5	3.3 ± 0.4	16.5 ± 1.2a
	Surfactant	7.7 ± 0.5b	4.1 ± 0.2b	0.9 ± 0.1a	41.2 ± 2.8b	137.0 ± 11.2b	31.5 ± 3.6	1.2 ± 0.1b	2.0 ± 0.5b	3.3 ± 0.2	2.7 ± 0.7	0.14 ± 0.0b
	Salt spray plus surfactant	8.0 ± 0.3b	4.7 ± 0.3b	0.7 ± 0.1a	56.2 ± 7.8b	197.2 ± 34.5b	35.8 ± 1.6	16.8 ± 1.2a	14.8 ± 1.2a	3.9 ± 0.5	3.4 ± 0.5	20.3 ± 1.2a
Significance		**	**	*	***	***	ns	***	***	ns	ns	**
Summer (II CP)	Control	21.6 ± 1.1a	18.5 ± 1.0a	0.2 ± 0.0	220.3 ± 12.5a	860.0 ± 45.0a	35.2 ± 2.0	3.2 ± 0.5c	9.2 ± 0.5c	4.6 ± 0.2c	4.3 ± 0.2b	0.0 ± 0.0c
	Salt spray	14.1 ± 0.6b	12.0 ± 0.7b	0.2 ± 0.0	167.5 ± 12.8b	624.5 ± 32.1b	25.6 ± 2.5	21.2 ± 2.6b	55.0 ± 0.6a	7.4 ± 0.8b	8.0 ± 0.4a	17.1 ± 1.2b
	Surfactant	20.0 ± 0.9a	17.3 ± 0.8a	0.2 ± 0.0	188.2 ± 8.4b	814.0 ± 24.7a	30.9 ± 3.5	1.5 ± 0.1C	4.5 ± 0.7d	3.2 ± 0.5d	4.2 ± 0.1b	0.0 ± 0.0c
	Salt spray plus surfactant	14.3 ± 0.3b	11.2 ± 0.8b	0.2 ± 0.0	179.0 ± 8.5b	691.9 ± 19.5b	33.2 ± 1.4	28.9 ± 2.6a	51.8 ± 0.8b	9.9 ± 0.5a	7.3 ± 0.6a	27.7 ± 1.2a
Significance		***	***	ns	*	***	ns	***	***	***	***	***

Significance of differences in parameters: ns = not significant; * $p < 0.05$; ** $p < 0.01$; *** $p < 0.001$. Data followed by a different letter were significantly different according to Tukey's test.

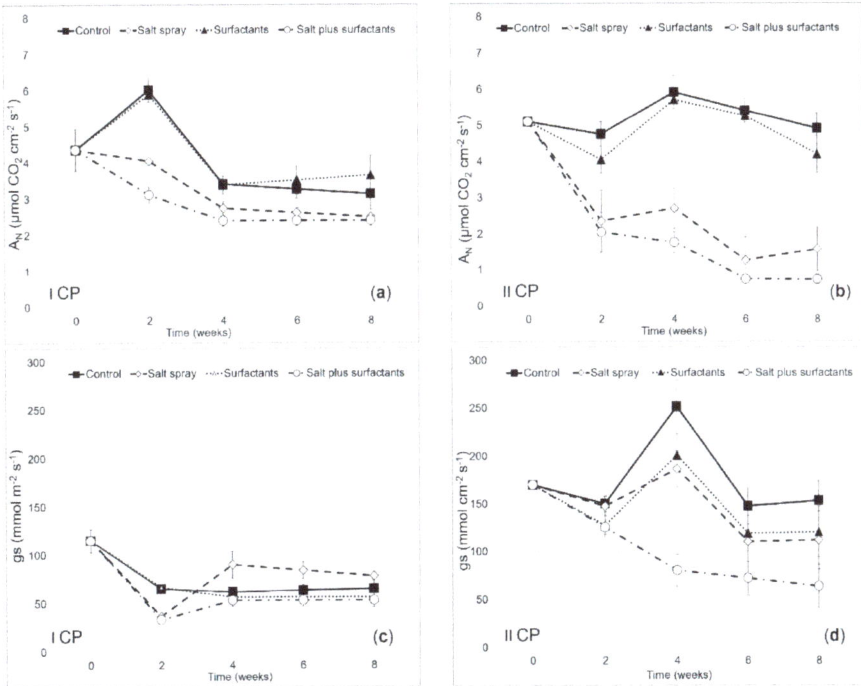

Figure 1. Net photosynthesis (A_N) (**a**,**b**) and leaf stomatal conductance (gs) (**c**,**d**) in Callistemon plants during I CP and II CP. Plants were subjected twice a week to nebulization treatments for 8 weeks, depending on their treatments: Control (deionized water), salt spray (solution simulating the composition of seawater), surfactants (sodium dodecylbenzene-sulfonate), and a solution with seawater and anionic surfactant. Mean values ± SE (n = 4).

The leaf damage percentages were significantly different among treatments. During I CP, the average leaf damage percentage was 18% in plants treated with salt spray and salt spray plus surfactants. During II CP, the maximum value (~27%) was observed in salt spray plus surfactants plants; no leaf damage was observed in plants treated with only surfactants in either cycle period (Table 2).

Values are the means ± SE of three replicate samples. n.s. = not significant; *, **, *** significant at $p < 0.05$, 0.01 and $p < 0.001$, respectively. The values in the same column followed by the same letter are not significantly different at $p \leq 0.05$ (Tukey's test). The gas exchange in both cycle periods was significantly modified as a result of the treatments (Figure 1). From the first measurement (2 weeks), *Callistemon* plants in I CP treated with salt spray alone or in combination with surfactants showed reductions of 33% and 50% in A_N, compared with the control plants; at the end of I CP, a reduction of 23% in this activity in salt spray and salt spray and surfactant plants was observed (Figure 1a). The stomatal conductance values during I CP showed no significant differences (Figure 1c).

During II CP, the reduction in photosynthesis activity was significant immediately after the second week, and this trend was maintained until the end of the experimental period. Under control conditions, the net photosynthesis was about 5 µmol CO_2 m^{-2} s^{-1}. Salt spray alone or in combination caused a progressive decrease in net photosynthesis assimilation; at the end of the cycle period, a reduction of ~60% was reached (Figure 1b). A similar response was registered in the stomatal conductance. During the warmer season, stomatal conductance showed higher values throughout the growth cycle, but a more intense reduction was seen at the end of the trial in the salt spray plus surfactants group (~60%) (Figure 1d), compared with the control plants.

The different values of chlorophyll *a* fluorescence (Fv/Fm ratio) after exposure to different treatments in different growth periods are presented in Figure 2a,b. The plants treated with salt spray and salt spray plus surfactant showed reductions in both cycle periods, compared with the control plants; the Fv/Fm achieved respective values of 0.67 and 0.71 in I CP, and 0.67 and 0.65 during II CP.

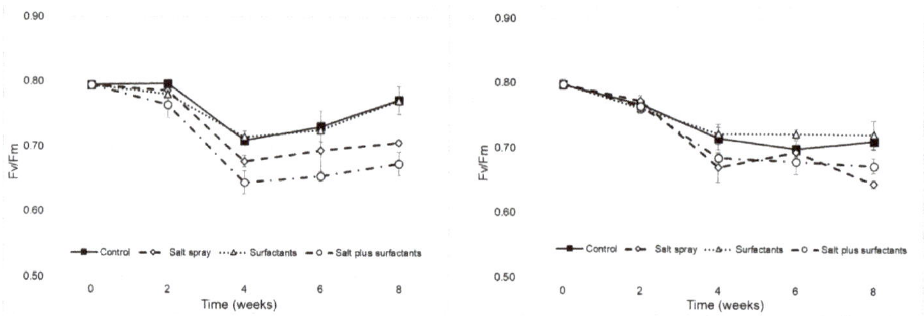

(a) (b)

Figure 2. Maximum quantum efficiency of the PSII (Fv/Fm) of *Callistemon* plants during I CP (**a**) and II CP (**b**) treated with distilled water (control), salt spray, surfactants, and salt plus surfactants. Values are means ± SE (n = 3).

In the final part of the experiment, in both cycle periods, the leaf proline content increased in the salt spray and salt plus surfactants treatment groups, compared with the control and surfactants plants, in both cycle periods, by 65% and 37% in I CP, and 75% and 63% in II CP (Figure 3).

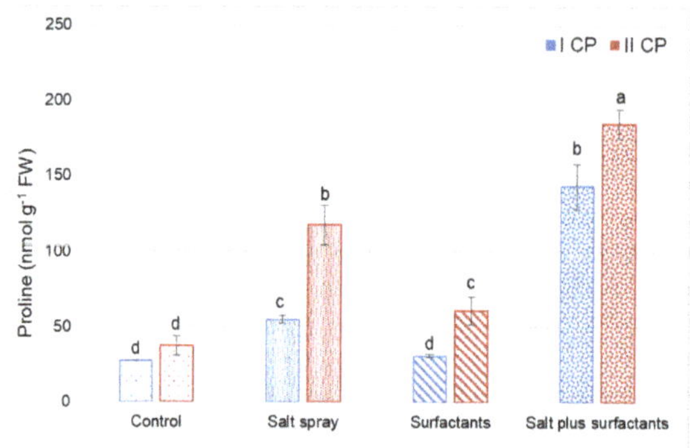

Figure 3. Proline contents in leaves of *Callistemon* during I CP (■) and II CP (■) treated with distilled water (control, □ and □), salt spray (□ and □), surfactants (□ and □), and salt plus surfactants (□ and □). Values are means ± SE (n = 3). Values with the same letter are not significantly different as determined by Tukey's test ($p < 0.05$).

The enzyme activities, SOD, CAT, and GPX, exhibited significant differences in the cycle periods, depending on treatments (Figure 4).

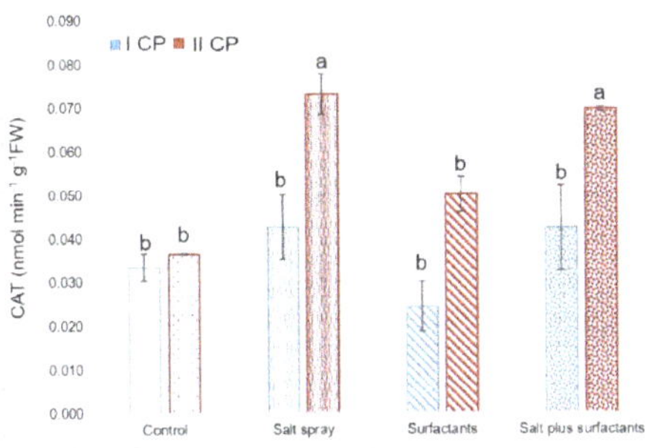

(a)

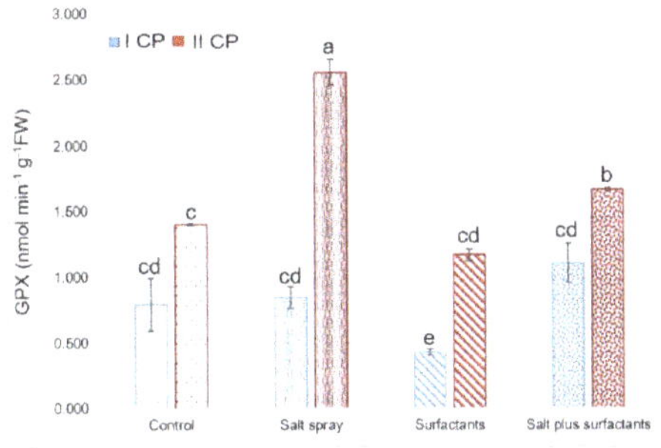

(b)

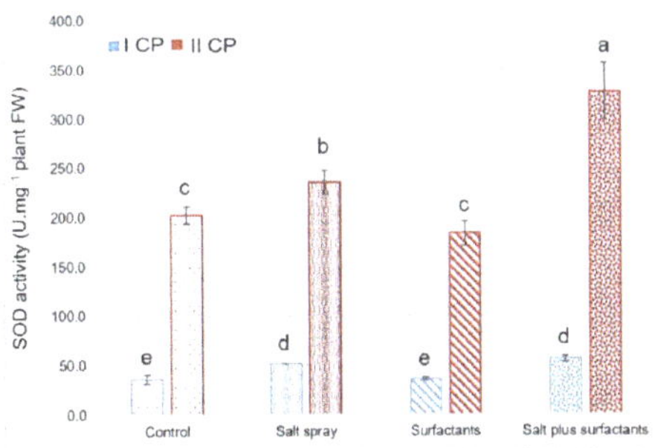

(c)

Figure 4. Effects of different treatments on catalase (CAT (**a**)), peroxidase (GPX (**b**)), and superoxide dismutase (SOD (**c**)) activity in *Callistemon* plants during I CP ( ) and II CP ( ) treated with distilled water (control, and), salt spray (and), surfactants (and), and salt plus surfactants (and). Values are means ± SE ($n = 3$). Values with the same letter are not significantly different by Tukey's test ($p < 0.05$).

The CAT activity at the end of I CP showed no significant differences. In II CP, however, the salt spray and the salt spray plus surfactants treatment groups showed significant increases in this enzyme by 47% compared with the control and surfactants treatments (Figure 4a).

The GPX activity during I CP was lower in the surfactants compared with the other treatments (Figure 4b). In the final part of II CP, however, the salt treatments increased the activity of this enzyme by 42% and 12% in the salt and salt plus surfactants treatments, respectively (Figure 4b).

In I CP, the SOD activity was much lower and showed its highest values in the salt treatments, with an increase of 20%. In the second cycle period, the observed values were much higher and were again even higher in the salt treatment groups, with increases of 21% and of 42% in the salt spray and salt plus surfactants groups, respectively (Figure 4c).

To visualize the associations among the different treatments in terms of the plants' morphological, physiological, and biochemical parameters in the two cycle periods, a principal component analysis was performed for all analyzed parameters (PCA; Figure 5a,b; Tables S1 and S2). The first two principal components (PC1 and PC2) accounted for 59% and 28% and 73% and 19% of the total variance during I CP and II CP, respectively, and these components were associated with eigenvalues higher than one.

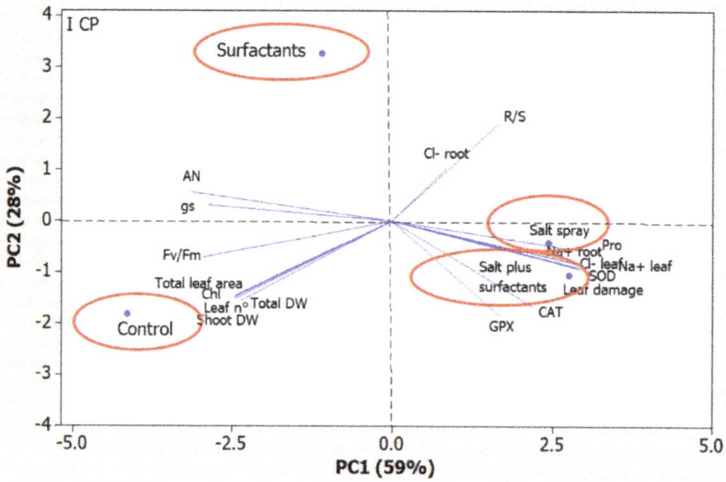

(a)

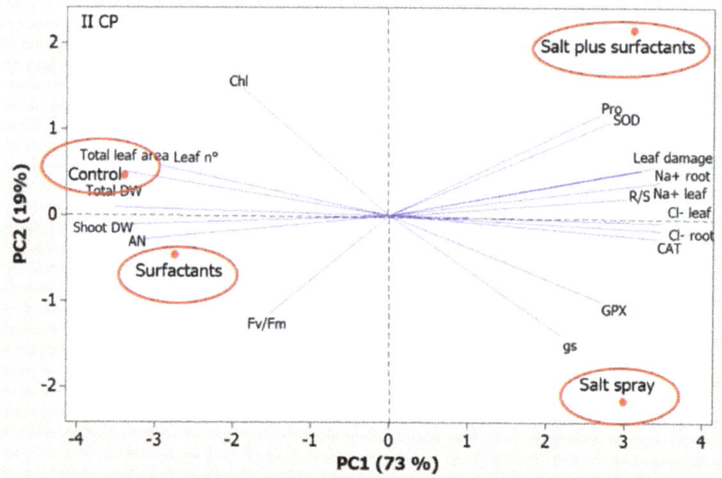

(b)

Figure 5. Principal component loading plot and scores of the PCA on the total and shoot dry biomass, R/S, leaf number, total leaf area, chlorophyll content (Chl), Na^+ leaf and root content, Cl^- leaf and root content, leaf damage, gas exchange (A_N and gs), Fv/Fm, proline (Pro), and enzyme activity (GPX, CAT, and SOD), for *Callistemon* plants undergoing different salt treatments (control, salt spray, surfactants, and salt plus surfactants) according to the first two principal components in I CP (**a**) and II CP (**b**).

During I CP, PCA revealed that the morphological and physiological parameters were directly affected by the control and surfactants treatments, whereas the mineral contents and the biochemical parameters were directly correlated with the salt and salt plus surfactants treatments (Figure 5a).

A similar trend was found at the end of II CP (Figure 5b).

4. Discussion

The negative influence of salinity on ornamental plants has been studied. Nevertheless, only a limited number of papers have provided information on the effects of salt spray alone or in combination with surfactants. The identification of plants tolerant to salt spray is particularly important in the selection of shrubs and trees to be used in green areas [9].

A large number of studies on salt stress have used NaCl as the salinizing agent [52–54]. However, the compositions of salts in seawater are rather different. Saline solutions containing a single salt may give rise to misleading and erroneous interpretations of plants' responses to salinity [55,56]. In our studies, we have added to deionized water a solution reproducing the composition of seawater [46] so as to simulate the effects of salt spray along coastal areas.

Salt spray can hinder plant growth because it causes drought stress, disrupts membranes and enzyme systems, reduces nutrient uptake, causes necrosis or leaf loss, and can lead to plant mortality [15].

In both cycle periods, the total and shoot dry biomass were reduced in the presence of salt spray, alone or in combination with surfactants. Fragmented cuticles, disaggregated chloroplasts, and nuclei disrupted stomata, collapsed cell walls, coarsely granulated cytoplasm, and disorganized phloem leading to reduced biomass were observed after the application of spray NaCl [57].

Salt treatments reduced the epigeous dry biomass, reduced the leaf number and total leaf area, and did so more extensively during II CP. This behavior confirms that the plants reduce their leaf area to overcome stress [58]. Plants implement different mechanisms to overcome abiotic stress (drought, saline stress), including a reduction in the total leaf area, which leads to a reduction in water losses through stomata closure, which is the principal defense strategy of species subject to osmotic stress [12].

Reductions in leaf area can therefore be a strategy of adaptation. The reduction in this parameter is generally related to a decrease in the total leaf number and size. In the presence of abiotic stress (drought and salt stress), plants reduce their leaf size so as to reduce the surface available for both the depositing of salt and the loss of water through transpiration [59]. Reduction in leaf number was observed in plants of *Scaevola sericea* [60] and *Crambe maritime* [61] sprayed with seawater. This agrees with the results of our study. The plants grown during II CP and treated with salt spray showed a greater reduction in leaf number. This result agrees with the experiments of Bussotti et al. [41], who found that the damage of surfactants was greater in the warmer season.

Aesthetic quality is the most important trait in ornamental plant species [62]. Visual quality is a practical tool used to evaluate the salt response of ornamental species [63]. Common responses in the presence of salt stress include leaf necrosis and reductions in chlorophyll [64]; greater sensitivity has been observed in older leaves [65]. In our study, leaf necrosis as a symptom of leaf damage was registered, and this ranged from minimal (surfactant) to moderate (salt spray and salt spray plus surfactants). Leaf damage could lead to a reduction in the photosynthetic rate, which might manifest in diminished growth [66].

A potential biomarker of salinity tolerance is indicated by the absorption and accumulation of ions in the leaves [2]. The Na^+ contents significantly increased during salt treatments in both cycle periods, although this was greater during the summer cycle, and the presence of surfactants intensified this effect further still. In this case, the presence of surfactants increased the phytotoxic effects, as reported by Sànchez-Blanco et al. [67]. High-rated absorptions of Na^+ and Cl^- in the shoots reduce growth rate due to ion toxicity. Excessive Na and Cl accumulation leads to molecular damage-causing growth arrest and plant death because these ions induce cytoplasmic toxicity [68].

Different studies suggest ornamental plants exposed to salt treatments show a reduction in net photosynthesis and stomatal conductance. In our work, salt treatments, alone or in combination with surfactants, caused a reduction in net photosynthesis as a consequence of the reduced CO_2 assimilation following stomatal closure, which particularly occurred during II CP.

The reduction in Fv/Fm is a clear signal that the PSII was affected by the presence of salt and that photoinhibition occurred. Generally, plants under normal conditions presented Fv/Fm ratios close to 0.8 [7], but in our case, the reduction in Fv/Fm was correlated with salt spray. The plants showed a reduction in this index in the presence of salt, and this reduction was greater when the salt was combined with a surfactant. For different plant species, regardless of whether the PSII damages occur in the leaves, this index is considered one of the most effective tools for the measurement of plant stress response, particularly before any morphological changes manifest [69]. The reduction in photosynthetic capacity is correlated with reductions in chlorophyll fluorescence in ornamental species; these cause physiological modifications that will lead to losses in sales for the growers [70].

One of the mechanisms adopted to defend against salt stress is proline accumulation, which can be used as a selection criterion for salt tolerance because an increase in this amino acid may be positively correlated with the level of salt tolerance [71]. In fact, modifications in soluble protein and proline content could regulate cellular redox potential [72] and help scavenge free radicals [73].

Increased ROS production, which damages biomolecules (e.g., lipids, proteins, and nucleic acids) and harms redox homeostasis, was observed in the presence of salt stress [74]. The increase in ROS activity causes an increase in oxidative stress due to ionic toxicity and osmotic imbalance and has negative effects on photosynthetic pigments [75]. In the presence of saline stress, plants produce reactive oxygen species (ROS), such as hydroxyl radical ($OH^{\bullet}$), singlet oxygen ($^1O^2$), superoxide radical ($O_2^{\bullet -}$), and hydrogen peroxide (H_2O_2), which have a toxic effect on plants [76]. Our results show that the enzyme activity was similar between the control and surfactant treatments but was significantly increased under the salt spray treatment. These results showed that *Callistemon* plants have significant salt tolerance, and as reported by Sklodowska et al. [77], a greater increase in SOD activity could help preserve membrane integrity and osmoregulation by helping regulate the level of superoxide radical in the presence of high saline levels.

To better understand the tolerance mechanisms related to salt stress and surfactants, it will be relevant to investigate and understand the morphological, physiological, and biochemical parameters of plants grown under stressed and not-stressed conditions (Supplementary Tables S1 and S2). To evaluate stress tolerance and the relationships among the different parameters analyzed in our study, an effective approach is the analysis of principal components (PCA) [12]. Our PCA results demonstrate that the variations in stress tolerance between *Callistemon* during I CP and II CP are linked to variations in biochemical and mineral contents.

5. Conclusions

Our trial has demonstrated the combined effects of salt spray and surfactant on the ornamental shrub species largely used in Mediterranean green areas near the sea. Salt spray, alone or in combination with surfactant action, influences plants' growth and aesthetic features, especially during warmer seasons when transpiration is more intense and causes stomatal opening. To overcome stressful conditions, plants undergo several changes, such as the activation of enzymes and the accumulation of proline and minerals (Na^+ and Cl^-). Interaction with surfactants amplifies these adverse effects.

The results of this trial help us to better understand the responses of plant species used in green areas near the coastline and will thus guide the choice of plant species.

Supplementary Materials: The following supporting information can be downloaded at: https://www.mdpi.com/article/10.3390/horticulturae8030261/s1, Figure S1: Air temperature (°C) and relative humidity (%) during the two cycle periods, Table S1: Pearson's correlation coefficients among morphometric, physiological, and biochemical parameters from *Callistemon* potted plants exposed to different treatments during I CP, Table S2: Pearson's correlation coefficients among morphometric, physiological, and biochemical parameters from *Callistemon* potted plants exposed to different treatments during II CP.

Author Contributions: Conceptualization, D.R. and S.T.; methodology, D.R. and S.T.; software, S.T.; validation, D.R.; formal analysis, S.T. and G.L.F.; investigation, S.T. and G.L.F.; resources, D.R.; data curation, S.T.; writing—original draft preparation, D.R. and S.T.; writing—review and editing, D.R. and S.T.; supervision, D.R. All authors have read and agreed to the published version of the manuscript.

Funding: This research received no external funding.

Institutional Review Board Statement: Not applicable.

Informed Consent Statement: Not applicable.

Data Availability Statement: All data for this study has been included in the tables, figures, and Supplemental Materials.

Conflicts of Interest: The authors declare no conflict of interest.

References

1. Farieri, E.; Toscano, S.; Ferrante, A.; Romano, D. Identification of ornamental shrubs tolerant to saline aerosol for coastal urban and peri-urban greening. *Urban For. Urban Green.* **2016**, *18*, 9–18. [CrossRef]
2. Ferrante, A.; Trivellini, A.; Malorgio, F.; Carmassi, G.; Vernieri, P.; Serra, G. Effect of seawater on leaves of six plant species potentially useful for ornamental purposes in coastal areas. *Sci. Hortic.* **2011**, *128*, 332–341. [CrossRef]
3. Sánchez-Blanco, M.J.; Rodríguez, P.; Morales, M.A.; Torrecillas, A. Contrasting physiological responses of dwarf sea-lavender and marguerite to simulated sea aerosol deposition. *J. Environ. Qual.* **2003**, *32*, 2238–2244. [CrossRef] [PubMed]
4. Greaver, T.L.; Sternberg, L.L.D.S. Linking marine resources to ecotonal shifts of water uptake by terrestrial dune vegetation. *Ecology* **2006**, *87*, 2389–2396. [CrossRef]
5. Franco, J.A.; Martínez-Sánchez, J.J.; Fernández, J.A.; Bañón, S. Selection and nursery production of ornamental plants for landscaping and xerogardening in semi-arid environments. *J. Hortic. Sci. Biotechnol.* **2006**, *81*, 3–17. [CrossRef]
6. Niu, G.; Rodriguez, D.S. Relative salt tolerance of selected herbaceous perennials and groundcovers. *Sci. Hortic.* **2006**, *110*, 352–358. [CrossRef]
7. Toscano, S.; Ferrante, A.; Romano, D. Response of Mediterranean ornamental plants to drought stress. *Horticulturae* **2019**, *5*, 6. [CrossRef]
8. Zollinger, N.; Koenig, R.; Cerny-Koenig, T.; Kjelgren, R. Relative salinity tolerance of intermountain western United States native herbaceous perennials. *HortScience* **2007**, *42*, 529–534. [CrossRef]
9. Toscano, S.; Branca, F.; Romano, D.; Ferrante, A. An evaluation of different parameters to screen ornamental shrubs for salt spray tolerance. *Biology* **2020**, *9*, 250. [CrossRef]
10. Wu, S.; Sun, Y.; Niu, G.; Pantoja, G.L.G.; Rocha, A.C. Responses of six Lamiaceae landscape species to saline water irrigation. *J. Environ. Hortic.* **2016**, *34*, 30–35. [CrossRef]
11. Marcum, K.B.; Pessarakli, M.; Kopec, D. Relative salinity tolerance of 21 turf-type desert salt grasses compared to bermudagrass. *HortScience* **2005**, *40*, 827–829. [CrossRef]
12. Toscano, S.; Ferrante, A.; Romano, D.; Tribulato, A. Interactive effects of drought and saline aerosol stress on morphological and physiological characteristics of two ornamental shrub species. *Horticulturae* **2021**, *7*, 517. [CrossRef]
13. Cheplick, G.P.; Demetri, H. Impact of saltwater spray and sand deposition on the coastal annual *Triplasis purpurea* (Poaceae). *Am. J. Bot.* **1999**, *86*, 703–710. [CrossRef] [PubMed]
14. Elhaak, M.A.; Migahid, M.M.; Wegmann, K. Ecophysiological studies on *Euphorbia paralias* under soil salinity and sea water spray treatments. *J. Arid Environ.* **1997**, *35*, 459–471. [CrossRef]
15. Scheiber, S.M.; Sandrock, D.; Alvarez, E.; Brennan, M.M. Effect of salt spray concentration on growth and appearance of 'Gracillimus' maiden grass and 'Hamelin' fountain grass. *HortTechnology* **2008**, *18*, 34–38. [CrossRef]
16. Rizzo, V.; Toscano, S.; Farieri, E.; Romano, D. Antioxidative defense mechanism in *Callistemon citrinus* (Curtis) Skeels and *Viburnum tinus* L. 'Lucidum' in response to seawater aerosol and surfactants. *J. Agric. Sci. Technol.* **2019**, *21*, 911–925.
17. Palmer, M.; Hatley, H. The role of surfactants in wastewater treatment: Impact, removal and future techniques: A critical review. *Water Res.* **2018**, *147*, 60–72. [CrossRef]
18. Savé, R. What is stress and how to deal with it in ornamental plants? *Acta Hortic.* **2009**, *813*, 241–254. [CrossRef]
19. Nicolotti, G.; Rettori, A.; Paoletti, E.; Gullino, M.L. Morphological and physiological damage by surfactant-polluted seaspray on *Pinus pinea* and *Pinus halepensis*. *Environ. Monit. Assess.* **2005**, *105*, 175–191. [CrossRef]
20. Bussotti, F.; Bottacci, A.; Grossoni, P.; Mori, B.; Tani, C. Cytological and structural changes in *Pinus pinea* L. needles following the application of anionic surfactant. *Plant Cell Environ.* **1997**, *20*, 513–520. [CrossRef]
21. Schreiber, L.; Bach, S.; Kirsch, T.; Knoll, D.; Schalz, K.; Riederer, M. A simple photometric device analysing cuticular transport physiology: Surfactant effect on permeability of isolated cuticular membranes of *Prunus laurocerasus* L. *J. Exp. Bot.* **1995**, *46*, 1915–1921. [CrossRef]

22. Raddi, S.; Cherubini, P.; Lauteri, M.; Magnani, F. The impact of sea erosion on coastal *Pinus pinea* stands: A diachronic analysis combining tree-rings and ecological markers. *For. Ecol. Manag.* **2009**, *257*, 773–781. [CrossRef]
23. Rettori, A.; Paoletti, E.; Nicolotti, G.; Gullino, M.L. Ecophysiological responses of Mediterranean pines to simulated sea aerosol polluted with an anionic surfactant: Prospects for biomonitoring. *Ann. For. Sci.* **2005**, *62*, 351–360. [CrossRef]
24. Bussotti, F.; Bottacci, A.; Grossoni, P.; Mori, B.; Tani, C. Anatomical and ultrastructural alterations in *Pinus pinea* L. needles treated with simulated sea aerosol. *Agric. Med.* **1995**, 148–155.
25. Tezara, W.; Martínez, D.; Rengifo, E.; Herrera, A. Photosynthetic responses of the tropical spiny shrubs *Lycium nodosum* (Solanaceae) to drought, soil salinity and saline spray. *Ann. Bot.* **2003**, *92*, 757–765. [CrossRef]
26. Sudhir, P.; Murthy, S.D.S. Effects of salt stress on basic processes of photosynthesis. *Photosynthetica* **2004**, *42*, 481–486. [CrossRef]
27. Parida, A.K.; Das, A.B. Salt tolerance and salinity effects on plants: A review. *Ecotoxicol. Environ. Saf.* **2005**, *60*, 324–349. [CrossRef]
28. Sharma, N.; Gupta, N.K.; Gupta, S.; Hasegawa, H. Effect of NaCl salinity on photosynthetic rate, transpiration rate, and oxidative stress tolerance in contrasting wheat genotypes. *Photosynthetica* **2005**, *43*, 609–613. [CrossRef]
29. Toscano, S.; Farieri, E.; Ferrante, A.; Romano, D. Physiological and biochemical responses in two ornamental shrubs to drought stress. *Front. Plant Sci.* **2016**, *7*, 645. [CrossRef] [PubMed]
30. Bowler, C.; Montagu, M.V.; Inze, D. Superoxide dismutase and stress tolerance. *Annu. Rev. Plant Biol.* **1992**, *43*, 83–116. [CrossRef]
31. Malan, C.; Greyling, M.M.; Gressel, J. Correlation between CuZn superoxide dismutase and glutathione reductase, and environmental and xenobiotic stress tolerance in maize inbreds. *Plant Sci.* **1990**, *69*, 157–166. [CrossRef]
32. Perl, A.; Perl-Treves, R.; Galili, S.; Aviv, D.; Shalgi, E.; Malkin, S.; Galun, E. Enhanced oxidative-stress defense in transgenic potato expressing tomato Cu, Zn superoxide dismutases. *Theor. Appl. Genet.* **1993**, *85*, 568–576. [CrossRef] [PubMed]
33. Alscher, R.G.; Erturk, N.; Heath, L.S. Role of superoxide dismutases (SODs) in controlling oxidative stress in plants. *J. Exp. Bot.* **2002**, *53*, 1331–1341. [CrossRef] [PubMed]
34. Ruley, A.T.; Sharma, N.C.; Sahi, S.V. Antioxidant defense in a lead accumulating plant, *Sesbania drummondii*. *Plant Physiol. Biochem.* **2004**, *42*, 899–906. [CrossRef] [PubMed]
35. Deeba, F.; Pandey, A.K.; Ranjan, S.; Mishra, A.; Singh, R.; Sharma, Y.K.; Shirke, P.A.; Pandey, V. Physiological and proteomic responses of cotton (*Gossypium herbaceum* L.) to drought stress. *Plant Physiol. Biochem.* **2012**, *53*, 6–18. [CrossRef] [PubMed]
36. Deng, Y.; Chen, S.; Chen, F.; Cheng, X.; Zhang, F. The embryo rescue derived intergeneric hybrid between chrysanthemum and *Ajania przewalskii* shows enhanced cold tolerance. *Plant Cell Rep.* **2011**, *30*, 2177–2186. [CrossRef] [PubMed]
37. Deng, Y.; Jiang, J.; Chen, S.; Huang, C.; Fang, W.; Chen, F. Drought tolerance of intergeneric hybrids between *Chrysanthemum morifolium* and *Ajania przewalskii*. *Sci. Hortic.* **2012**, *148*, 17–22. [CrossRef]
38. Mereu, S.; Gerosa, G.; Marzuoli, R.; Fusaro, L.; Salvatori, E.; Finco, A.; Spano, D.; Manes, F. Gas exchange and JIP-test parameters of two Mediterranean maquis species are affected by sea spray and ozone interaction. *Environ. Exp. Bot.* **2011**, *73*, 80–88. [CrossRef]
39. Niinemets, Ü. Responses of forest trees to single and multiple environmental stresses from seedlings to mature plants: Past stress history, stress interactions, tolerance and acclimation. *For. Ecol. Manag.* **2010**, *260*, 1623–1639. [CrossRef]
40. Gonthier, P.; Nicolotti, G.; Rettori, A.; Paoletti, E.; Gullino, M.L. Testing *Nerium oleander* as a biomonitor for surfactant polluted marine aerosol. *Int. J. Environ. Health Res.* **2010**, *4*, 1–10.
41. Bussotti, F.; Grossoni, P.; Pantani, F. The role of marine salt and surfactants in the decline of Tyrrhenian coastal vegetation in Italy. *Ann. For. Sci.* **1995**, *52*, 251–261. [CrossRef]
42. Lippi, G.; Serra, G.; Vernieri, P.; Tognoni, F. Response of potted *Callistemon* species to high salinity. *Acta Hortic.* **2003**, *609*, 247–250. [CrossRef]
43. Mugnai, S.; Ferrante, A.; Petrognani, L.; Serra, G.; Vernieri, P. Stress-induced variation in leaf gas exchange and chlorophyll *a* fluorescence in Callistemon plants. *Res. J. Biol. Sci.* **2009**, *4*, 913–921.
44. Vernieri, P.; Mugnai, S.; Borghesi, E.; Petrognani, L.; Serra, G. Non-chemical growth control of potted *Callistemon laevis* Anon. *Agr. Med.* **2006**, *136*, 85.
45. Álvarez, S.; Sánchez-Blanco, M.J. Comparison of individual and combined effects of salinity and deficit irrigation on physiological, nutritional and ornamental aspects of tolerance in *Callistemon laevis* plants. *J. Plant Physiol.* **2015**, *185*, 65–74. [CrossRef]
46. Elshatshat, S.A. The effect of simulated seawater on water permeability of isolated leaf cuticular layers. *Int. J. Agric. Biol.* **2010**, *12*, 150–152.
47. Moran, R.; Porath, D. Chlorophyll determination in intact tissues using N,NDimethylformamide. *Plant Physiol.* **1980**, *65*, 78–79. [CrossRef]
48. Bates, L.S.; Waldren, R.P.; Teare, I.D. Rapid determination of free proline for water-stress studies. *Plant Soil* **1973**, *39*, 205–207. [CrossRef]
49. Bian, S.; Jiang, Y. Reactive oxygen species, antioxidant enzyme activities and gene expression patterns in leaves and roots of Kentucky bluegrass in response to drought stress and recovery. *Sci. Hortic.* **2009**, *120*, 264–270. [CrossRef]
50. Aebi, H. Catalase In Vitro. *Meth. Enzymol.* **1984**, *105*, 121–130. [CrossRef]
51. Giannopolitis, C.N.; Ries, S.K. Superoxide dismutases: I. Occurrence in higher plants. *Plant Physiol.* **1977**, *59*, 309–314. [CrossRef] [PubMed]
52. Eom, S.H.; Setter, T.L.; DiTommaso, A.; Weston, L.A. Differential growth response to salt stress among selected ornamentals. *J. Plant Nutr.* **2007**, *30*, 1109–1126. [CrossRef]

53. Marosz, A. Effect of soil salinity on nutrient uptake, growth, and decorative value of four ground cover shrubs. *J. Plant Nutr.* **2004**, *27*, 977–989. [CrossRef]
54. Wahome, P.K. Mechanisms of salt (NaCl) stress tolerance in horticultural crops—A mini review. *Acta Hortic.* **2003**, *609*, 127–131. [CrossRef]
55. Carter, C.T.; Grieve, C.M. Salt tolerance of floriculture crops. In *Ecophysiology of High Salinity Tolerant Plants*; Khān, M.A., Khan, M.A., Weber, D.J., Eds.; Springer: Dordrecht, The Netherlands, 2008; pp. 279–287.
56. Grattan, S.R.; Grieve, C.M. Salinity-mineral nutrient relations in horticultural crops. *Sci. Hortic.* **1998**, *78*, 127–157. [CrossRef]
57. Touchette, B.W.; Rhodes, K.L.; Smith, G.A.; Poole, M. Salt spray induces osmotic adjustment and tissue rigidity in smooth cordgrass, *Spartina alterniflora* (Loisel.). *Estuaries Coasts* **2009**, *32*, 917–925. [CrossRef]
58. Alarcón, J.J.; Morales, M.A.; Torrecillas, A.; Sánchez-Blanco, M.J. Growth, water relations and accumulation of organic and inorganic solutes in the halophyte *Limonium latifolium* cv. Avignon and its interspecific hybrid *Limonium caspia* × *Limonium latifolium* cv. Beltlaard during salt stress. *J. Plant Physiol.* **1999**, *154*, 795–801. [CrossRef]
59. Morant-Manceau, A.; Pradier, E.; Tremblin, G. Osmotic adjustment, gas exchanges and chlorophyll fluorescence of a hexaploid triticale and its parental species under salt stress. *J. Plant Physiol.* **2004**, *161*, 25–33. [CrossRef]
60. Goldstein, G.; Drake, D.R.; Alpha, C.; Melcher, P.; Heraux, J.; Azocar, A. Growth and photosynthetic responses of *Scaevola sericea*, a Hawaiian coastal shrub, to substrate salinity and salt spray. *Int. J. Plant Sci.* **1996**, *157*, 171–179. [CrossRef]
61. De Vos, A.C.; Broekman, R.; Groot, M.P.; Rozema, J. Ecophysiological response of *Crambe maritima* to airborne and soil-borne salinity. *Ann. Bot.* **2010**, *105*, 925–937. [CrossRef]
62. Bernstein, L.; Francois, L.E.; Clark, R.A. Salt tolerance of ornamental shrubs and ground covers. *Am. Soc. Hortic. Sci. J.* **1972**, *97*, 550–556.
63. Niu, G.; Cabrera, R.I. Growth and physiological responses of landscape plants to saline water irrigation: A review. *HortScience* **2010**, *45*, 1605–1609. [CrossRef]
64. Ashraf, M. Biotechnological approach of improving plant salt tolerance using antioxidants as markers. *Biotechnol. Adv.* **2009**, *27*, 84–93. [CrossRef] [PubMed]
65. Parvaiz, A.; Satyawati, S. Salt stress and phyto-biochemical responses of plants-a review. *Plant Soil Environ.* **2008**, *54*, 89. [CrossRef]
66. Griffiths, M.E.; Orians, C.M. Salt spray effects on forest succession in rare coastal sandplain heathlands: Evidence from field surveys and *Pinus rigida* transplant experiments. *J. Torrey Bot. Soc.* **2004**, *131*, 23–31. [CrossRef]
67. Sánchez-Blanco, M.J.; Ferrández, T.; Navarro, A.; Bañon, S.; Alarcón, J.J. Effects of irrigation and air humidity preconditioning on water relations, growth and survival of *Rosmarinus officinalis* plants during and after transplanting. *J. Plant Physiol.* **2004**, *161*, 1133–1142. [CrossRef]
68. Flowers, T.J.; Munns, R.; Colmer, T.D. Sodium chloride toxicity and the cellular basis of salt tolerance in halophytes. *Ann. Bot.* **2015**, *115*, 419–431. [CrossRef]
69. Waldhoff, D.; Furch, B.; Junk, W.J. Fluorescence parameters, chlorophyll concentration, and anatomical features as indicators for flood adaptation of an abundant tree species in Central Amazonia: *Symmeria paniculata*. *Environ. Exp. Bot.* **2002**, *48*, 225–235. [CrossRef]
70. García-Caparrós, P.; Lao, M.T. The effects of salt stress on ornamental plants and integrative cultivation practices. *Sci. Hortic.* **2018**, *240*, 430–439. [CrossRef]
71. Kaur, G.; Asthir, B. Proline: A key player in plant abiotic stress tolerance. *Biol. Plant.* **2015**, *59*, 609–619. [CrossRef]
72. Verbruggen, N.; Hermans, C. Proline accumulation in plants: A review. *Amino Acids* **2008**, *35*, 753–759. [CrossRef] [PubMed]
73. Gupta, B.; Huang, B. Mechanism of salinity tolerance in plants: Physiological, biochemical, and molecular characterization. *Int. J. Genom.* **2014**, *2014*, 701596. [CrossRef] [PubMed]
74. Kundu, P.; Gill, R.; Ahlawat, S.; Anjum, N.A.; Sharma, K.K.; Ansari, A.A.; Hasanuzzaman, M.; Ramakrishna, A.; Chauhan, N.; Tuteja, N.; et al. Targeting the redox regulatory mechanisms for abiotic stress tolerance in crops. In *Biochemical, Physiological and Molecular Avenues for Combating Abiotic Stress Tolerance in Plants*; Wani, S.H., Ed.; Academic Press Elsevier: London, UK, 2018; pp. 151–220. [CrossRef]
75. Bernstein, L. Osmotic adjustment of plants to saline media. II. Dynamic phase. *Am. J. Bot.* **1963**, *50*, 360–370. [CrossRef]
76. De Gara, L.; Foyer, C.H. Ying and Yang interplay between reactive oxygen and reactive nitrogen species controls cell functions. *Plant Cell Environ.* **2017**, *40*, 459–461. [CrossRef]
77. Skłodowska, M.; Gapińska, M.; Gajewska, E.; Gabara, B. Tocopherol content and enzymatic antioxidant activities in chloroplasts from NaCl-stressed tomato plants. *Acta Physiol. Plant.* **2009**, *31*, 393–400. [CrossRef]

MDPI
St. Alban-Anlage 66
4052 Basel
Switzerland
Tel. +41 61 683 77 34
Fax +41 61 302 89 18
www.mdpi.com

Horticulturae Editorial Office
E-mail: horticulturae@mdpi.com
www.mdpi.com/journal/horticulturae

www.ingramcontent.com/pod-product-compliance
Lightning Source LLC
Chambersburg PA
CBHW061548210926
51850CB00009B/2043